Modeling the Internet and the Web

Modeling the Internet and the Web

Probabilistic Methods and Algorithms

Pierre Baldi
School of Information and Computer Science,
University of California, Irvine, USA

Paolo Frasconi
Department of Systems and Computer Science,
University of Florence, Italy

Padhraic Smyth
School of Information and Computer Science,
University of California, Irvine, USA

WILEY

Published by   John Wiley & Sons Ltd, The Atrium, Southern Gate, Chichester,
                    West Sussex PO19 8SQ, England
                    Phone   (+44) 1243 779777

Email (for orders and customer service enquiries): cs-books@wiley.co.uk
Visit our Home Page on www.wileyeurope.com or www.wiley.com

This publication is designed to provide accurate and authoritative information in regard to the subject matter
covered. It is sold on the understanding that the Publisher is not engaged in rendering professional services.
If professional advice or other expert assistance is required, the services of a competent professional should
be sought.

Other Wiley Editorial Offices

John Wiley & Sons Inc., 111 River Street, Hoboken, NJ 07030, USA

Jossey-Bass, 989 Market Street, San Francisco, CA 94103-1741, USA

Wiley-VCH Verlag GmbH, Boschstr. 12, D-69469 Weinheim, Germany

John Wiley & Sons Australia Ltd, 33 Park Road, Milton, Queensland 4064, Australia

John Wiley & Sons (Asia) Pte Ltd, 2 Clementi Loop #02-01, Jin Xing Distripark, Singapore 129809

John Wiley & Sons Canada Ltd, 22 Worcester Road, Etobicoke, Ontario, Canada M9W 1L1

Wiley also publishes its books in a variety of electronic formats. Some content that appears in print may
not be available in electronic books.

British Library Cataloguing in Publication Data

A catalogue record for this book is available from the British Library

ISBN 0-470-84906-1

Typeset in 10/12pt Times by T&T Productions Ltd, London.
Printed and bound in Great Britain by Biddles Ltd, Guildford, Surrey.
This book is printed on acid-free paper responsibly manufactured from sustainable forestry
in which at least two trees are planted for each one used for paper production.

*To Ezio and José (P.B.), to Neda (P.F.) and
to Seosamh and Bríd Áine (P.S.)*

Contents

Preface   xiii

1   Mathematical Background   1

1.1   Probability and Learning from a Bayesian Perspective   1
1.2   Parameter Estimation from Data   4
       1.2.1   Basic principles   4
       1.2.2   A simple die example   6
1.3   Mixture Models and the Expectation Maximization Algorithm   10
1.4   Graphical Models   13
       1.4.1   Bayesian networks   13
       1.4.2   Belief propagation   15
       1.4.3   Learning directed graphical models from data   16
1.5   Classification   17
1.6   Clustering   20
1.7   Power-Law Distributions   22
       1.7.1   Definition   22
       1.7.2   Scale-free properties (80/20 rule)   24
       1.7.3   Applications to Languages: Zipf's and Heaps' Laws   24
       1.7.4   Origin of power-law distributions and Fermi's model   26
1.8   Exercises   27

2   Basic WWW Technologies   29

2.1   Web Documents   30
       2.1.1   SGML and HTML   30
       2.1.2   General structure of an HTML document   31
       2.1.3   Links   32
2.2   Resource Identifiers: URI, URL, and URN   33
2.3   Protocols   36
       2.3.1   Reference models and TCP/IP   36
       2.3.2   The domain name system   37
       2.3.3   The Hypertext Transfer Protocol   38
       2.3.4   Programming examples   40

	2.4	Log Files	41
	2.5	Search Engines	44
		2.5.1 Overview	44
		2.5.2 Coverage	45
		2.5.3 Basic crawling	46
	2.6	Exercises	49

3   Web Graphs   **51**

	3.1	Internet and Web Graphs	51
		3.1.1 Power-law size	53
		3.1.2 Power-law connectivity	53
		3.1.3 Small-world networks	56
		3.1.4 Power law of PageRank	57
		3.1.5 The bow-tie structure	58
	3.2	Generative Models for the Web Graph and Other Networks	60
		3.2.1 Web page growth	60
		3.2.2 Lattice perturbation models: between order and disorder	61
		3.2.3 Preferential attachment models, or the rich get richer	63
		3.2.4 Copy models	66
		3.2.5 PageRank models	67
	3.3	Applications	68
		3.3.1 Distributed search algorithms	68
		3.3.2 Subgraph patterns and communities	70
		3.3.3 Robustness and vulnerability	72
	3.4	Notes and Additional Technical References	73
	3.5	Exercises	74

4   Text Analysis   **77**

	4.1	Indexing	77
		4.1.1 Basic concepts	77
		4.1.2 Compression techniques	79
	4.2	Lexical Processing	80
		4.2.1 Tokenization	80
		4.2.2 Text conflation and vocabulary reduction	82
	4.3	Content-Based Ranking	82
		4.3.1 The vector-space model	82
		4.3.2 Document similarity	83
		4.3.3 Retrieval and evaluation measures	85
	4.4	Probabilistic Retrieval	86
	4.5	Latent Semantic Analysis	88
		4.5.1 LSI and text documents	89
		4.5.2 Probabilistic LSA	89
	4.6	Text Categorization	93

	4.6.1	k nearest neighbors	93
	4.6.2	The Naive Bayes classifier	94
	4.6.3	Support vector classifiers	97
	4.6.4	Feature selection	102
	4.6.5	Measures of performance	104
	4.6.6	Applications	106
	4.6.7	Supervised learning with unlabeled data	111
4.7	Exploiting Hyperlinks		114
	4.7.1	Co-training	114
	4.7.2	Relational learning	115
4.8	Document Clustering		116
	4.8.1	Background and examples	116
	4.8.2	Clustering algorithms for documents	117
	4.8.3	Related approaches	119
4.9	Information Extraction		120
4.10	Exercises		122

5 Link Analysis — 125

5.1	Early Approaches to Link Analysis	126
5.2	Nonnegative Matrices and Dominant Eigenvectors	128
5.3	Hubs and Authorities: HITS	131
5.4	PageRank	134
5.5	Stability	138
	5.5.1 Stability of HITS	139
	5.5.2 Stability of PageRank	139
5.6	Probabilistic Link Analysis	140
	5.6.1 SALSA	140
	5.6.2 PHITS	142
5.7	Limitations of Link Analysis	143

6 Advanced Crawling Techniques — 149

6.1	Selective Crawling	149
6.2	Focused Crawling	152
	6.2.1 Focused crawling by relevance prediction	152
	6.2.2 Context graphs	154
	6.2.3 Reinforcement learning	155
	6.2.4 Related intelligent Web agents	157
6.3	Distributed Crawling	158
6.4	Web Dynamics	160
	6.4.1 Lifetime and aging of documents	161
	6.4.2 Other measures of recency	167
	6.4.3 Recency and synchronization policies	167

7   Modeling and Understanding Human Behavior on the Web   171

7.1   Introduction   171
7.2   Web Data and Measurement Issues   172
       7.2.1   Background   172
       7.2.2   Server-side data   174
       7.2.3   Client-side data   177
7.3   Empirical Client-Side Studies of Browsing Behavior   179
       7.3.1   Early studies from 1995 to 1997   180
       7.3.2   The Cockburn and McKenzie study from 2002   181
7.4   Probabilistic Models of Browsing Behavior   184
       7.4.1   Markov models for page prediction   184
       7.4.2   Fitting Markov models to observed page-request data   186
       7.4.3   Bayesian parameter estimation for Markov models   187
       7.4.4   Predicting page requests with Markov models   189
       7.4.5   Modeling runlengths within states   193
       7.4.6   Modeling session lengths   194
       7.4.7   A decision-theoretic surfing model   198
       7.4.8   Predicting page requests using additional variables   199
7.5   Modeling and Understanding Search Engine Querying   201
       7.5.1   Empirical studies of search behavior   202
       7.5.2   Models for search strategies   207
7.6   Exercises   208

8   Commerce on the Web: Models and Applications   211

8.1   Introduction   211
8.2   Customer Data on the Web   212
8.3   Automated Recommender Systems   212
       8.3.1   Evaluating recommender systems   214
       8.3.2   Nearest-neighbor collaborative filtering   215
       8.3.3   Model-based collaborative filtering   218
       8.3.4   Model-based combining of votes and content   223
8.4   Networks and Recommendations   224
       8.4.1   Email-based product recommendations   224
       8.4.2   A diffusion model   226
8.5   Web Path Analysis for Purchase Prediction   228
8.6   Exercises   232

Appendix A   Mathematical Complements   235

A.1   Graph Theory   235
       A.1.1   Basic definitions   235
       A.1.2   Connectivity   236
       A.1.3   Random graphs   236

A.2   Distributions   237
       A.2.1   Expectation, variance, and covariance   237
       A.2.2   Discrete distributions   237
       A.2.3   Continuous distributions   238
       A.2.4   Weibull distribution   240
       A.2.5   Exponential family   240
       A.2.6   Extreme value distribution   241
A.3   Singular Value Decomposition   241
A.4   Markov Chains   243
A.5   Information Theory   243
       A.5.1   Mathematical background   244
       A.5.2   Information, surprise, and relevance   247

Appendix B   List of Main Symbols and Abbreviations   253

References   257

Index   277

Preface

Since its early ARPANET inception during the Cold War, the Internet has grown by a staggering nine orders of magnitude. Today, the Internet and the World Wide Web pervade our lives, having fundamentally altered the way we seek, exchange, distribute, and process information. The Internet has become a powerful social force, transforming communication, entertainment, commerce, politics, medicine, science, and more. It mediates an ever growing fraction of human knowledge, forming both the largest library and the largest marketplace on planet Earth.

Unlike the invention of earlier media such as the press, photography, or even the radio, which created specialized passive media, the Internet and the Web impact all information, converting it to a uniform digital format of bits and packets. In addition, the Internet and the Web form a dynamic medium, allowing software applications to control, search, modify, and filter information without human intervention. For example, email messages can carry programs that affect the behavior of the receiving computer. This active medium also promotes human intervention in sharing, updating, linking, embellishing, critiquing, corrupting, etc., information to a degree that far exceeds what could be achieved with printed documents.

In common usage, the words 'Internet' and 'Web' (or World Wide Web or WWW) are often used interchangeably. Although they are intimately related, there are of course some nuances which we have tried to respect. 'Internet', in particular, is the more general term and implicitly includes physical aspects of the underlying networks as well as mechanisms such as email and peer-to-peer activities that are not directly associated with the Web. The term 'Web', on the other hand, is associated with the information stored and available on the Internet. It is also a term that points to other complex networks of information, such as webs of scientific citations, social relations, or even protein interactions. In this sense, it is fair to say that a predominant fraction of our book is about the Web and the information aspects of the Internet. We use 'Web' every time we refer to the World Wide Web and 'web' when we refer to a broader class of networks or other kinds of networks, i.e. web of citations.

As the Internet and the Web continue to expand at an exponential rate, it also evolves in terms of the devices and processors connected to it, e.g. wireless devices and appliances. Ever more human domains and activities are ensnared by the Web, thus creating challenging problems of ownership, security, and privacy. For instance, we are quite far from having solved the security, privacy, and authentication problems that would allow us to hold national Internet elections.

As scientists, the Web has also become a tool we use on a daily basis for tasks ranging from the mundane to the intractable, to search and disseminate information, to exchange views and collaborate, to post job listings, to retrieve and quote (by Uniform Resource Locator (URL)) bibliographic information, to build Web servers, and even to compute. There is hardly a branch of computer science that is not affected by the Internet: not only the most obvious areas such as networking and protocols, but also security and cryptography; scientific computing; human interfaces, graphics, and visualization; information retrieval, data mining, machine learning, language/text modeling and artificial intelligence, to name just a few.

What is perhaps less obvious and central to this book is that not only have the Web and the Internet become essential tools of scientific enterprise, but they have also themselves become the objects of active scientific investigation. And not only for computer scientists and engineers, but also for mathematicians, economists, social scientists, and even biologists.

There are many reasons why the Internet and the Web are exciting, albeit young, topics for scientific investigation. These reasons go beyond the need to improve the underlying technology and to harness the Web for commercial applications. Because the Internet and the Web can be viewed as dynamic constellations of interconnected processors and Web pages, respectively, they can be monitored in many ways and at many different levels of granularity, ranging from packet traffic, to user behavior, to the graphical structure of Web pages and their hyperlinks. These measurements provide new types of large-scale data sets that can be scientifically analyzed and 'mined' at different levels. Thus researchers enjoy unprecedented opportunities to, for instance:

- gather, communicate, and exchange ideas, documents, and information;

- monitor a large dynamic network with billions of nodes and one order of magnitude more connections;

- gather large training sets of textual or activity data, for the purposes of modeling and predicting the behavior of millions of users;

- analyze and understand interests and relationships within society.

The Web, for instance, can be viewed as an example of a very large distributed and dynamic system with billions of pages resulting from the uncoordinated actions of millions of individuals. After all, anyone can post a Web page on the Internet and link it to any other page. In spite of this complete lack of central control, the graphical structure of the Web is far from random and possesses emergent properties shared with other complex graphs found in social, technological, and biological systems. Examples of properties include the power-law distribution of vertex connectivities and the small-world property – any two Web pages are usually only a few clicks away from each other. Similarly, predictable patterns of congestion (e.g. traffic jams) have also been observed in Internet traffic. While the exploitation of these regularities may be beneficial to providers and consumers, their mere existence and discovery has become a topic of basic research.

Why Probabilistic Modeling?

By its very nature, a very large distributed, decentralized, self-organized, and evolving system necessarily yields uncertain and incomplete measurements and data. Probability and statistics are the fundamental mathematical tools that allow us to model, reason and proceed with inference in uncertain environments. Not only are probabilistic methods needed to deal with noisy measurements, but many of the underlying phenomena, including the dynamic evolution of the Internet and the Web, are themselves probabilistic in nature. As in the systems studied in statistical mechanics, regularities may emerge from the more or less random interactions of myriads of small factors. Aggregation can only be captured probabilistically. Furthermore, and not unlike biological systems, the Internet is a very high-dimensional system, where measurement of all relevant variables becomes impossible. Most variables remain hidden and must be 'factored out' by probabilistic methods.

There is one more important reason why probabilistic modeling is central to this book. At a fundamental level the Web is concerned with information retrieval and the semantics, or meaning, of that information. While the modeling of semantics remains largely an open research problem, probabilistic methods have achieved remarkable successes and are widely used in information retrieval, machine translation, and more. Although these probabilistic methods bypass or fake semantic understanding, they are, for instance, at the core of the search engines we use every day. As it happens, the Internet and the Web themselves have greatly aided the development of such methods by making available large corpora of data from which statistical regularities can be extracted.

Thus, probabilistic methods pervasively apply to diverse areas of Internet and Web modeling and analysis, such as network traffic, graphical structure, information retrieval engines, and customer behavior.

Audience and Prerequisites

Our aim has been to write an interdisciplinary textbook about the Internet both to fill a specific niche at the undergraduate and graduate level and to serve as a reference for researchers and practitioners from academia, industry, and government. Thus, it is aimed at a relatively broad audience including both students and more advanced researchers, with diverse backgrounds. We have tried to provide a succinct and self-contained description of the main concepts and methods in a manner accessible to computer scientists and engineers and also to those whose primary background is in other disciplines touched by the Internet and the Web. We hope that the book will be of interest to students, postdoctoral fellows, faculty members and researchers from a variety of disciplines including Computer Science, Engineering, Statistics, Applied Mathematics, Economics and Business, and Social Sciences.

The topic is quite broad. On the surface the Web could appear to be a limited sub-discipline of computer science, but in reality it is impossible for a single researcher to have an in-depth knowledge and understanding of all the areas of science and technology touched by the Internet and the Web. While we do not claim to cover all aspects of the Internet – for instance, we do not look in any detail at the physical layer – we do try to cover the most important aspects of the Web at the *information level* and provide pointers to the reader for topics that are left out. We propose a unified treatment based on mathematical and probabilistic modeling that emphasizes the unity of the field, as well as its connections to other areas such as machine learning, data mining, graph theory, information retrieval, and bioinformatics.

The prerequisites include an understanding of several basic mathematical concepts at an undergraduate level, including probabilistic concepts and methods, basic calculus, and matrix algebra, as well as elementary concepts in data structures and algorithms. Some additional knowledge of graph theory and combinatorics is helpful but not required. Mathematical proofs are usually short and mathematical details that can be found in the cited literature are sometimes left out in favor of a more intuitive treatment. We expect the typical reader to be able to gather complementary information from the references, as needed. For instance we refer to, but do not provide the details of, the algorithm for finding the shortest path between two vertices in a graph, since this is readily found in other textbooks.

We have included many concrete examples, such as examples of pseudocode and analyses of specific data sets, as well as exercises of varying difficulty at the end of each chapter. Some are meant to encourage basic thinking about the Internet and the Web, and the corresponding probabilistic models. Other exercises are more suited for class projects and require computer measurements and simulations, such as constructing the graph of pages and hyperlinks associated with one's own institution. These are complemented by more mathematical exercises of varying levels of difficulty.

While the book can be used in a course about the Web, or as complementary reading material in, for instance, an information retrieval class, we are also in the process of using it to teach a course on the application of probability, statistics, and information theory in computer science by providing a unified theme, set of methods, and a variety of 'close-to-home' examples and problems aimed at developing both mathematical intuition and computer simulation skills.

Content and General Outline of the Book

We have strived to write a comprehensive but reasonably concise introductory book that is self-contained and summarizes a wide range of results that are scattered throughout the literature. A portion of the book is built on material taken from articles we have written over the years, as well as talks, courses, and tutorials. Our main focus is not on the history of a rapidly evolving field, but rather on what we believe are the primary relevant methods and algorithms, and a general way of thinking about modeling of the Web that we hope will prove useful.

Chapter 1 covers in succinct form most of the mathematical background needed for the following chapters and can be skipped by those with a good familiarity with its material. It contains an introduction to basic concepts in probabilistic modeling and machine learning – from the Bayesian framework and the theory of graphical models to mixtures, classification, and clustering – these are all used throughout various chapters and form a substantial part of the 'glue' of this book.

Chapter 2 provides an introduction to the Internet and the Web and the foundations of the WWW technologies that are necessary to understand the rest of the book, including the structure of Web documents, the basics of Internet protocols, Web server log files, and so forth. Server log files, for instance, are important to thoroughly understand the analysis of human behavior on the Web in Chapter 7. The chapter also deals with the basic principles of Web crawlers. Web crawling is essential to gather information about the Web and in this sense is a prerequisite for the study of the Web graph in Chapter 3.

Chapter 3 studies the Internet and the Web as large graphs. It describes, models, and analyzes the power-law distribution of Web sizes, connectivity, PageRank, and the 'small-world' properties of the underlying graphs. Applications of graphical properties, for instance to improve search engines, are also covered in this chapter and further studied in later chapters.

Chapter 4 deals with text analysis in terms of indexing, content-based ranking, latent semantic analysis, and text categorization, providing the basic components (together with link analysis) for understanding how to efficiently retrieve information over the Web.

Chapter 5 builds upon the graphical results of Chapter 4 and deals with link analysis, inferring page relevance from patterns of connectivity, Web communities, and the stability and evolution of these concepts with time.

Chapter 6 covers advanced crawling techniques – selective, focused, and distributed crawling and Web dynamics. It is essential material in order to understand how to build a new search engine for instance.

Chapter 7 studies human behavior on the Web. In particular it builds and studies several probabilistic models of human browsing behavior and also analyzes the statistical properties of search engine queries.

Finally, Chapter 8 covers various aspects of commerce on the Web, including analysis of customer Web data, automated recommendation systems, and Web path analysis for purchase prediction.

Appendix A contains a number of technical sections that are important for reference and for a thorough understanding of the material, including an informal introduction to basic concepts in graph theory, a list of standard probability densities, a short section on Singular Value Decomposition, a short section on Markov chains, and a brief, critical, overview of information theory.

What Is New and What Is Omitted

On several occasions we present new material, or old material but from a somewhat new perspective. Examples include the notion of surprise in Appendix A, as well as a simple model for power-law distributions originally due to Enrico Fermi that seems to have been forgotten, which is described in Chapter 2. The material in this book and its treatment reflect our personal biases. Many relevant topics had to be omitted in order to stay within reasonable size limits. In particular, the book contains little material about the physical layer, about any hardware, or about Internet protocols. Other important topics that we would have liked to cover but had to be left out include the aspects of the Web related to security and cryptography, human interfaces and design. We do cover many aspects of text analysis and information retrieval, but not all of them, since a more exhaustive treatment of any of these topics would require a book by itself. Thus, in short, the main focus of the book is on the information aspects of the Web, its emerging properties, and some of its applications.

Notation

In terms of notation, most of the symbols used are listed at the end of the book, in Appendix B. A symbol such as 'D' represents the data, regardless of the amount or complexity. Boldface letters are usually reserved for matrices and vectors. Capital letters are typically used for matrices and random variables, lowercase letters for scalars and random variable realizations. Greek letters such as θ typically denote the parameters of a model. Throughout the book P and E are used for 'probability' and 'expectation'. If X is a random variable, we often write $P(x)$ for $P(X = x)$, or sometimes just $P(X)$ if no confusion is possible. $E[X]$, var$[X]$, and cov$[X, Y]$, respectively, denote the expectation, variance, and covariance associated with the random variables X and Y with respect to the probability distributions $P(X)$ and $P(X, Y)$.

We use the standard notation $f(n) = o(g(n))$ to denote a function $f(n)$ that satisfies $f(n)/g(n) \to 0$ as $n \to \infty$, and $f(n) = O(g(n))$ when there exists a constant $C > 0$ such that $f(n) \leqslant Cg(n)$ when $n \to \infty$. Similarly, we use $f(n) = \Omega(g(n))$ to denote a function $f(n)$ such that asymptotically there are two constants C_1 and C_2 with $C_1 g(n) \leqslant f(n) \leqslant C_2 g(n)$. Calligraphic style is reserved for particular functions, such as error or energy ($\mathcal{E}$), entropy and relative entropy ($\mathcal{H}$). Finally, we often deal with quantities characterized by many indices. Within a given context, only the most relevant indices are indicated.

Acknowledgements

Over the years, this book has been supported directly or indirectly by grants and awards from the US National Science Foundation, the National Institutes of Health,

NASA and the Jet Propulsion Laboratory, the Department of Energy and Lawrence Livermore National Laboratory, IBM Research, Microsoft Research, Sun Microsystems, HNC Software, the University of California MICRO Program, and a Laurel Wilkening Faculty Innovation Award. Part of the book was written while P.F. was visiting the School of Information and Computer Science (ICS) at UCI, with partial funding provided by the University of California. We also would like to acknowledge the general support we have received from the Institute for Genomics and Bioinformatics (IGB) at UCI and the California Institute for Telecommunications and Information Technology (Cal(IT)^2). Within IGB, special thanks go the staff, in particular Suzanne Knight, Janet Ko, Michele McCrea, and Ann Marie Walker. We would like to acknowledge feedback and general support from many members of Cal(IT)^2 and thank in particular its directors Bill Parker, Larry Smarr, Peter Rentzepis, and Ramesh Rao, and staff members Catherine Hammond, Ashley Larsen, Doug Ramsey, Stuart Ross, and Stephanie Sides. We thank a number of colleagues for discussions and feedback on various aspects of the Web and probabilistic modeling: David Eppstein, Chen Li, Sharad Mehrotra, and Mike Pazzani at UC Irvine, as well as Albert-László Barabási, Nicoló Cesa-Bianchi, Monica Bianchini, C. Lee Giles, Marco Gori, David Heckerman, David Madigan, Marco Maggini, Heikki Mannila, Chris Meek, Amnon Meyers, Ion Muslea, Franco Scarselli, Giovanni Soda, and Steven Scott. We thank all of the people who have helped with simulations or provided feedback on the various versions of this manuscript, especially our students Gianluca Pollastri, Alessandro Vullo, Igor Cadez, Jianlin Chen, Xianping Ge, Joshua O'Madadhain, Scott White, Alessio Ceroni, Fabrizio Costa, Michelangelo Diligenti, Sauro Menchetti, and Andrea Passerini. We also acknowledge Jean-Pierre Nadal, who brought to our attention the Fermi model of power laws. We thank Xinglian Yie and David O'Hallaron for providing their data on search engine queries in Chapter 8. We also thank the staff from John Wiley & Sons, Ltd, in particular Senior Editor Sian Jones and Robert Calver, and Emma Dain at T&T Productions Ltd. Finally, we acknowledge our families and friends for their support in the writing of this book.

Pierre Baldi, Paolo Frasconi and Padhraic Smyth
October 2002, Irvine, CA

1

Mathematical Background

In this chapter we review a number of basic concepts in probabilistic modeling and data analysis that are used throughout the book, including parameter estimation, mixture models, graphical models, classification, clustering, and power-law distributions. Each of these topics is worthy of an entire chapter (or even a whole book) by itself, so our treatment is necessarily brief. Nonetheless, the goal of this chapter is to provide the introductory foundations for models and techniques that will be widely used in the remainder of the book. Readers who are already familiar with these concepts, or who want to avoid mathematical details during a first reading, can safely skip to the next chapter. More specific technical mathematical concepts or mathematical complements that concern only a specific chapter rather than the entire book are given in Appendix A.

1.1 Probability and Learning from a Bayesian Perspective

Throughout this book we will make frequent use of probabilistic models to characterize various phenomena related to the Web. Both theory and experience have shown that probability is by far the most useful framework currently available for modeling uncertainty. Probability allows us to reason in a coherent manner about events and make inferences about such events given observed data. More specifically, an event e is a proposition or statement about the world at large. For example, let e be the proposition that 'the number of Web pages in existence on 1 January 2003 was greater than five billion'. A well-defined proposition e is either true or false – by some reasonable definition of what constitutes a Web page, the total number that existed in January 2003 was either greater than five billion or not. There is, however, considerable uncertainty about what this number was back in January 2003 since, as we will discuss later in Chapters 2 and 3, accurately estimating the size of the Web is a quite challenging problem. Consequently there is uncertainty about whether the proposition e is true or not.

Modeling the Internet and the Web   P. Baldi, P. Frasconi and P. Smyth
© 2003 P. Baldi, P. Frasconi and P. Smyth   ISBN: 0-470-84906-1

A probability, $P(e)$, can be viewed as a number that reflects our uncertainty about whether e is true or false in the real world, given whatever information we have available. This is known as the 'degree of belief' or Bayesian interpretation of probability (see, for instance, Berger 1985; Box and Tiao 1992; Cox 1964; Gelman *et al.* 1995; Jaynes 2003) and is the one that we will use by default throughout this text. In fact, to be more precise, we should use a conditional probability $P(e \mid \mathcal{I})$ in general to represent degree of belief, where $\mathcal{I}$ is the background information on which our belief is based. For simplicity of notation we will often omit this conditioning on $\mathcal{I}$, but it may be useful to keep in mind that everywhere you see a $P(e)$ for some proposition e, there is usually some implicit background information $\mathcal{I}$ that is known or assumed to be true.

The Bayesian interpretation of a probability $P(e)$ is a generalization of the more classic interpretation of a probability as the relative frequency of successes to total trials, estimated over an infinite number of hypothetical repeated trials (the so-called 'frequentist' interpretation). The Bayesian interpretation is more useful in general, since it allows us to make statements about propositions such as 'the number of Web pages in existence' where a repeated trials interpretation would not necessarily apply.

It can be shown that, under a small set of reasonable axioms, degrees of belief can be represented by real numbers and that when rescaled to the [0, 1] interval these degrees of confidence must obey the rules of probability and, in particular, Bayes' theorem (Cox 1964; Jaynes 1986, 2003; Savage 1972). This is reassuring, since it means that the standard rules of probability still apply whether we are using the degree of belief interpretation or the frequentist interpretation. In other words, the rules for manipulating probabilities such as conditioning or the law of total probability remain the same no matter what semantics we attach to the probabilities.

The Bayesian approach also allows us to think about probability as being a dynamic entity that is updated as more data arrive – as we receive more data we may naturally change our degree of belief in certain propositions given these new data. Thus, for example, we will frequently refer to terms such as $P(e \mid D)$ where D is some data. In fact, by Bayes' theorem,

$$P(e \mid D) = \frac{P(D \mid e)P(e)}{P(D)}. \tag{1.1}$$

The interpretation of each of the terms in this equation is worth discussing. $P(e)$ is your belief in the event e before you see any data at all, referred to as your *prior probability* for e or prior degree of belief in e. For example, letting e again be the statement that 'the number of Web pages in existence on 1 January 2003 was greater than five billion', $P(e)$ reflects your degree of belief that this statement is true. Suppose you now receive some data D which is the number of pages indexed by various search engines as of 1 January 2003. To a reasonable approximation we can view these numbers as lower bounds on the true number and let's say for the sake of argument that all the numbers are considerably less than five billion. $P(e \mid D)$ now reflects your updated *posterior* belief in e given the observed data and it can be calculated by using Bayes' theorem via

Equation (1.1). The right-hand side of Equation (1.1) includes the prior, so naturally enough the posterior is proportional to the prior.

The right-hand side also includes $P(D \mid e)$, which is known as the *likelihood* of the data, i.e. the probability of the data under the assumption that e is true. To calculate the likelihood we must have a probabilistic model that connects the proposition e we are interested in with the observed data D – this is the essence of probabilistic learning. For our Web page example, this could be a model that puts a probability distribution on the number of Web pages that each search engine may find if the conditioning event is true, i.e. if there are in fact more than five billion Web pages in existence. This could be a complex model of how the search engines actually work, taking into account all the various reasons that many pages will not be found, or it might be a very simple approximate model that says that each search engine has some conditional distribution on the number of pages that will be indexed, as a function of the total number that exist. Appendix A provides examples of several standard probability models – these are in essence the 'building blocks' for probabilistic modeling and can be used as components either in likelihood models $P(D \mid e)$ or as priors $P(e)$.

Continuing with Equation (1.1), the likelihood expression reflects how likely the observed data are, given e and given some model connecting e and the data. If $P(D \mid e)$ is very low, this means that the model is assigning a low probability to the observed data. This might happen, for example, if the search engines hypothetically all reported numbers of indexed pages in the range of a few million rather than in the billion range. Of course we have to factor in the alternative hypothesis, $\bar{e}$, here and we must ensure that both $P(e) + P(\bar{e}) = 1$ and $P(e \mid D) + P(\bar{e} \mid D) = 1$ to satisfy the basic axioms of probability. The 'normalization' constant in the denominator of Equation (1.1) can be calculated by noting that $P(D) = P(D \mid e)P(e) + P(D \mid \bar{e})P(\bar{e})$. It is easy to see that $P(e \mid D)$ depends both on the prior and the likelihood in terms of 'competing' with the alternative hypothesis $\bar{e}$ – the larger they are relative to the prior for $\bar{e}$ and the likelihood for $\bar{e}$, then the larger our posterior belief in e will be.

Because probabilities can be very small quantities and addition is often easier to work with than multiplication, it is common to take logarithms of both sides, so that

$$\log P(e \mid D) = \log P(D \mid e) + \log P(e) - \log P(D). \tag{1.2}$$

To apply Equation (1.1) or (1.2) to any class of models, we only need to specify the prior $P(e)$ and the data likelihood $P(D \mid e)$.

Having updated our degree of belief in e, from $P(e)$ to $P(e \mid D)$, we can continue this process and incorporate more data as they become available. For example, we might later obtain more data on the size of the Web from a different study – call this second data set D_2. We can use Bayes' rule to write

$$P(e \mid D, D_2) = \frac{P(D_2 \mid e, D)P(e \mid D)}{P(D_2 \mid D)}. \tag{1.3}$$

Comparing Equations (1.3) and (1.2) we see that the old posterior $P(e \mid D)$ plays the role of the new prior when data set D_2 arrives.

The use of priors is a strength of the Bayesian approach, since it allows the incorporation of prior knowledge and constraints into the modeling process. In general, the effects of priors diminish as the number of data points increases. Formally, this is because the log-likelihood $\log P(D \mid e)$ typically increases linearly with the number of data points in D, while the prior $\log P(e)$ remains constant. Finally, and most importantly, the effects of different priors, as well as different models and model classes, can be assessed within the Bayesian framework by comparing the corresponding probabilities.

The computation of the likelihood is of course model dependent and is not addressed here in its full generality. Later in the chapter we will briefly look at a variety of graphical model and mixture model techniques that act as 'components' in a 'flexible toolbox' for the construction of different types of likelihood functions.

1.2   Parameter Estimation from Data

1.2.1   Basic principles

We can now turn to the main type of inference that will be used throughout this book, namely estimation of parameters θ under the assumption of a particular functional form for a model M. For example, if our model M is the Gaussian distribution, then we have two parameters: the mean and standard deviation of this Gaussian.

In what follows below we will refer to priors, likelihoods, and posteriors relating to sets of parameters θ. Typically our parameters θ are a set of real-valued numbers. Thus, both the prior $P(\theta)$ and the posterior $P(\theta \mid D)$ are defining probability density functions over this set of real-valued numbers. For example, our prior density might assert that a particular parameter (such as the standard deviation) must be positive, or that the mean is likely to lie within a certain range on the real line. From a Bayesian viewpoint our prior and posterior reflect our degree of belief in terms of where θ lies, given the relevant evidence. For example, for a single parameter θ, $P(\theta > 0 \mid D)$ is the posterior probability that θ is greater than zero, given the evidence provided by the data. For simplicity of notation we will not always explicitly include the conditioning term for the model M, $P(\theta \mid M)$ or $P(\theta \mid D, M)$, but instead write expressions such as $P(\theta)$ which implicitly assume the presence of a model M with some particular functional form.

The general objective of parameter estimation is to find or approximate the 'best' set of parameters for a model – that is, to find the set of parameters θ maximizing the posterior $P(\theta \mid D)$, or $\log P(\theta \mid D)$. This is called maximum *a posteriori* (MAP) estimation.

In order to deal with positive quantities, we can minimize $-\log P(\theta \mid D)$:

$$\mathcal{E}(\theta) = -\log P(\theta \mid D) = -\log P(D \mid \theta) - \log P(\theta) + \log P(D). \qquad (1.4)$$

From an optimization standpoint, the logarithm of the prior $P(\theta)$ plays the role of a regularizer, that is, of an additional penalty term that can be used to enforce additional constraints such as smoothness. Note that the term $P(D)$ in (1.4) plays the role of a normalizing constant that does not depend on the parameters θ, and is therefore irrelevant for this optimization. If the prior $P(\theta)$ is uniform over parameter space, then the problem reduces to finding the maximum of $P(D \mid \theta)$, or $\log P(D \mid \theta)$. This is known as maximum-likelihood (ML) estimation.

A substantial portion of this book and of machine-learning practice in general is based on MAP estimation, that is, the minimization of

$$\mathcal{E}(\theta) = -\log P(D \mid \theta) - \log P(\theta), \tag{1.5}$$

or the simpler ML estimation procedure, that is, the minimization of

$$\mathcal{E}(\theta) = -\log P(D \mid \theta). \tag{1.6}$$

In many useful and interesting models the function being optimized is complex and its modes cannot be found analytically. Thus, one must resort to iterative and possibly stochastic methods such as gradient descent, expectation maximization (EM), or simulated annealing. In addition, one may also have to settle for approximate or sub-optimal solutions, since finding global optima of these functions may be computationally infeasible. Finally it is worth mentioning that mean posterior (MP) estimation is also used, and can be more robust than the mode (MAP) in certain respects. The MP is found by estimating θ by its expectation $E[\theta]$ with respect to the posterior $P(\theta \mid D)$, rather than the mode of this distribution.

Whereas finding the optimal model, i.e. the optimal set of parameters, is common practice, it is essential to note that this is really useful only if the distribution $P(\theta \mid D)$ is sharply peaked around a unique optimum. In situations characterized by a high degree of uncertainty and relatively small amounts of available data, this is often not the case. Thus, a full Bayesian approach focuses on the function $P(\theta \mid D)$ over the entire parameter space rather than at a single point. Discussion of the full Bayesian approach is somewhat beyond the scope of this book (see Gelman $et\ al.$ (1995) for a comprehensive introductory treatment). In most of the cases we will consider, the simpler ML or MAP estimation approaches are sufficient to yield useful results – this is particularly true in the case of Web data sets which are often large enough that the posterior distribution can be reasonably expected to be concentrated relatively tightly about the posterior mode.

The reader ought also to be aware that whatever criterion is used to measure the discrepancy between a model and data (often described in terms of an error or energy function), such a criterion can always be defined in terms of an underlying probabilistic model that is amenable to Bayesian analysis. Indeed, if the fit of a model $M = M(\theta)$ with parameters θ is measured by some error function $f(\theta, D) \geqslant 0$ to be minimized, one can always define the associated likelihood as

$$P(D \mid M(\theta)) = \frac{e^{-f(\theta, D)}}{Z}, \tag{1.7}$$

where $Z = \int_\theta e^{-f(\theta, D)}\, d\theta$ is a normalizing factor (the 'partition function' in statistical mechanics) that ensures the probabilities integrate to unity. As a result, minimizing the error function is equivalent to ML estimation or, more generally, MAP estimation, since Z is a constant and maximizing $\log P(D \mid M(\theta))$ is the same as minimizing $f(\theta, D)$ as a function of θ. For example, when the sum of squared differences is used as an error function (a rather common practice), this implies an underlying Gaussian model on the errors. Thus, the Bayesian point of view clarifies the probabilistic assumptions that underlie any criteria for matching models with data.

1.2.2 A simple die example

Consider an m-ary alphabet of symbols, $A = \{a_1, \ldots, a_m\}$. Assume that we have a data set D that consists of a set of n observations from this alphabet, where $D = X_1, \ldots, X_n$. We can convert the raw data D into counts for each symbol $\{n_1, \ldots, n_m\}$. For example, we could write a program to crawl the Web to find Web pages containing English text and 'parse' the resulting set of pages to remove HTML and other non-English words. In this case the alphabet A would consist of a set of English words and the counts would be the number of times each word was observed.

The simplest probabilistic model for such a 'bag of words' is a memoryless model consisting of a die with m sides, one for each symbol in the alphabet, and where we assume that the set D has been generated by n successive independent tosses of this die. Because the tosses are independent and there is a unique underlying die, for likelihood considerations it does not matter whether the data were generated from many Web pages or from a single Web page. Nor does it even matter what the order of the words on any page is, since we are not modeling the sequential order of the words in our very simple model. In Chapter 4 we will see that this corresponds to the widely used 'bag of words' model for text analysis.

Our model M has m parameters π_i, $1 \leqslant i \leqslant m$, corresponding to the probability of producing each symbol, with $\sum_i \pi_i = 1$. Assuming independent repeated draws from this model M, this defines a multinomial model with a likelihood given by

$$P(D \mid \pi) = \prod_{i=1}^{m} \pi_i^{n_i}, \qquad (1.8)$$

where n_i is the number of times symbol i appears in D. The negative log-posterior is then

$$-\log P(\pi \mid D) = -\sum_i n_i \log \pi_i - \log P(\pi) + \log P(D). \qquad (1.9)$$

If we assume a uniform prior distribution over the parameters π, then the MAP parameter estimation problem is identical to the ML parameter estimation problem and can be solved by optimizing the Lagrangian

$$\mathcal{L} = -\sum_i n_i \log \pi_i - \lambda \left(1 - \sum_i \pi_i \right) \qquad (1.10)$$

associated with the negative log-likelihood and augmented by the normalizing constraint (see in addition the exercises at the end of Chapter 7). Setting the partial derivatives $\partial \mathcal{L}/\partial \pi_i$ to zero immediately yields $\pi_i = n_i/\lambda$. Using the normalization constraint gives $\lambda = n$, so that finally, as expected, we get the estimates

$$\pi_i^{ML} = \frac{n_i}{n} \quad \text{for all } a_i \in A. \tag{1.11}$$

Note that the value of the negative log-likelihood per letter, for the optimal parameter set π^{ML}, approaches the entropy (see Appendix A) $\mathcal{H}(\pi^{ML})$ of π^{ML} as $n \to \infty$:

$$\lim_{n \to \infty} -\frac{1}{n} \sum_i n_i \log \frac{n_i}{n} = -\sum_i \pi_i^{ML} \log \pi_i^{ML} = \mathcal{H}(\pi^{ML}). \tag{1.12}$$

Another way of looking at these results is to say that, except for a constant entropy term, the negative log-likelihood is essentially the relative entropy between the fixed die probabilities p_i and the observed frequencies n_i/n. In Chapter 7 we will see how this idea can be used to quantify how well various sequential models can predict which Web page a Web surfer will request next.

The observed frequency estimate $\pi_i^{ML} = n_i/n$ is of course intuitive when n is large. The strong law of large numbers tells us that for large enough values of n, the observed frequency will almost surely be very close to the true value of π_i. But what happens if n is small relative to m and some of the symbols (or words) are never observed in D? Do we want to set the corresponding probability to zero? In general this is not a good idea, since it would lead us to assign probability zero to any new data set containing one or more symbols that were not observed in D.

This is a problem that is most elegantly solved by introducing a Dirichlet prior on the space of parameters (Berger 1985; MacKay and Peto 1995a). This approach is used again in Chapter 7 for modeling Web surfing patterns with Markov chains and the basic definitions are described in Appendix A. A Dirichlet distribution on a probability vector $\pi = (\pi_1, \ldots, \pi_m)$ with parameters α and $q = (q_1, \ldots, q_m)$ has the form

$$\mathcal{D}_{\alpha q}(\pi) = \frac{\Gamma(\alpha)}{\prod_i \Gamma(\alpha q_i)} \prod_{i=1}^{m} \pi_i^{\alpha q_i - 1} = \prod_{i=1}^{m} \frac{\pi_i^{\alpha q_i - 1}}{Z(i)}, \tag{1.13}$$

with $\alpha, \pi_i, q_i \geq 0$ and $\sum \pi_i = \sum q_i = 1$. Alternatively, it can be parameterized by a single vector $\boldsymbol{\alpha}$, with $\alpha_i = \alpha q_i$. When $m = 2$, it is also called a Beta distribution (Figure 1.1). For a Dirichlet distribution, $E(\pi_i) = q_i$, $\text{var}(\pi_i) = q_i(1 - q_i)/(\alpha + 1)$, and $\text{cov}(\pi_i \pi_j) = -q_i q_j/(\alpha + 1)$. Thus, q is the mean of the distribution, and α determines how peaked the distribution is around its mean. Dirichlet priors are the natural conjugate priors for multinomial distributions. In general, we say that a prior is conjugate with respect to a likelihood when the functional form of the posterior is identical to the functional form of the prior. Indeed, the likelihood in Equation (1.8) and the Dirichlet prior have the same functional form. Therefore the posterior associated with the product of the likelihood with the prior has also the same functional form and must be a Dirichlet distribution itself.

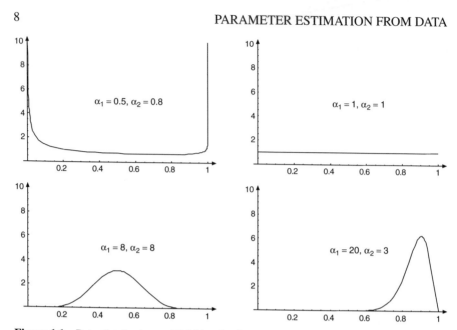

Figure 1.1   Beta distribution or Dirichlet distribution with $m = 2$. Different shapes can be obtained by varying the parameters α_1 and α_2. For instance, $\alpha_1 = \alpha_2 = 1$ corresponds to the uniform distribution. Likewise, $\alpha_1 = \alpha_2 \neq 1$ corresponds to a bell-shaped distribution centered on 0.5, with height and width controlled by $\alpha_1 + \alpha_2$.

Indeed, with a Dirichlet prior $\mathcal{D}_{\alpha q}(\boldsymbol{\pi})$ the negative log-posterior becomes

$$- \log P(\boldsymbol{\pi} \mid D) = - \sum_i [n_i + \alpha q_i - 1] \log \pi_i + \log Z + \log P(D). \qquad (1.14)$$

Z is the normalization constant of the Dirichlet distribution and does not depend on the parameters π_i. Thus the MAP optimization problem is very similar to the one previously solved, except that the counts n_i are replaced by $n_i + \alpha q_i - 1$. We immediately get the estimates

$$\pi_i^{\text{MAP}} = \frac{n_i + \alpha q_i - 1}{n + \alpha - m} \quad \text{for all } a_i \in a, \qquad (1.15)$$

provided this estimate is positive. In particular, the effect of the Dirichlet prior is equivalent to adding pseudocounts to the observed counts. When q is uniform, we say that the Dirichlet prior is *symmetric*. Notice that the uniform distribution over $\boldsymbol{\pi}$ is a special case of a symmetric Dirichlet prior, with $q_i = 1/\alpha = 1/m$. It is also clear from (1.14) that the posterior distribution $P(M \mid D)$ is a Dirichlet distribution $\mathcal{D}_{\beta r}$ with $\beta = n + \alpha$ and $r_i = (n_i + \alpha q_i)/(n + \alpha)$.

The expectation of the posterior is the vector r_i, which is slightly different from the MAP estimate (1.8). This suggests using an alternative estimate for π_i, the predictive

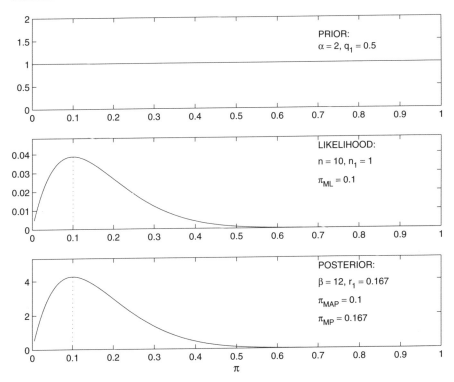

Figure 1.2 An illustration of parameter estimation with $m = 2$ and a 'flat' prior.

distribution or MP (mean posterior) estimate

$$\pi_i^{MP} = \frac{n_i + \alpha q_i}{n + \alpha}. \tag{1.16}$$

This is often a better choice. Here, in particular, the MP estimate minimizes the expected relative entropy distance (see Appendix A) $f(\pi^{MP}) = E(\mathcal{H}(\pi, \pi^{MP}))$, where the expectation is taken with respect to the posterior $P(\pi \mid D)$.

Figure 1.2 illustrates a simple example of parameter estimation for this model for an alphabet with only $m = 2$ symbols. In this example we have a 'flat' Beta prior ($\alpha = 0.5$) and our data consist of $n = 10$ data points in total, with only $n_1 = 1$ observations for the first of the two words. Since the prior is flat, the posterior Beta distribution has the same shape as the likelihood function. Figure 1.3 shows the same inference problem with the same data, but with a different prior – now the prior is 'stronger' and favors a parameter π that is around 0.5. In this case the likelihood and the posterior have different shapes and the posterior in effect lies 'between' the shapes of the likelihood and the prior.

The simple die model, where there is no 'memory' between the events, is also called a Markov model of order zero. Naturally if we could somehow model the sequential dependence between symbols, assuming such dependence exists, we could get a

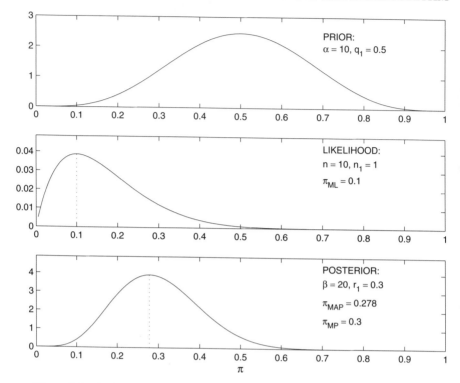

Figure 1.3   An illustration of parameter estimation with $m = 2$ and a 'nonflat' prior.

better model and make more accurate predictions on future data. A standard class of models for sequence data is the finite-state Markov chain, described in Appendix A. The application of Markov chains to the problem of inferring the importance of Web pages from the Web graph structure is described in the context of the PageRank algorithm in Chapter 5. A second application of Markov chains to the problem of predicting the page that a Web surfer will request next is treated in detail in Chapter 7.

1.3   Mixture Models and the Expectation Maximization Algorithm

One way to build complex probabilistic models out of simpler models is the concept of a mixture. In mixture models, a complex distribution P is parameterized as a linear convex combination of simpler or canonical distributions in the form

$$P = \sum_{l=1}^{K} \lambda_l P_l, \tag{1.17}$$

where the $\lambda_i \geq 0$ are called the mixture coefficients and satisfy $\sum_i \lambda_i = 1$. The distributions P_i are called the components of the mixture and have their own parameters (means, standard deviations, etc.). Mixture distributions provide a flexible way for modeling complex distributions, combining together simple building-blocks, such as Gaussian distributions. A review of mixture models can be found in Everitt (1984), Titterington *et al.* (1985), McLachlan and Peel (2000), and Hand *et al.* (2001). Mixture models are used, for instance, in clustering problems, where each component in the mixture corresponds to a different cluster from which the data can be generated and the mixture coefficients represent the frequency of each cluster.

To be more precise, imagine a data set $D = (d_1, \ldots, d_n)$ and an underlying mixture model with K components of the form

$$P(d_i \mid M) = \sum_{l=1}^{K} P(M_l) P(d_i \mid M_l) = \sum_{l=1}^{K} \lambda_l P(d_i \mid M_l), \qquad (1.18)$$

where $\lambda_l \geq 0$, $\sum_l \lambda_l = 1$, and M_l is the model for mixture l. Assuming that the data points are conditionally independent given the model, we have

$$P(D \mid M) = \prod_{i=i}^{n} P(d_i \mid M).$$

The Lagrangian associated with the log-likelihood and with the normalization constraints on the mixing coefficients is given by

$$\mathcal{L} = \sum_{i=1}^{n} \log \left(\sum_{l=1}^{K} \lambda_l P(d_i \mid M_l) \right) + \mu \left(1 - \sum_{l=1}^{K} \lambda_l \right) \qquad (1.19)$$

with the corresponding critical equation

$$\frac{\partial \mathcal{L}}{\partial \lambda_l} = \sum_{i=1}^{n} \frac{P(d_i \mid M_l)}{P(d_i)} - \mu = 0. \qquad (1.20)$$

Multiplying each critical equation by λ_l and summing over l immediately yields the value of the Lagrange multiplier $\mu = n$. Multiplying the critical equation again by $P(M_l) = \lambda_l$, and using Bayes' theorem in the form

$$P(M_l \mid d_i) = P(d_i \mid M_l) \frac{P(M_l)}{P(d_i)} \qquad (1.21)$$

yields

$$\lambda_l^{\text{ML}} = \frac{1}{n} \sum_{i=1}^{n} P(M_l \mid d_i). \qquad (1.22)$$

Thus, the ML estimate of the mixing coefficients for class l is the sample mean of the conditional probabilities that d_i comes from model l. Observe that we could have estimated these mixing coefficients using the MAP framework and, for example, a Dirichlet prior on the mixing coefficients.

Consider now that each model M_l has its own vector of parameters $\boldsymbol{\theta}_l$. Differentiating the Lagrangian with respect to each parameter of each component θ_{lj} gives

$$\frac{\partial \mathcal{L}}{\partial \theta_{lj}} = \sum_{i=1}^{n} \frac{\lambda_l}{P(d_i)} \frac{\partial P(d_i \mid M_l)}{\partial \theta_{lj}}. \tag{1.23}$$

Substituting Equation (1.21) into Equation (1.23) finally provides the critical equation

$$\sum_{i=1}^{n} P(M_l \mid d_i) \frac{\partial \log P(d_i \mid M_l)}{\partial \theta_{lj}} = 0 \tag{1.24}$$

for each l and j. The ML equations for estimating the parameters are weighted averages of the ML equations

$$\frac{\partial \log P(d_i \mid M_l)}{\partial \theta_{lj}} = 0 \tag{1.25}$$

arising from each point separately. As in Equation (1.22), the weights are the probabilities of d_i being generated from model l.

The ML Equations (1.22) and (1.24) can be used iteratively to search for ML estimates. This is a special case of a more general algorithm known as the Expectation Maximization (EM) algorithm (Dempster *et al.* 1977). In its most general form the EM algorithm is used for inference in the presence of missing data given a probabilistic model for both the observed and missing data. For mixture models, the missing data are considered to be a set of n labels for the n data points, where each label indicates which of the K components generated the corresponding data point.

The EM algorithm proceeds in an iterative manner from a starting point. The starting point could be, for example, a randomly chosen setting of the model parameters. Each subsequent iteration consists of an E step and M step. In the E step, the membership probabilities $p(M_l \mid d_i)$ of each data point are estimated for each mixture component. The M step is equivalent to K separate estimation problems with each data point contributing to the log-likelihood associated with each of the K components with a weight given by the estimated membership probabilities. Variations of the M step are possible depending, for instance, on whether the parameters θ_{lj} are estimated by gradient descent or by solving Equation (1.24) exactly.

A different flavor of this basic EM algorithm can be derived depending on whether the membership probabilities $P(M_l \mid d_i)$ are estimated in hard (binary) or soft (actual posterior probabilities) fashion during the E step. In a clustering context, the hard version of this algorithm is also known as 'K-means', which we discuss later in the section on clustering.

Another generalization occurs when we can specify priors on the parameters of the mixture model. In this case we can generalize EM to the MAP setting by using MAP equations for parameter estimates in the M step of the EM algorithm, rather than ML equations.

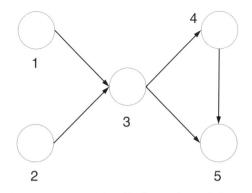

Figure 1.4 A simple Bayesian network with five nodes and five random variables and the global factorization property, $P(X_1, X_2, X_3, X_4, X_5) = P(X_1)P(X_2)P(X_3 \mid X_1, X_2)$ $P(X_4 \mid X_3)P(X_5 \mid X_3, X_4)$, associated with the Markov independence assumptions. For instance, conditioned on X_3, X_5 is independent of X_1 and X_2.

1.4 Graphical Models

Probabilistic modeling in complex domains leads to high-dimensional probability distributions over data, model parameters, and other hidden variables. These high-dimensional probability distributions are typically intractable to compute with. The theory of graphical models (Lauritzen 1996; Whittaker 1990) provides a principled graph-theoretic approach to modeling or approximating high-dimensional distributions in terms of simpler, more 'local', lower-dimensional distributions. Probabilistic graphical models can be developed using both directed and undirected graphs, each with different probabilistic semantics. Here, for simplicity, we concentrate on the directed approach which corresponds to the Bayesian or belief network class of models (Buntine 1996; Charniak 1991; Frey 1998; Heckerman 1997; Jensen 1996; Jordan 1999; Pearl 1988; Whittaker 1990).

1.4.1 Bayesian networks

A Bayesian network model M consists of a set of random variables $X_1, \ldots, X_n$ and an underlying directed acyclic graph (DAG) $G = (V, E)$, such that each random variable is uniquely associated with a vertex of the DAG. Thus, in what follows we will use variables and nodes interchangeably. The parameters θ of the model are the numbers that specify the local conditional probability distributions $P(X_i \mid X_{\text{pa}[i]})$, $1 \leqslant i \leqslant n$, where $X_{\text{pa}[i]}$ denotes the parents of node i in the graph (Figure 1.4). The fundamental property of the model M is that the global probability distribution must be equal to the product of the local conditional distributions

$$P(X_1, \ldots, X_n) = \prod_i P(X_i \mid X_{\text{pa}[i]}). \tag{1.26}$$

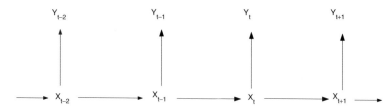

Figure 1.5 The underlying Bayesian network structure for both hidden Markov models and Kalman filter models. The two independence assumptions underlying both models are that (1) the current state X_t only depends on the past state X_{t-1} and (2) the current state Y_t only depends on the current state X_t. In the hidden Markov model the state variables X_t are discrete-valued, while in the Kalman filter model the state variables are continuous.

The local conditional probabilities can be specified in terms of lookup tables (for categorical variables). This is often impractical, due to the size of the tables, requiring in general $O(k^{p+1})$ values if all the variables take k values and have p parents. A number of more compact but also less general representations are often used, such as noisy OR models (Pearl 1988) or neural-network-style representations such as sigmoidal belief networks (Neal 1992). In these neural-network representations, the local conditional probabilities are defined by local connection weights and sigmoidal functions for the binary case, or normalized exponentials for the general multivariate case. Another useful representation is a decision tree approximation (Chickering *et al.* 1997), which will be discussed in more detail in Chapter 8 in the context of probabilistic models for recommender systems.

It is easy to see why the graph must be acyclic. This is because in general it is not possible to consistently define the joint probability of the variables in a cycle from the product of the local conditioning probabilities. That is, in general, the product $P(X_2 \mid X_1)P(X_3 \mid X_2)P(X_1 \mid X_3)$ does not consistently define a distribution on X_1, X_2, X_3.

The direction of the edges can represent causality or time course if this interpretation is natural, or the direction can be chosen more for convenience if an obvious causal ordering of the variables is not present.

The factorization in Equation (1.26) is a generalization of the factorization in simple Markov chain models, and it is equivalent to any one of a set of independence properties which generalize the independence properties of first-order Markov chains, where 'the future depends on the past only through the present' (see Appendix A). For instance, conditional on its parents, a variable X_i is independent of all other nodes, except for its descendants. Another equivalent statement is that, conditional on a set of nodes I, X_i is independent of X_j if and only if i and j are *d-separated*, that is, if there is no *d*-connecting path from i to j with respect to I (Charniak 1991; Pearl 1988).

A variety of well-known probabilistic models can be represented as Bayesian networks, including finite-state Markov models, mixture models, hidden Markov models, Kalman filter models, and so forth (Baldi and Brunak 2001; Bengio and Frasconi 1995; Ghahramani and Jordan 1997; Jordan *et al.* 1997; Smyth *et al.* 1997). Representing these models as Bayesian networks provides a unifying language and framework for what would otherwise appear to be rather different models. For example, both hidden Markov and Kalman filter models share the same underlying Bayesian network structure as depicted in Figure 1.5.

1.4.2 Belief propagation

A fundamental operation in Bayesian networks is the propagation of evidence, which consists of updating the probabilities of sets of variables conditioned on other variables whose values are observed or assumed. An example would be calculating $P(X_1 \mid x_4, x_{42})$, where the x-values represent specific values of the observed variables and where we could have a model involving (say) 50 variables in total – thus, there are 47 other variables whose values we are not interested in here and whose values must be averaged over (or 'marginalized out') if these values are unknown. This process is also known as belief propagation or inference. Belief propagation is NP-complete in the general case (Cooper 1990). But for singly connected graphs (no more than one path between any two nodes in the underlying undirected graph), propagation can be executed in time linear in n, the number of nodes, using a simple message-passing approach (Aji and McEliece 2000; Pearl 1988). In the general case, all known exact algorithms for multiply connected networks rely on the construction of an equivalent singly connected network, the junction tree. The junction tree is constructed by clustering the original variables according to the cliques of the corresponding triangulated moral graph as described in Pearl (1988), Lauritzen and Spiegelhalter (1988), and Shachter (1988) with refinements in Jensen et al. (1990), and Dechter (1999). A similar algorithm for the estimation of the most probable configuration of a set of variables, given observed or assumed values of other variables, is given in Dawid (1992). Shachter et al. (1994) show that all of the known exact inference algorithms are equivalent in some sense to the algorithms in Jensen et al. (1990) and Dawid (1992).

In practice, belief propagation is often tractable in sparsely connected graphs. Although it is known to be NP-complete in the general case, approximate propagation algorithms have been derived using a variety of methods including Monte Carlo methods such as Gibbs sampling (Gilks et al. 1994; York 1992), mean field methods, and variational methods (Ghahramani 1998; Jaakkola and Jordan 1997; Saul and Jordan 1996). An important observation, supported by both empirical evidence and results in coding theory, is that the simple message-passing algorithm of Pearl (1988) yields reasonably good approximations for certain classes of networks in the multiply connected case (see McEliece et al. 1997 for details). More recent results on belief propagation, and approximation methods for graphical models with cycles, can be found in Weiss (2000), Yedidia et al. (2000), Aji and McEliece (2000), Kask and Dechter (1999), McEliece and Yildirim (2002) and references therein.

Table 1.1   Four basic Bayesian network learning problems depending on whether the structure of the network is known in advance or not, and whether there are hidden (unobserved) variables or not.

	No hidden	Hidden
Known structure	easy	doable (via EM)
Unknown structure	doable	hard

1.4.3   Learning directed graphical models from data

There are several levels of learning in graphical models in general and Bayesian networks in particular. These range from learning the entire graph structure (the edges in the model) to learning the local conditional distributions when the structure is known. To a first approximation, four different situations can be considered, depending on whether the structure of the network is known or not, and whether the network contains unobserved data, such as hidden nodes (variables), which are completely unobserved in the model (Table 1.1).

When the structure is known and there are no hidden variables, the problem is a relatively straightforward statistical question of estimating probabilities from observed frequencies. The die example in this chapter is an example of this situation, and we discussed how ML, MAP, and MP ideas can be applied to this problem. At the other end of the spectrum, learning both the structure and parameters of a network that contains unobserved variables can be a very difficult task. Reasonable approaches exist in the two remaining intermediary cases. When the structure is known but contains hidden variables, algorithms such as EM can be applied, as we have described earlier for mixture models. Considerably more details and other pointers can be found in Buntine (1996), Heckerman (1997, 1998), and Jordan (1999).

When the structure is unknown, but no hidden variables are assumed, a variety of search algorithms can be formulated to search for the structure (and parameters) that optimize some particular objective function. Typically these algorithms operate by greedily searching the space of possible structures, adding and deleting edges to effect the greatest change in the objective function. When the complexity (number of parameters) of a model M is allowed to vary, using the likelihood as the objective function will inevitably lead to the model with the greatest number of parameters, since it is this model that can fit the training data D the best. In the case of searching through a space of Bayesian networks, this means that the highest likelihood network will always be the one with the maximal number of edges. While such a network may provide a good fit to the training data, it may in effect 'overfit' the data. Thus, it may not generalize well to new data and will often be outperformed by more parsimonious (sparse) networks. Rather than selecting the model that maximizes the likelihood $P(D \mid M)$, the Bayesian approach is to select the model with the maximum posterior probability given the data, $P(M \mid D)$, where we average over parameter uncertainty

$$P(M \mid D) \propto \int_{\theta} P(D \mid \theta, M) P(\theta \mid M) \, d\theta,$$

where $P(D \mid \theta, M)$ is the likelihood and $P(\theta \mid M)$ is the prior (here we explicitly include both θ and M). There is an implicit penalty in effect here. By averaging over parameter space, instead of picking the most likely parameter, more complex models will in effect be penalized for having a higher-dimensional parameter space and will only 'win' if their predictive power on the training data (measured via the likelihood) can overcome the inherent penalty that arises from having to integrate over a higher-dimensional space. Heckerman (1998) provides a full discussion of how this Bayesian estimation approach can be used to automatically construct Bayesian networks from data.

1.5 Classification

Classification consists of learning a mapping that can classify a set of measurements on an object, such as a d-dimensional vector of attributes, into one of a finite number K of classes or categories. This mapping is referred to as a *classifier* and it is typically learned from training data. Each training instance (or data point) consists of two parts: an input part x and an associated output class 'target' c, where $c \in \{1, 2, \ldots, K\}$. The classifier mapping can be thought of as a function of the inputs, $g(x)$, which maps any possible input x into an output class c.

For example, email filtering can be formulated as a classification problem. In this case the input x is a set of attributes defined on an email message, and $c = 1$ if and only if the message is spam. We can construct a training data set of pairs of email messages and labels:

$$D = \{[x_1, c_1], \ldots, [x_n, c_n]\}.$$

The class labels in the training data c can be obtained by having a human manually label each email message as spam or non-spam.

The goal of classification learning is to take a training data set D and estimate the parameters of a classification function $g(x)$. Typically we seek the best function g, from some set of candidate functions, that minimizes an empirical loss function, namely

$$\mathcal{E} = \sum_{i=1}^{n} l(c_i, g(x_i)),$$

where $l(c_i, g(x_i))$ is defined as the loss that is incurred when our predicted class label is $g(x_i)$, given input x_i, and the true class label is c_i. A widely used loss function for classification is the so-called 0-1 loss function, where $l(a, b)$ is zero if $a = b$ and one otherwise, or in other words we incur a loss of zero when our prediction matches the true class label and a loss of one otherwise. Other loss functions may be more appropriate in certain situations. For example, in email filtering we might want to assign a higher penalty or loss to classifying as spam an email message that is really non-spam, versus the other type of error, namely classifying an email message as non-spam that is really spam.

Generally speaking, there are two broad approaches to classification, the probabilistic approach and the discriminative or decision-boundary approach. In the probabilistic approach we can learn a probability distribution $P(x \mid c)$ for each of the K classes, as well as the marginal probability for each class $P(c)$. This can be done straightforwardly by dividing the training data D into K different subsets according to class labels, assuming some functional form for $P(x \mid c)$ for each class, and then using ML or Bayesian estimation techniques to estimate the parameters of $P(x \mid c)$ for each of the K classes. Once these are known we can then use Bayes' rule to calculate the posterior probability of each of the classes, given an input x:

$$P(c = k \mid x) = \frac{P(x \mid c = k)P(c = k)}{\sum_{j=1}^{K} P(x \mid c = j)P(c = j)}, \quad 1 \leqslant k \leqslant K. \qquad (1.27)$$

To make a class label prediction for a new input that is not in the training data x, we calculate $P(c = k \mid x)$ for each of the K classes. If we are using the 0-1 loss function, then the optimal decision is to choose the most likely class, i.e.

$$\hat{c} = \arg\max_{k}\{P(c = k \mid x)\}.$$

An example of this general approach is the so-called Naive Bayes classifier, to be discussed in more detail in Chapter 4 in the context of document classification, where the assumption is made that each of the individual attributes in x are conditionally independent given the class label

$$P(x \mid c = k) = \prod_{j=1}^{m} p(x_j \mid c = k), \quad 1 \leqslant k \leqslant K,$$

where x_j is the jth attribute and m is the total number of attributes in the input x.

A limitation of this general approach is that by modeling $P(x \mid c = k)$ directly, we may be doing much more work than is necessary to discriminate between the classes. For example, say the number of attributes is $m = 100$ but only two of these attributes carry any discriminatory information – the other 98 are irrelevant from the point of view of making a classification decision. A good classifier would ignore these 98 features. Yet the full probabilistic approach we have prescribed here will build a full 100-dimensional distribution model to solve this problem. Another way to state this is that by using Bayes' rule, we are solving the problem somewhat indirectly: we are using a generative model of the inputs and then 'inverting' this via Bayes' rule to get $P(c = k \mid x)$.

A probabilistic solution to this problem is to instead focus on learning the posterior (conditional) probabilities $P(c = k \mid x)$ *directly*, and to bypass Bayes' rule. Conceptually, this is somewhat more difficult to do than the previous approach, since the training data provide class labels but do not typically provide 'examples' of values of $P(c = k \mid x)$ directly. One well-known approach in this category is to assume a logistic functional form for $P(c \mid x)$,

$$P(c = 1 \mid x) = \frac{1}{1 + e^{-w^T x - w_0}},$$

where for simplicity we assume a two-class problem and where w is a weight vector of dimension d, w^T is the transpose of this vector, and $w^T x$ is the scalar inner product of w and x. Equivalently, we can represent this equation in 'log-odds' form

$$\log \frac{P(c = 1 \mid x)}{1 - P(c = 1 \mid x)} = w^T x + w_0,$$

where now the role of the weights in the vector w is clearer: a large positive (negative) weight w_j for attribute x_j means that, as x_j gets larger, the probability of class c_1 increases (decreases), assuming all other attribute values are fixed. Estimation of the weights or parameters of this logistic model from labeled data D can be carried out using iterative algorithms that maximize ML or Bayesian objective functions. Multilayer perceptrons, or neural networks, can also be interpreted as logistic models where multiple logistic functions are combined in various layers.

The other alternative to the probabilistic approach is to simply seek a function f that optimizes the relevant empirical loss function, with no direct reference to probability models. If x is a d-dimensional vector where each attribute is numerical, these models can often be interpreted as explicitly searching for (and defining) decision regions in the d-dimensional input space x. There are many non-probabilistic classification methods available, including perceptrons, support vector machines and kernel approaches, and classification trees. In Chapter 4, in the context of document classification, we will discuss one such method, support vector machines, in detail.

The advantage of this approach to classification is that it seeks to directly maximize the chosen loss function for the problem, and no more. In this sense, if the loss function is well defined, the direct approach can in a certain sense be optimal. However, in many cases the loss function is not known precisely ahead of time, or it may be desirable to have posterior class probabilities available for various reasons, such as for ranking or for passing probabilistic information on to another decision making algorithm. For example, if we are classifying documents, we might not want the classifier to make any decision on documents whose maximum class probability is considered too low (e.g. less than 0.9), but instead to pass such documents to a human decision-maker for closer scrutiny and a final decision.

Finally, it is important to note that our ultimate goal in building a classifier is to be able to do well on predicting the class labels of *new* items, not just the items in the training data. For example, consider two classifiers where the first one is very simple with only d parameters, one per attribute, and the second is very complex with 100 parameters per attribute. Further assume that the functional form of the second model includes the first one as a special case. Clearly the second model can in theory always do as well, if not better than, the first model, in terms of fitting to the training data. But the second model might be hopelessly overfitting the data and on new unseen data it might produce *less accurate* predictions than the simpler model.

In this sense, minimizing the empirical loss function on the training data is only a surrogate for what we would really like to minimize, namely the expected loss over all future unseen data points. Of course this future loss is impossible to know. Nonetheless there is a large body of work in machine learning and statistics on various methods that try to estimate how well we will do on future data using only the available training data D and that then use this information to construct classifiers that generalize more accurately. A full discussion of these techniques is well beyond the scope of this book, but an excellent treatment for the interested reader can be found in Hastie *et al.* (2001) and Devroye *et al.* (1996).

1.6 Clustering

Clustering is very similar to classification, except that we are not provided with class labels in the training data. For this reason classification is often referred to as *supervised* learning and clustering as *unsupervised* learning. Clustering essentially means that we are trying to find 'natural' classes that are suggested by the data. The problem is that this is a rather vague prescription and as a result there are many different ways to define what precisely is meant by a cluster, the quality of a particular clustering of the data, and algorithms to optimize a particular cluster 'quality' function or objective function given a set of data. As a consequence there are a vast number of different clustering algorithms available. In this section we briefly introduce two of the more well-known methodologies and refer the reader to other sources for more complete discussions (Hand *et al.* 2001; Hastie *et al.* 2001).

One of the simplest and best known clustering algorithms is the K-means algorithm. In a typical implementation the number of clusters is fixed *a priori* to some value K. K representative points or centers are initially chosen for each cluster more or less at random. These points are also called centroids or prototypes. Then at each step:

(i) each point in the data is assigned to the cluster associated with the closest representative;

(ii) after the assignment, new representative points are computed for instance by averaging or by taking the center of gravity of each computed cluster;

(iii) the two procedures above are repeated until the system converges or fluctuations remain small.

Notice that the K-means algorithm requires choosing the number of clusters, being able to compute a distance or similarity between points, and computing a representative for each cluster given its members.

From this general version of K-means one can define different algorithmic variations, depending on how the initial centroids are chosen, how symmetries are broken, whether points are assigned to clusters in a hard or soft way, and so forth. A good implementation ought to run the algorithm multiple times with different initial conditions, since any individual run may converge to a local extremum of the objective function.

We can interpret the K-means algorithm as a special case of the EM algorithm for mixture models described earlier in this chapter. More specifically, the description of K-means given above corresponds to a 'hard' version of EM for mixture models, where the membership probabilities are either zero or one, each point being assigned to only one cluster. It is well known that the center of gravity of a set of points minimizes its average quadratic distance to any fixed point. Therefore, in the case of a mixture of spherical Gaussians, the M step of the K-means algorithm described above maximizes the corresponding quadratic log-likelihood and provides an ML estimate for the center of each Gaussian component.

When the objective function corresponds to an underlying probabilistic mixture model (Everitt 1984; McLachlan and Peel 2000; Titterington *et al.* 1985), K-means is an approximation to the classical EM algorithm (Baldi and Brunak 2001; Dempster *et al.* 1977), and as such it typically converges toward a solution that is at least a local ML or maximum posterior solution. A classical case is when Euclidean distances are used in conjunction with a mixture of Gaussian models.

We can also use the EM algorithm for mixture models in its original form with probabilistic membership weights to perform clustering. This is sometimes referred to as *model-based* clustering. The general procedure is again to fix K in advance, select specific parametric models for each of the K clusters, and then use EM to learn the parameters of this mixture model. The resulting component models provide a generative probabilistic model for the data in each cluster. For example, if the data points are real-valued vectors in a d-dimensional space, then each component model can be a d-dimensional Gaussian or any other appropriate multivariate distribution. More generally, we can use mixtures of graphical models where each mixture component is encoded as a particular graphical model. A particularly simple and widely used graphical model is the so-called 'Naive Bayes' model, described in the previous section on classification, where the attributes are assumed to be conditionally independent given the component.

A useful feature of model-based clustering (and mixture models in general) is that the data need not be in vector form to define a clustering algorithm. For example, we can define mixtures of Markov chains to cluster sequences, where the underlying model assumes that the observed data are being generated by K different Markov chains and the goal is to learn these Markov models without knowing which sequence came from which model. In Chapter 7 we will describe this idea in more detail in an application involving clustering of Web users based on their Web page requests.

The main limitation of model-based clustering is the requirement to specify a parametric functional form for the probability distribution of each component – in some applications this may be difficult to do. However, if we are able to assume some functional form for each cluster, the gains can be substantial. For example, we can include a component in the model to act as a background cluster to 'absorb' data points that are outliers or that do not fit any of the other components very well. We can also estimate the posterior distribution $P(K \mid D)$ for different numbers of clusters K to see what value of K is most likely given the data (see Fraley and Raftery (2002) for a variety of such extensions). Another drawback of the standard mixture model approach to clustering, and indeed of many other approaches to clustering, is the assumption that each data point was generated by one and only one cluster. In clustering text documents on the Web, for example, we might want to allow documents to belong to multiple clusters, e.g. 'sports' and 'finance'. Recent proposals by Hofmann (2001) and Blei *et al.* (2002a) develop probabilistic models to specifically address this issue – we will discuss such models in Chapter 4 for modeling text documents and in Chapter 8 in the context of recommender algorithms.

1.7   Power-Law Distributions

1.7.1   Definition

Power-law distributions arise naturally in a variety of domains. In the case of the Web, power-law distributions appear to be a ubiquitous phenomenon and have been reported, for instance, for

- Web page sizes;

- Web page connectivity;

- Web connected components' size;

- Web page access statistics;

- Web browsing behavior.

Each one of these examples will be studied in great detail in the coming chapters.

Continuous or discrete power-law distributions are characterized by the fact that for large values of the independent variable x the distribution decays polynomially as $x^{-\gamma}$, with $\gamma > 1$. This is in sharp contrast with many other standard distributions, such as exponential, Poisson, Binomial, Gaussian, and Gamma, where the asymptotic decay of the distribution is exponential. This is intuitively summarized by saying that in a power-law distribution, rare events are not so rare.

Power-law distributions on ranked data are often called Zipf distributions, or Pareto–Zipf distributions. In language applications, for instance, the distribution of words ranked by their frequency in a large corpus of text is invariably a power-law distribution, also known as Zipf's Law. There is a vast literature on power-law distributions and the related log-normal distribution (e.g. Mihail and Papadimitriou 2002; Mitzenmacher 2002).

A discrete power-law distribution with coefficient $\gamma > 1$ is a distribution of the form

$$P(X = k) = Ck^{-\gamma} \qquad (1.28)$$

for $k = 1, 2, \ldots$. The corresponding density in the continuous case is

$$f(x) = Cx^{-\gamma} \tag{1.29}$$

for $x \in [1, +\infty)$. In many real life situations associated with power-law distributions, the distribution for small values of k or x may deviate from the expressions in Equations (1.28) and (1.29). Thus, a more flexible definition is to say that Equations (1.28) and (1.29) describe the behavior for sufficiently large values of x or k.

In a log–log plot, the signature of a power-law distribution is a line with a slope determined by the coefficient γ. This provides a simple means for testing power-law behavior and for estimating the exponent γ. An example of ML estimation of γ is provided in Chapter 7.

In both the discrete and continuous cases, moments of order $m \geqslant 0$ are finite if and only if $\gamma > m + 1$. In particular, the expectation is finite if and only if $\gamma > 2$, and the variance is finite if and only if $\gamma > 3$. In the discrete case,

$$E[X^m] = \sum_{k=1}^{\infty} Ck^{m-\gamma} = C\zeta(\gamma - m), \tag{1.30}$$

where ζ is Riemann's zeta function ($\zeta(s) = \sum_k 1/k^s$). In the continuous case,

$$E[X^m] = \int_1^{\infty} Cx^{m-\gamma}\, dx = \frac{C}{\gamma - m - 1}. \tag{1.31}$$

In particular, $C = \gamma - 1$, $E[X] = \gamma - 1/(\gamma - 2)$ and

$$\operatorname{var}[X] = \frac{\gamma - 1}{\gamma - 3} - \left(\frac{\gamma - 1}{\gamma - 2}\right)^2.$$

A simple consequence is that in a power-law distribution the average behavior is not the most frequent or the most typical, in sharp contrast with what is observed with, for instance, a Gaussian distribution. Furthermore, the total mass of the points to the right of the average is greater than the total mass of points to the left. Thus, in a random sample, the majority of points are to the right of the average.

Another interesting property of power-law distributions is that the cumulative tail, that is the area under the tail of the density, also has a power-law behavior with exponent $\gamma - 1$. This is obvious since $\int_x^{\infty} u^{-\gamma}\, du = x^{-\gamma+1}/(\gamma - 1)$.

There is also a natural connection between power-law distributions and log-normal distributions (see Appendix A). X has a log-normal density if $\log X$ has a normal density with mean μ and variance σ^2. This implies that the density of X is

$$f(x) = \frac{1}{\sqrt{2\pi}\sigma x} e^{-(\log x - \mu)^2/2\sigma^2} \tag{1.32}$$

and thus $\log f(x) = C - \log x - (\log x - \mu)^2/2\sigma^2$. In the range where σ is large compared with $|\log x - \mu|$, we have $\log f(x) \approx C - \log x$, which corresponds to a straight line. Very basic generative models can lead to log-normal or to power-law distributions depending on small variations and therefore it is not surprising that in a number of situations both models have been proposed to capture the same phenomena, such as the distributions of computer file sizes (see, for instance, Perline 1996 and Mitzenmacher 2002).

1.7.2   Scale-free properties (80/20 rule)

Power-law distributions result in scale-free properties in the sense that, for $1 \leqslant a \leqslant b$,

$$\frac{\sum_b^\infty C/i^\gamma}{\sum_a^\infty C/i^\gamma} \approx \frac{\int_b^\infty Cx^{-\gamma}\,dx}{\int_a^\infty Cx^{-\gamma}\,dx} = \left(\frac{b}{a}\right)^{1-\gamma}. \tag{1.33}$$

In other words, the ratio of the sums depends only on the ratio b/a and not on the absolute scale of a or b – things look the same at all length scales. This scale-free property is also known in more folkloristic terms as the 80/20 rule, which basically states that '80% of your business comes from 20% of your clients'. Of course it is not the particular value $80/20 = 4/1$ ratio that matters, since this may vary from one phenomenon to the next, it is the existence of such a ratio across different scales, or 'business sizes' (and business types). This property is also referred to as 'self-scaling' or 'self-similar', since the ratio of the areas associated with a and b remains constant, regardless of their actual values or frequencies. One important consequence of this behavior is that observations made at one scale can be used to infer behavior at other scales.

1.7.3   Applications to Languages: Zipf's and Heaps' Laws

As briefly mentioned above, a power-law distribution on ranked data is also called a Zipf Law. If, for instance, one plots the frequency rank of words contained in a large corpus of text data versus the number of occurrences or actual frequencies, one invariably obtains a power-law distribution, with exponent close to one (but see Gelbukh and Sidorov 2001). An example is plotted in Figure 1.6. Under simple assumptions, this is also observed in random text generated by a Markov model (Li 1992).

A second related law for textual data is Heaps' Law. Heaps' Law is the observation that asymptotically, a corpus of text containing N words typically contains on the order of CN^β distinct words, with $0 < \beta < 1$. Empirically, β was found to be between 0.4 and 0.6 by some authors (Araújo et al. 1997; Moura et al. 1997). However, the value of the exponent varies somewhat with the language, the type of data set, and so forth. Using the Web data set from the WEB $\rightarrow$ KB (World Wide Knowledge Base) project at Carnegie Mellon University, consisting of several thousand Web pages collected from Computer Science departments of various universities, we find a characteristic exponent equal to 0.76 (Figure 1.7).

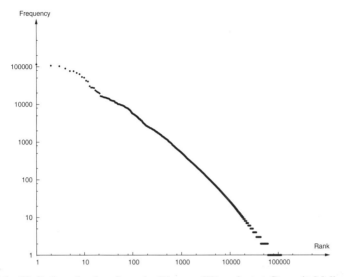

Figure 1.6 Zipf's Law for data from the WEB → KB project at Carnegie Mellon University consisting of 8282 Web pages collected from Computer Science departments of various universities. Tokenization was carried out by removing all HTML tags and by replacing each non-alphanumeric character by a space. Log–log plot of word ranks versus frequencies.

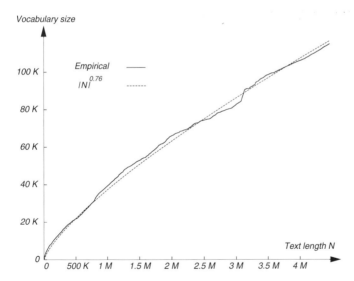

Figure 1.7 Heaps' Law for data from WEB → KB project at Carnegie Mellon University, consisting of 8282 Web pages collected from Computer Science departments of various universities. Tokenization was carried out by removing all HTML tags and by replacing each non-alphanumeric character by a space. Plot of total number of words (text length) versus the number of distinct words (vocabulary size). The characteristic exponent for this Web data set is 0.76.

1.7.4   Origin of power-law distributions and Fermi's model

It is in general easy to detect power-law behavior in data: after all a power-law curve yields a straight line in a log–log plot. Explaining the origin of the power-law behavior, however, is a different story. Because power laws are so ubiquitous, several attempts have been made to try to reduce their origin to a single, or perhaps a few, 'canonical' models. This is not completely unreasonable in the same way that the Gaussian distribution is often explained by some kind of central limit theorem, i.e. it is the distribution that emerges from the sum of a large number of more or less independent small effects.

As in the case of the Gaussian, however, it does not seem likely that a single model can account for all occurrences of power-law distributions, and certainly not with any precision, since it is easy to introduce small perturbations in a power-law distribution in a large number of different ways. In Chapter 4, we will look at a simple generative model that tries to provide an explanation for the power law observed in the connectivity of the Web. Here we describe a very simple model that yields power-law distributions of objects submitted to two exponential laws. The model was originally proposed by Fermi (1949) while modeling particles and cosmic radiation.

Imagine that particles are being produced continuously at a constant rate and that particles 'age', so that the current age distribution is an exponential of the form $\lambda e^{-\lambda t}$. As it ages, a particle gains energy exponentially, so that the energy of a particle of age t is $w(t) = Ce^{\alpha t}$. By combining the distribution of particle ages with the energy associated with each age, it is easy to find the distribution π of energies at time t. In particular,

$$P(w_0 \leqslant w \leqslant w_0 + dw) = P(w_0 \leqslant Ce^{\alpha t} \leqslant w_0 + dw)$$
$$= \int_{w_0}^{w_0 + dw} \pi(w) \, dw = \pi(w_0) \, dw. \qquad (1.34)$$

This is equivalent to

$$P\left(\frac{1}{\alpha} \log \frac{w_0}{C} \leqslant t \leqslant \frac{1}{\alpha} \log \frac{w_0 + dw}{C}\right) = \left(\frac{w_0}{C}\right)^{-\lambda/\alpha} - \left(\frac{w_0 + dw}{C}\right)^{-\lambda/\alpha} \qquad (1.35)$$

by using the exponential distribution of t. This finally yields the density

$$\pi(w) = \frac{\lambda}{\alpha} \left(\frac{w}{C}\right)^{-\lambda/\alpha} \frac{1}{w}, \qquad (1.36)$$

with power-law exponent $\gamma = 1 + \lambda/\alpha$. Thus a power-law density results naturally from the competition of two exponential phenomena: a positive exponential controlling the growth of the energy and a negative exponential controlling the age distribution.

A different generative model of power-law behavior based on the notion of preferential attachment will be given in Chapter 3. A third and more indirect model based on information theory is due to Mandelbrot (see Mandelbrot 1977 and references therein; Carlson and Doyle 1999; Zhu *et al.* 2001) and is given in one of the exercises below.

1.8   Exercises

Exercise 1.1. Prove that Equations (1.15) and (1.16) are correct.

Exercise 1.2. In the die model, compute the posterior when the prior is a *mixture* of Dirichlet distributions. Can you infer a general principle?

Exercise 1.3. The die model with a single Dirichlet prior is simple enough that one can proceed with 'higher levels' of Bayesian inference, beyond the computation of the posterior or its mode. For example, explicitly compute the normalizing factor (also called the 'evidence') $P(D)$. Use the evidence to derive a strategy for optimizing the parameters of the Dirichlet prior.

Exercise 1.4. Consider a mixture model with a Dirichlet prior on the mixing coefficients. Study the MAP and MP estimates for the mixing coefficients.

Exercise 1.5. Compute the differential entropy (see Appendix A) of the continuous power-law distribution $f(x) = Cx^{-\gamma}$, $x \geqslant 1$.

Exercise 1.6. Simulate a first-order Markov language (by selecting an alphabet containing the 'space' symbol, defining a die probability vector over the alphabet, tossing the die many times, and computing word statistics) whether Zipf's and Heaps' Laws are true or not in this case. Experiment with the size of the alphabet, the letter probabilities, the order of the Markov model, and the length of the generated sequence.

Exercise 1.7. Find a large text data set (for instance from WEB $\rightarrow$ KB) and compute the corresponding Zipf and Heaps curves, e.g. word rank versus word frequency and text size versus vocabulary, or the respective logarithms.

Exercise 1.8. Study the connection between Zipf's and Heaps' Laws. In particular, is Heaps' Law a consequence of Zipf's Law?

Exercise 1.9. Collect statistics about the size of cities, or computer files, and determine whether there is an underlying power-law distribution. Provide an explanation for your observation.

Exercise 1.10. Mandelbrot's information theoretic generative model for power-law distributions considers random text generated using a set of M different words with associated probabilities $p_1, \ldots, p_M$, ranked in decreasing order, and associated 'transmission costs' $C_1, \ldots, C_M$. A possible cost function is to take C_i proportional to $\log i$. Why?
Now suppose that the goal is to design the probabilities p_i in order to optimize the average amount of information per unit transmission cost. The average cost per word is $E[C] = \sum_i p_i C_i$ and the average information, or entropy (see Appendix A), per word is $\mathcal{H}(\boldsymbol{p}) = -\sum_i p_i \log p_i$. Compute the values of the p_i to maximize the quantity $\mathcal{H}/C$ and show that, with logarithmic costs, it corresponds to a power law.

Exercise 1.11. Show that the product of two independent log-normal random variables is also log-normal. Use this simple fact to derive a generative model for log-normal distributions.

2

Basic WWW Technologies

Providing a complete description of the technology by which the Web is implemented is beyond the scope of this book. However, a basic understanding of Web technology is essential for the rest of this book and for implementing Web algorithms.

In simple terms, the World Wide Web is implemented by combining three essential components: resources, resource identifiers, and transfer protocols. Resources are defined in a very general way as conceptual mappings to concrete or abstract entities (Berners-Lee *et al.* 1998). In the most common case these entities correspond to, but are not limited to, Web pages and other kinds of files, chosen from a large assortment of proprietary and public domain formats. Note the importance of defining resources in a conceptual rather than physical way. For example, we may say that the site of the World Wide Web Consortium (W3C) is a resource. This conceptual entity does not change over time, at least in the short term. The concrete entity, on the other hand, could include sequences of bits that represent Web pages, for example stored as files in some computer disk. This level of detail, however, is unnecessary. Moreover, the actual files may change at any time without affecting the conceptual entity.

Resource identifiers are string of characters representing generalized addresses that may also implicitly contain precise instructions for accessing the identified resource. For example, the string `http://www.w3.org/` is used to identify the W3C website. Protocols are essentially conventions that regulate the communication between two connected entities such as a Web user agent (e.g. a browser) and a server. The key property of the Web is that any resource may contain several references to other resources (encoded as identifiers), and it is these hyperlinks that give rise to an interconnected network. Protocols provide a convenient way for an agent to navigate this network by following the referenced resources.

As pointed out in the Preface, people often tend to confuse the Internet and the Web. On one hand, because the common use of Web services involves only a subset of Internet protocols at the application level (see below), the Internet encompasses something larger than just the Web. On the other hand, according to the above definition, several 'webs' may be conceived, not just those that use Internet protocols for fast communication. For example, the whole body of scientific literature forms a web through the mechanism of citations, but the access protocol, in general, still implies a

Modeling the Internet and the Web   P. Baldi, P. Frasconi and P. Smyth
© 2003 P. Baldi, P. Frasconi and P. Smyth   ISBN: 0-470-84906-1

visit to the library, whether an electronic or a 'brick and mortar' one. Hence, in turn, under an extended set of protocols the Web encompasses more than just the Internet.

2.1 Web Documents

A vast proportion of existing Web resources are HTML documents. HTML (Hyper-Text Markup Language) is a non-proprietary format that provides basic tagging support for structuring and laying out text and images as we commonly experience them when using a Web browser. Briefly, the language allows one to specify the formatting of headings, paragraphs, lists, tables, and so forth, to create fillable forms for electronic transactions, to include procedural scripts, and to combine text with several types of multimedia objects such as video or audio clips. The original language was developed by Tim Berners-Lee at CERN but was extended in several ways during the 1990s until the present version 4.01 (Raggett *et al.* 1999).

2.1.1 SGML and HTML

HTML is actually one application of the Standard Generalized Markup Language (SGML), an international standard that defines how descriptive markup should be embedded in a document (ISO 1986). Application in this context has a special meaning: SGML is in fact a *metalanguage*, i.e. a formalism that permits us to specify languages, and HTML is a language that can be specified in SGML. In general, markup refers to any extra information that is distinguished from the document contents and that characterizes the structure and/or the style of the document. *Descriptive* markup is used to describe different *elements* of a document, like chapters, sections, and paragraphs in a book. For example, LaTeX, the typesetting system that was employed to prepare this book, extensively uses descriptive elements for defining the structure of a document, although it is not derived from SGML. *Procedural* markup, on the other hand, is a lower-level specification of the sequence of processing steps that are necessary to render the document in its final form. Common WYSIWYG ('what you see is what you get') applications for document editing tend to stress procedural markup, i.e. specification of how to display the elements, while the main focus of descriptive markup in SGML is about the structure of the document, i.e. specification of what the elements are. More details on SGML can be found in Sperberg-McQueen and Burnard (2002) and Raggett *et al.* (1999). The reader should also be aware that, in 1996, a subset of SGML, called the eXtensible Markup Language (XML), was developed to simplify the distribution and the processing of SGML documents on the Web. XML is released as a W3C recommendation.

As an SGML application, HTML needs to specify a valid character set, and the syntax and semantics of tags. In SGML, valid *element types* and their relations are formally specified in a special set of declarations called the Document Type Definition (DTD). As a loose analogy, the relationship between an SGML application and its DTD is similar to the relationship between a programming language (like C or Java) and its formal grammar. In other words, the DTD of HTML specifies the metalevel of a generic HTML document. For example, the following declaration in the DTD of HTML 4.01 specifies that an unordered list is a valid element and that it should consist of one or more list items,

```
<!ELEMENT UL - - (LI)+          -- unordered list -->
```

where ELEMENT is a keyword that introduces a new element type, UL is the new element type, which represents an unordered list being defined and LI is the element type that represents a list item. Here, metasymbols (), for grouping and +, for repetition, are similar to the corresponding metasymbols in regular expressions. In particular, + means that the preceding symbol (i.e. LI) should appear one or more times. The declaration of an element in a document is generally obtained by enclosing some content within a pair of matching *tags*. A start tag is realized by the *element name* enclosed in angle brackets, as in <LI>. The end tag is distinguished by adding a slash before the element name, as in </LI>. The semantics in this case says that any text enclosed between <LI> and </LI> should be treated as a list item. Note that, besides enriching the document with structure, markup may affect the appearance of the document in the window of a visual browser (e.g. the text of a list item should be rendered with a certain right indentation and preceded by a bullet).

Some elements may be enriched with attributes. For example, the following is a portion of the declaration of the element IMG, used to embed inline images in HTML documents:

```
<!ELEMENT IMG - O EMPTY      -- Embedded image -->
<!ATTLIST IMG
    src     %URI;  #REQUIRED -- URI of image to embed --
    alt     %Text; #REQUIRED -- short description --
    ...
```

The attribute src is a required resource identifier (see Section 2.2 below) that allows us to access the file containing the image to be displayed. In the simplest case this is just a local filename. The attribute alt is text that should be used in place of the image, for example when the document is rendered in a terminal-based browser. An actual image would be then inserted as

```
<img src="Web.png" alt="a Web">
```

This is an example of an element that must not use a closing tag (as implied by the keywords O and EMPTY in the element declaration).

2.1.2   General structure of an HTML document

HTML 4.01 is a language with a fairly large set of elements which, for lack of space, cannot be described here in complete detail. The interested reader is urged to consult the W3C recommendation of 24 December 1999 (Raggett *et al.* 1999) for a complete specification. Here we limit our discussion to the analysis of the general structure of a document and we present a rather simple example.

A valid HTML document is comprised of the following three parts.

Version information. This is a declarative section that specifies which DTD is used in the document (see lines 1–2 in Figure 2.1). The W3C recommends three possible DTDs: strict, transitional, or frameset. Strict contains all the elements that are not deprecated (i.e. outdated by new elements) and are not related to frames, transitional allows deprecated elements, and frameset contains all of the above plus frames.

Header. This is a declarative section that contains *metadata*, i.e. data about the document. This section is enclosed within <head> and </head> in Figure 2.1 (lines 4–8). In our example, line 5 specifies the character set (ISO-8859-1, aka Latin-1) and line 6 describes the authors. Metadata in this case are specified by setting attributes of the element meta. The title is instead an element (line 7).

Body. This is the actual document content. This element is enclosed within <body> and </body> in Figure 2.1 (lines 9–18).

In Figure 2.1 we see some other examples of HTML elements. <h1> and <h2> are used for headings, <ul> consists of an unordered list, whose (bulleted) items are tagged by <li>, and so on.

2.1.3   Links

The <A> element (see, for example, lines 15 and 16) is used in HTML to implement *links*, an essential feature of hypertexts. As we will see later in this book, a link is rather similar to an edge in a directed graph. It connects two objects referred to as *anchors*. In our example, the source anchors are the elements of the document delimited by <A> and </A>. In the World Wide Web, the target anchor is a resource that may be physically stored in the same server as the source document, or may be in a different server, possibly located in another country or continent (see Figure 2.2). The target resource is identified by means of a special string called the Uniform Resource Identifier (URI). If the attribute href of the element <A> is set, its value is the URI of the target anchor. In our example there are two URIs, namely the two strings http://www.w3.org/ and toc.html.

Note that links can be implemented in several other ways. For HTML documents, source anchors can be associated with images, forms, or active elements implemented using scripting languages or applets. But linking is not limited to HTML documents, since many document formats now support mechanisms for encoding URIs.

In 2000, HTML was refomulated as an XML application. The resulting language was called XHTML and released as a W3C recommendation. Details can be found at http://www.w3.org/TR/html/.

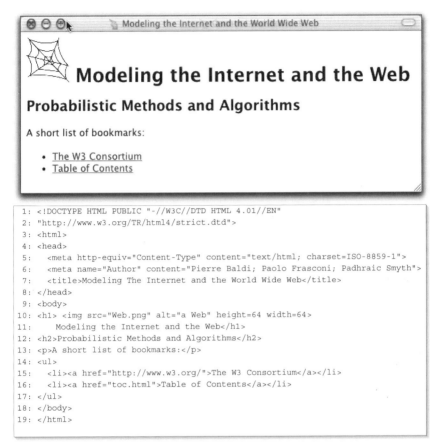

```
 1: <!DOCTYPE HTML PUBLIC "-//W3C//DTD HTML 4.01//EN"
 2: "http://www.w3.org/TR/html4/strict.dtd">
 3: <html>
 4: <head>
 5:   <meta http-equiv="Content-Type" content="text/html; charset=ISO-8859-1">
 6:   <meta name="Author" content="Pierre Baldi; Paolo Frasconi; Padhraic Smyth">
 7:   <title>Modeling The Internet and the World Wide Web</title>
 8: </head>
 9: <body>
10: <h1> <img src="Web.png" alt="a Web" height=64 width=64>
11:     Modeling the Internet and the Web</h1>
12: <h2>Probabilistic Methods and Algorithms</h2>
13: <p>A short list of bookmarks:</p>
14: <ul>
15:   <li><a href="http://www.w3.org/">The W3 Consortium</a></li>
16:   <li><a href="toc.html">Table of Contents</a></li>
17: </ul>
18: </body>
19: </html>
```

Figure 2.1   Simple HTML document (bottom) demonstrating the use of
a few common tags. The rendered document is shown on the top.

2.2   Resource Identifiers: URI, URL, and URN

Resources are referred to by means of strings with a controlled syntax and semantics, called the resource identifiers. Uniform Resource Identifiers (URI) are the most general set of identifiers. Uniformity refers to the generality of these identifiers and to the property that the interpretation of each string should be unambiguous, regardless of the context. The precise definition of URIs has evolved from the informational document released as RFC 1630 (Berners-Lee 1994) to the draft standard RFC 2396 (Berners-Lee *et al.* 1998) that describes in detail the general grammar of URIs. Since the early formalizations, the design of URIs has focused on the concepts of extensibility and completeness. Extensibility means that the specification should leave room for adding new naming schemes in the future, perhaps for types of services that have not been invented yet. Completeness means that it must be possible to encode every possible naming scheme. This need is justified by the definition of

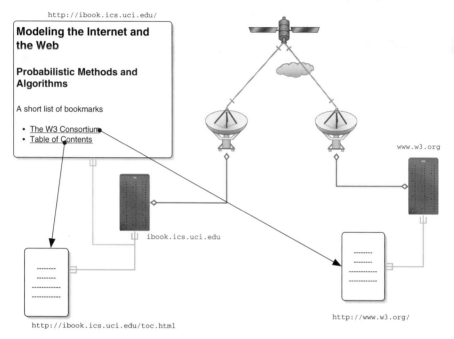

Figure 2.2   Anchors and Web links.

resources as conceptual mappings to entities. This definition is extremely general and not limited to electronic documents that can be accessed in a networked environment. Everything that has an identity can be considered as a resource, including persons or abstract entities. Indeed, URIs have been defined that identify telephone or fax numbers (e.g. `fax:+39-055-2222222`), short message services in GSM mobile phones (e.g. `sms:+393301234567`), or television broadcasts (e.g. `tv:rai.it`).

URIs actually include two overlapping subsets of identifiers: Uniform Resource Locators (URLs) and Uniform Resource Names (URNs). URLs are identifiers that explicitly encode details about the algorithm used to access the resource. For example, the URI `ftp://ftp.ietf.org/rfc/rfc2396.txt` is a URL because it specifies that the document containing the text of RFC 2396 can be accessed via file transfer protocol from the ftp server of the Internet Engineering Task Force. URNs, on the other hand, are those URIs that must persist and remain unique even if the resource becomes unavailable.

In order to maximize the ability of transcribing them, URIs are strings over a minimal subset of the 7-bit ASCII character set. This allows us, for example, to advertise URIs via traditional media such as radio or printed newspapers. A small number of characters is reserved for special purposes. For example, the percent sign (`%`) is used as an escape character forcing the next two characters to be interpreted as hex digits. In this way a space can be represented in a URI by the substring `%20` and a tilde by %7E. Similarly, the question mark ? is reserved in URIs that represent queries to delimit the identifier of a queryable object and the query itself.

An *absolute* URI consists of two portions separated by a colon: a scheme specifier followed by a second portion whose syntax and semantics depend on the particular scheme. In BNF (Backus–Naur Form), the general syntax of absolute URIs is expressed as

⟨*absoluteURI*⟩   ::= ⟨*scheme*⟩ : (⟨*hierarchicalPart*⟩ | ⟨*opaquePart*⟩)

As an example, three common schemes for identifying Web resources are http, https and ftp. Incidentally, the names of these schemes correspond to the TCP/IP application level protocols (see Section 2.3.1) that are used to access the resource. In these three cases, the second part of the URI is hierarchical and consists of a double slash // followed by a so-called *authority* component (e.g. a host name) and optionally by a *path* and/or a *query*. For example, the URI http://bsd.slashdot.org/article.pl?sid= 02/08/16/0041250 contains all three components: the authority is bsd. slashdot.org, the absolute path is /article.pl, and the query is the substring following the question mark. More generally, the authority can be omitted and we have

⟨*hierarchicalPart*⟩   ::= (⟨*networkPath*⟩ | ⟨*absolutePath*⟩) [? ⟨*query*⟩]

⟨*networkPath*⟩   ::= / / ⟨*authority*⟩ [⟨*absolutePath*⟩]

⟨*absolutePath*⟩   ::= / ⟨*pathSegments*⟩

The authority can be either a server with an IP, or a reference to a registry of a naming authority:

⟨*authority*⟩   ::= ⟨*server*⟩ | ⟨*registryName*⟩

For Internet URIs we are interested in the first case. While in the simplest case a server is specified by its full hostname, in general we must allow the possibility of password-protected sites and servers that do not listen to the default TCP/IP port for a given protocol. Thus

⟨*server*⟩   ::= [[⟨*userinfo*⟩ @] ⟨*hostport*⟩]

⟨*hostport*⟩   ::= ⟨*host*⟩ [: ⟨*port*⟩]

⟨*host*⟩   ::= ⟨*hostname*⟩ | ⟨*IPv4address*⟩

For example, the string more:less@www.gnacz.org:8081 specifies that the resource can be accessed on the nonstandard port 8081 of the host www.gnacz.org and that the access requires logging in as user more with password less. The host can also be specified using the numerical IP address (see Section 2.3.1):

⟨*IPv4address*⟩   ::= 1*⟨*digit*⟩ . 1*⟨*digit*⟩ . 1*⟨*digit*⟩ . 1*⟨*digit*⟩

By website we generally mean the collection of resources that share the same authority in their URI. Anchors in actual HTML pages may also contain *relative* URIs. For example, the absolute URI `http://www.gnacz.org/foo/bar.html` may be simply referenced to as `/foo/bar.html` in any source anchor contained in a page belonging to the site `www.gnacz.org`.

⟨*URI-reference*⟩   ::= [⟨*absoluteURI*⟩ | ⟨*relativeURI*⟩] [# ⟨*fragment*⟩]

⟨*relativeURI*⟩   ::= (⟨*networkPath*⟩ | ⟨*absolutePath*⟩
                        | ⟨*relativePath*⟩) [? ⟨*query*⟩]

The full grammar specification for URIs can be found in Berners-Lee *et al.* (1998).

2.3   Protocols

Protocols describe how messages are encoded and specify how messages should be exchanged. Protocols that regulate the functioning of computer networks are layered in a hierarchical fashion. The lower layer in the hierarchy is typically related to the physical communication mechanisms (e.g. it may be concerned with optical fibers or wireless communication). Higher levels are related to the functioning of specific applications such as email or file transfer. A hierarchical organization of protocols allows one to employ different levels of abstraction when describing and implementing the software and the hardware components of a networked environment. For example, a host in the Internet is characterized by a 32 bit address, independently of whether the physical connection goes through a home DSL cable, through an office Ethernet LAN, or through an airport wireless network. Similarly, email protocols are a lingua franca spoken by clients and servers. Sending email just involves connecting to a server and using this language. All of the details about how the information is actually transmitted to the recipient are hidden by the hierarchical mechanism of encapsulation by which higher level messages are embedded into a format understood by the lower level protocols.

2.3.1   Reference models and TCP/IP

A rather famous model of protocol hierarchy was proposed in the early 1980s by the International Organization for Standardization (ISO) and was called the Open System Interconnect (OSI), shortly known as the ISO/OSI reference model (Day 1995; Day and Zimmerman 1983). It consists of the following seven layers.

Physical: concerned with the transmission of electrical or optical (possibly also acoustic) signals.

Data link: provides error control and divides data into frames.

Network: provides routing of packets from source to destination.

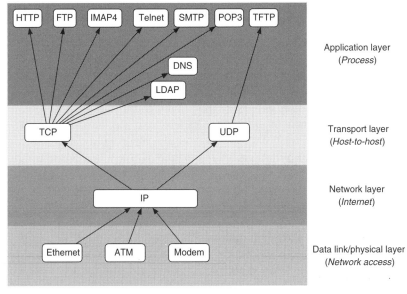

Figure 2.3 Layers in the TCP/IP reference model.

Transport: provides end-to-end reliability (guaranteeing all packets reach destination).

Session: primitive functions that coordinate the dialoging of applications.

Presentation: concerned about the transfer syntax used by applications.

Application: applications such as file transfer, email, browsing, etc.

In a sense, the value of the ISO model is more theoretical than practical, since different protocols are actually implemented in the Internet. The two most important ones are the Transmission Control Protocol (TCP) and the Internet Protocol (IP). While both TCP and IP are acronyms for two specific protocols, the term TCP/IP commonly refers to a large suite of different protocols that have been implemented since the late 1960s, forming one of the core technologies underpinning the Internet.

2.3.2   The domain name system

Each physical device actively connected to the Internet must have a unique address that, according to the IPv4 protocol, is a 32 bit address commonly represented as a string of four decimal numbers separated by dots, as in 128.200.84.11 (see Deering and Hinden (1998) for a specification of the more recent IPv6). IP addresses can have one or more domain names attached. For example, 128.200.222.100 corresponds to the domain name `ftp.uci.edu`. Names are obviously more understandable to humans than numbers. However, whenever a user agent needs to open a connection to an Internet host for which the name is known (e.g. `ftp.uci.edu`), it first must determine the IP number. The mapping from names to numbers, or *resolution*, is realized by the domain name service (DNS).

Names are organized hierarchically. The rightmost component of a name corresponds to the highest level in the hierarchy and is known as the top-level domain (TLD). Some historical TLDs include `.com`, `.edu`, `.org`, and `.gov`. Other TLDs are, for example, associated with ISO 3166 country codes, such as `.de`, `.fr`, `.it`, or `.uk`. The management of the domain name system and the root server system (that handles resolution for TLDs), as well as the allocation of the IP address space, is currently the responsibility of the Internet Corporation for Assigned Names and Numbers (ICANN). Within each TLD, several subdomains are registered (e.g. under the management of a local naming authority) and the process can be repeated recursively.

Resolution is carried out by consulting a hierarchical distributed database. We illustrate the mechanism by using an example. Suppose the user agent running on `lnx32.abcd.net` needs to translate the name `ftp.uni-gnat.edu`. The agent may first query a local database that holds the most commonly used hostnames (this is normally stored in the file `/etc/hosts` in Unix-like systems). If the search is unsuccessful, or no local database is maintained, a *recursive* query is typically sent to a local DNS server, say `dns.abcd.net`. This server may be able to provide an answer by consulting its cache. This happens if the same query has been answered before and the result has been stored in the server's cache. If not, then `dns.abcd.net` will need to consult other DNS servers, becoming itself a client. For example, it might be configured to send a non-recursive query to the DNS server of its Internet service provider, say `dns.abcdprov.com`. Having received a non-recursive query, if `dns.abcdprov.com` cannot provide an answer, it does not forward the query to another server but rather it returns to `dns.abcd.net` a referral to another DNS server that might be able to answer. Suppose it returns a referral to a root server. `dns.abcd.net` will now repeat the query to the root server, which replies with a referral to the DNS server of the `.edu` domain. The iteration continues and `dns.abcd.net` sends the query to the server for the `.edu`, which suggests that `dns.uni-gnat.edu` is the authority for the zone to which `ftp.uni-gnat.edu` belongs. Eventually, `dns.abcd.net` receives the answer from `dns.uni-gnat.edu` and returns it to the original client `lnx32.abcd.net`.

2.3.3 The Hypertext Transfer Protocol

The World Wide Web is largely based on the Hypertext Transfer Protocol (HTTP), specifying the format of messages exchanged by a client, called in this case the *user agent* (it could be a user browser or a crawler), and a server, i.e. the application program that accepts connections and provides the requested services. Communication is established via a TCP connection to the default port (number 80) of the server machine, although the URI can specify a different port number. In the simplest

Table 2.1   Methods in HTTP 1.1.

Method name	Description
GET	Retrieve an entity identified by a request URI (e.g. fetch a Web page or a file from the server)
HEAD	Identical to GET except that the server must not return a message body in response (i.e. returns only the header)
POST	Append the enclosed entity. The supplied URI will handle the entity (e.g. used to post a message to a newsgroup or to append a record to a database)
PUT	Store an enclosed entity under the supplied URI (e.g. store a Web page or a file with the server in order to create or modify a site)
DELETE	Delete the resource identified by the request URI
TRACE	Server echocs the request for debugging purposes
OPTIONS	Server returns a list of supported methods (e.g. to query the server's capabilities)
CONNECT	Reserved for dynamic proxies

case the user agent directly connects to the server but, more generally, intermediary agents – like proxies, tunnels, or gateways – can actually be present.

Messages are exchanged in ASCII format and pertain to one of several possible *methods*, as detailed in Table 2.1. HTTP is essentially a request/response protocol. A method essentially corresponds to a request from the user agent to the server. The server responds to the request with a response message.

The most common method is GET, which allows the user agent to fetch an HTML Web page, or another document such as an image or a file, from the server. Its usage is illustrated in Figure 2.4. In this example, the response of the server is simply the HTML code of the Web page associated with the requested URI (`http://www.ics.uci.edu/`). In HTTP 1.0 the communication would be released upon completion of the request, since HTTP 1.1 (Fielding *et al.* 1999; Krishnamurthy *et al.* 1999) connections are persistent, i.e. the link remains active after a request. Persistence offers considerable savings in overhead when several requests are sent to the server, for instance while downloading a page with several images.

While some entities are simply documents stored in the server's file system, it is also possible to have requests for entities that are actively generated by the server. This is normally the case for pages served as a result of user queries: the server's software could query a database management system to retrieve the requested information, create on-the-fly HTML code for display, and return the resulting ASCII stream as an HTTP response. The collection of documents 'behind' query forms is generally referred to as the hidden (or invisible) Web (Bergman 2000).

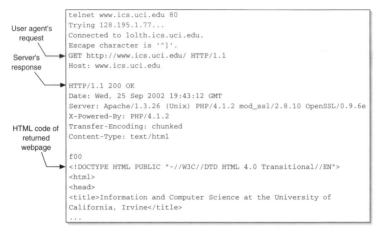

```
telnet www.ics.uci.edu 80
Trying 128.195.1.77...
Connected to lolth.ics.uci.edu.
Escape character is '^]'.
GET http://www.ics.uci.edu/ HTTP/1.1
Host: www.ics.uci.edu

HTTP/1.1 200 OK
Date: Wed, 25 Sep 2002 19:43:12 GMT
Server: Apache/1.3.26 (Unix) PHP/4.1.2 mod_ssl/2.8.10 OpenSSL/0.9.6e
X-Powered-By: PHP/4.1.2
Transfer-Encoding: chunked
Content-Type: text/html

f00
<!DOCTYPE HTML PUBLIC "-//W3C//DTD HTML 4.0 Transitional//EN">
<html>
<head>
<title>Information and Computer Science at the University of
California, Irvine</title>
...
```

Figure 2.4   Example of the use of the GET method in an HTTP 1.1 session.

2.3.4   Programming examples

The example shown in Figure 2.4 can be reproduced on almost any platform using a
terminal or a telnet application and is simply intended to demonstrate the mechanics of
the protocol. Figure 2.5, however, shows a simple programming example that requires
some basic knowledge of the Perl programming language (Wall *et al.* 1996). The
program fetches the header returned by the HTTP server and can be easily modified
to collect interesting pieces of information, such as the server software being run
(third line of output) or the timestamp (fourth line of output) that indicates the last
modification time of the page (in Chapter 6 we discuss how this information can
be useful to estimate the age distribution of Web pages). Lines 1–8 in Figure 2.5
are simply directives and variable declarations. A communication channel (socket)
is established in lines 10–12 by invoking the constructor of the INET class with
three arguments: the host (either specified by an IP address or a domain name), the
connection port (80) and a connection timeout in seconds. If a connection cannot be
established, the program is terminated in line 13. Line 15 prints a string to the socket
(that behaves like a character output device) issuing the HEAD request to the server,
using HTTP 1.0 (the commented line below shows how to use HTTP 1.1). Finally,
the loop in lines 18–20 is used to read and print (to the standard output), one line at
a time, the response returned by the server.

A similar program can be used to sample the IP address space. This technique,
explained in more detail in Chapter 3, can be used to estimate global properties of the
Web (e.g. the number of active sites). For example we may be interested in determining
the fraction of the addressable IPv4 space ($2^{32} = 4\,294\,967\,296$ hosts in total) with
running Web servers. Figure 2.6 shows a basic Perl script for sampling IPv4 addresses,
testing the presence of an HTTP server running on port 80. The program counts a
success if a socket can be established and, in addition, the server returns the status

```
1   #!/usr/bin/perl
2   # getHeader.pl
3   use strict;
4   use IO::Socket qw(:DEFAULT :crlf);
5   my $host = shift || "www.unifi.it";
6   my $path = shift || "/";
7   my $timeout= shift || 2;
8   my $socket;
9   # Create a socket for standard HTTP connection (port 80)
10  my $socket = IO::Socket::INET->new(PeerAddr => $host,
11                     PeerPort => 'http(80)',
12                     Timeout => $timeout)
13    or die "Cannot connect\n";
14  # Sends a HEAD request by printing on the socket
15  print $socket "HEAD $path HTTP/1.0",CRLF,CRLF;
16  #print $socket "HEAD $path HTTP/1.1",CRLF,"host: $host",CRLF,CRLF;
17  my $head;
18  while ($head = <$socket>) {
19      print $head;
20  }

% ./getHeader.pl
HTTP/1.1 200 OK
Date: Tue, 03 Dec 2002 23:59:59 GMT
Server: Apache/1.3.27 (Unix) PHP/4.2.3
Last-Modified: Mon, 04 Nov 2002 08:16:25 GMT
ETag: "f209-f91-3dc62cd9"
Accept-Ranges: bytes
Content-Length: 3985
Connection: close
Content-Type: text/html
```

Figure 2.5 Sample Perl script demonstrating the usage of the HEAD method in HTTP 1.0 and 1.1. A sample terminal output obtained by running the script is shown at the bottom.

code 200, which indicates a successful connection (Fielding *et al.* 1999). Although useful as a starting point, this basic program has several limitations (see Exercise 2.4).

2.4   Log Files

When users access a website, all transactions between the user's browser and the Web server software are logged in ASCII format in *server log files*. The main log file is usually known as a server transfer log or access log and typically contains fields such as the IP address from which a user's request originated, the time of the request, the method of the request (see Table 2.1), a numerical code called the *status code*, indicating the response from the server, and the size in bytes of the transaction. There are several other logs in addition to the server transfer log. These include the *referrer log*, which contains information about the Web page from which the request originated, indicating if the user came to the site via a hyperlink for example. The *agent log* contains information about the browser software making the request and can also indicate whether requests are originating from a 'robot' or 'spider' program if the requesting software identifies itself as such. The *error log* records error information for any requests that result in errors, such as a request for a Web page that cannot

```
 1  #!/usr/bin/perl
 2  # simpleSampleIP.pl
 3  use strict;
 4  use IO::Socket qw(:DEFAULT :crlf);
 5  my $trials = shift || 42949; # number of hosts to probe (default 42949)
 6  my $timeout = shift || 1;    # connection timeout (default 1 sec)
 7  my $found=0;                 # number of valid IP found
 8  my $host;                    # generic IP address to be probed
 9  my $socket;
10  for (my $i=1; $i<$trials; $i++) {
11      $host=sprintf("%d.%d.%d.%d",
12          int(rand(255)), int(rand(255)),
13          int(rand(255)), int(rand(255)));
14      print "trial $i: $host ... ";
15      select((select(STDOUT), $| = 1)[0]); # flush STDOUT
16      # Create a socket for standard HTTP connection (port 80)
17      my $failed=0;
18      my $socket = IO::Socket::INET->new(PeerAddr => $host,
19                          PeerPort => 'http(80)',
20                          Timeout => $timeout)
21          or $failed=1;
22      if (!$failed) {
23          print $socket "HEAD / HTTP/1.0",CRLF,CRLF;
24          my $header = <$socket>;    # read the header
25          my ($p, $r, $m) = split(/ /,$header); # extract status code
27          print "$p $r $m";
28          if ($r==200) { # accept only OK status
29                  $found++;
30          }
31      }
32      print " ($found) \n";
33  }
34  print "$found out of $trials IP accept connections on port 80\n";
```

Figure 2.6   Basic Perl script for sampling the IPv4 address space.

Table 2.2   An example of an entry in the www.ics.uci.edu Web server log file.

Field name	Value
Requested URL	/~frost/ics171/
Remote IP	68.5.116.149
Remote login name	—
Remote user	—
Time	[16/Nov/2002:23:51:05 -0800]
Method	GET /~frost/ics171/~HTTP 1.1
Status code	200
Bytes sent	9274
Referrer	—
User agent	Mozilla/5.0...Windows NT 5.0...

be found (code 404) or a 'server unavailable' response (code 503) (see Fielding *et al.* (1999) for details). Server-side logging software can be configured so that these various log files are combined into a single 'server log file' that records all of these fields for each request – we will assume this to be the case in what follows.

Table 2.3   A second example of an entry in the www.ics.uci.edu Web server log file.

Field Name	Value
Requested URL	/pub/ietf/http/hypermail/1996q1/0264.html
Remote IP	65.65.193.164
Remote login name	—
Remote user	—
Time	[16/Nov/2002:23:51:23 -0800]
Method	GET /pub/ietf/http/hypermail/ 1996q1/0264.html HTTP 1.0
Status code	302
Bytes sent	315
Referrer	http://www.google.com/ search?q=%22maximum+length+of+a+url %22&hl=en&lr=&ie=UTF-8&oe=UTF-8 &start=10&sa=N
User agent	Mozilla/5.0...Windows NT 5.0...

Table 2.2 shows an example of a single transaction from the www.ics.uci.edu server log file in November 2002. The first field is the requested URL, followed by the Remote IP address that generated the page request. The login and user fields are blank (indicated by '—'), since this is a publicly visible Web page that does not require any specific login procedure. The time field records the local time at the level of seconds. Being able to tell when a request arrived at this granularity is quite valuable when reconstructing a sequence of page requests for an individual IP address from a large log – we will discuss how such 'sessions' are created in more detail in Chapter 7. The method field corresponds to one of the HTTP methods listed in Table 2.1 – here the method is 'GET', a request for a specific page. The response of the server is recorded in the status code field, which in this case is 200, indicating 'success'. The next field records the total number of bytes sent from the server to the requester as a result of this transaction, if the request is successful. The referrer field contains information about where the user came from before they requested the current page. Frequently, as in this case, this referrer information is unavailable. This can happen, for example, when a user types a URL directly into the browser rather than following a hyperlink. However, when the referrer information is available, it can be very useful from a data analysis point of view, since it indicates how users are arriving at the site. The final field in the record is user agent, which in this case is a long string of information about the type of browser software being used to generate the request – only part of this string is displayed here.

Table 2.3 shows another example of a record from the log file of www.ics. uci.edu, with a timestamp that is 18 s after the previous example – in fact, there were several other requests in the intervening 18 s. This particular record is interesting because, unlike the other example, it does contain information in the referrer field. In particular, the referrer field tells us that the user that generated this request came via the Google search engine (`www.google.com`) by issuing the query 'maximum+length+of+a+url' to Google and then following a hyperlink in the displayed results to the `www.ics.uci.edu` website. The particular page being requested here is a page from the mailing list archives of the Internet Engineering Task Force (IETF) HTTP Working Group and is being provided via ftp.

2.5   Search Engines

2.5.1   Overview

We now introduce the basic design for a Web search engine. A user that browses the Web starting manually from a set of entry points, such as personal bookmarks or hierarchical structures such as the Open Directory Project (ODP, `http://dmoz.org/`), is unlikely to find answers to a general query in a reasonable amount of time. In particular, although links reported in hierarchies such as the ODP tend to be accurate thanks to human editing, only a very limited coverage of the Web contents is possible. According to a recent Pew Internet Project Report (2002), more than 25% of Internet users locate information on the Web with the help of search engines, and more than 80% of Internet users have used a search engine at least once. Search engines allow users to enter a query sentence (either using natural language or Boolean combinations of keywords) and receive as output an ordered list of URLs, ranked according to their *relevance* to the query.

To achieve these goals, modern search engines are complex aggregates of several pieces of software. In Figure 2.7 we show a very simplified architecture that nonetheless clarifies the role of some of the major components commonly found in real world implementations. *Crawlers*, also known as spiders, or Web robots, are computer programs that autonomously navigate the Web and download documents. Copies of the documents fetched by the crawler are then stored into an internal repository and subsequently indexed (see Section 4.1) to allow efficient searching. The link repository is essentially a representation of the Web as a graph. This component has become extremely important for improving the quality of ranking (to be further discussed in Chapter 5). The search engine processes a user's query, returning a set of matching documents that are eventually ranked according to criteria programmed into the scoring module.

The design of a search engines may vary significantly, depending on many factors. The most important distinction is probably between general-purpose engines (that are meant to provide general answers to the largest possible set of users) and thematic or specialized engines that focus on specific topics or types of documents.

Ideally, general-purpose engines should allow users to search the largest possible portion of the Web. However, various technological limitations impose trade-offs among conflicting objectives. Two of the most important measures of the performance for a search engine are coverage and recency. *Coverage* is the fraction of the visible Web that is searchable by the engine. *Recency* (or freshness) is the fraction of search-

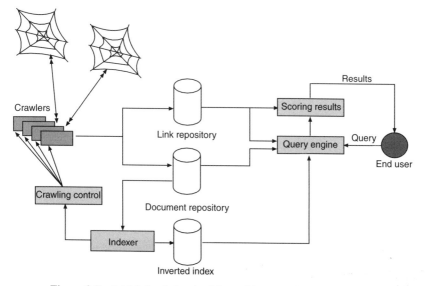

Figure 2.7   A high-level sketch of the architecture of a search engine.

able pages (or URLs) that are up-to-date in the internal repository data pages stored by the engine. One approach for maintaining a fresh repository consists of crawling the Web periodically in order to replace the contents of the index with updated versions of documents and possibly to add newly created documents. It is important to stress that fetching a large portion of the Web can take several days because of bandwidth limitations and that during the elapsed time between consecutive crawls the index is not up to date. As a result, achieving high performance in terms of both coverage and recency is a major challenge for general-purpose search engines, as earlier noted by Lawrence and Giles (1998b). This issue is explained in more detail in the following.

2.5.2   Coverage

Although several companies claimed by the mid 1990s that their search engines were able to offer an almost complete coverage of Web content, it soon became clear that only a relatively small fraction of existing Web pages could be fetched by a single search engine. Lawrence and Giles (1998b, 1999) describe two experiments aimed at measuring the performance of search engines in terms of coverage and freshness. They use the following approach, known as *overlap analysis*, for estimating the size of the indexable Web (see also Bharat and Broder 1998). Let W denote the set of Web pages and let $W_a \subset W$, $W_b \subset W$ be the pages crawled by two independent engines, a and b. How big are W_a and W_b with respect to W? Suppose we can uniformly sample documents and test for membership in the two sets. Denote by $P(W_a)$ and $P(W_b)$ the probabilities that a page has been crawled by a or b, respectively. We know that

$$P(W_a \cap W_b \mid W_b) = \frac{|W_a \cap W_b|}{|W_b|}. \tag{2.1}$$

Now if the two crawls are assumed to be independent, the left-hand side of Equation (2.1) reduces to $P(W_a)$, the coverage of engine a. So coverage can be easily obtained from the size of the intersection of the two crawls. This quantity, however, cannot be computed exactly unless one has access to the data stored in their repositories. In their first study, Lawrence and Giles (1998b) used a controlled set of 575 queries in order to sample pages and count the number of times both engines returned the same page. Assuming the resulting estimate $P(W_a)$ is correct, it is straightforward to infer the size of the entire Web as $|W_a|/P(W_a)$. Using this approach, it was estimated that the Web contained at least 320 million pages in November 1997, that about 60% of the Web was covered by the six major engines at that time, and that the maximum coverage from a single engine was only one third of the Web. Actually, since this approach is based on observations, it yields an estimate of the size of the so-called 'visible' Web, excluding (for example) pages that are behind forms, password protected, etc.

The experiment was repeated in February 1999, under slightly different conditions and using 11 search engines (Lawrence and Giles 1999). This later study found an estimated 42% collective coverage and a maximum 16% coverage by a single engine, while the estimated size of the Web grew to roughly 800 million pages. At the time of writing (November 2002), some of the major companies claim they can return results for over three billion pages, while it is widely believed that the Web now contains several billion pages. Fienberg *et al.* (1999) propose an interesting Bayesian approach to the same problem in a later paper.

A strategy for improving coverage is the construction of *meta-search engines* that dispatch a user query to several conventional engines simultaneously, collect the result and merge them using some fusion policy before returning a list to the user. SavvySearch (Dreilinger and Howe 1997) selects the search engine to which queries should be dispatched. Inquirus (Lawrence and Giles 1998a) improves the general meta-search approach by retrieving the pages returned by the search engines and using a fusion policy that depends on the page contents. A Bayesian approach to the meta-search problem is described in Aslam and Montague (2001).

2.5.3 Basic crawling

The algorithm SIMPLE-CRAWLER listed below is a simplified version of a basic general-purpose crawling process. The algorithm essentially performs a graph visit (Cormen *et al.* 2001). It starts from a collection S_0 of URLs (which are sometimes referred to as the *seeds*) and stores the documents and links encountered into two repositories, referenced by D and E, respectively. It maintains a list, Q, internally that contains the fringe of the portion of the graph visited so far. The function DEQUEUE extracts and returns one element out of the list. If Q is managed according to a first-in first-out (FIFO) policy, then the crawler performs a breadth-first search (BFS). Assuming that the purpose is to download all pages reachable from S_0, this *crawling strategy* may be reasonable, at least as a simple strategy to start with. However, we will see later that the URLs in Q may be sorted according to different criteria in order to focus the crawl in particular directions so as to improve the quality of the collected documents with respect to an assigned goal. In line 4, we invoke the function FETCH, which downloads the document addressed by the URL u. The fetched document is then stored in the repository D. In line 6, the function PARSE extracts all of the URLs contained in the document (the set of children of u) and puts them in a temporary list L. All the elements of L are in turn inserted into the crawling queue Q, unless they have already been crawled or they are already enqueued for crawling. The latter test is necessary to avoid loops, as the Web is not acyclic. Compared to standard graph visit algorithms, storing reached and fetched URLs replaces traditional node coloring. Note that in line 8 we also store the edges of the Web graph. We will see later how recent scoring algorithms can take advantage of this information.

$$\text{SIMPLE-CRAWLER}(S_0, D, E)$$

```
 1  Q ← S₀
 2  while Q ≠ ∅
 3  do u ← DEQUEUE(Q)
 4      d(u) ← FETCH(u)
 5      STORE(D, (d(u), u))
 6      L ← PARSE(d(u))
 7      for each v in L
 8      do STORE(E, (u, v))
 9          if ¬(v ∈ D ∨ v ∈ Q)
10              then ENQUEUE(Q, v).
```

While the above scheme is conceptually acceptable, in a real world application several practical modifications are also necessary.

- Under the assumption of a static Web, the contents of Q and D completely characterize the *state* of the crawler, in the sense that the process could be restarted with no information loss if Q and D are preserved. Because of their size and because of the need of persistent memory (to avoid data loss in the case of a crash), these objects must be stored on disk in practical implementations.

- In practice, it will be always necessary to halt SIMPLE-CRAWLER before the loop in lines 2–10 terminates with an empty Q, simply because storage resources are exhausted or because the time slot assigned to the crawling process has expired. In this case we are left with both *fetched* and *discovered* URLs (stored in D and Q, respectively). The contents of pages whose URLs were just discovered is unknown. Nonetheless, it is still possible to associate the text in the corresponding anchors with these URLs and return them as query results. Indeed, several search engines can (and do) return links to documents they have never downloaded.

- The time required to complete the downloading of a document is unknown because of many factors, including connection delays and network congestions. Moreover, the bandwidth available at the crawling site(s) may be significantly larger than at the location where the document to be downloaded is located. As a result, a single crawling process that just waits for the completion of each download before moving to the next one would be a poor design choice. Running concurrent fetching threads is thus the normal solution to maximize the exploitation of available bandwidth. Note that a multithreaded implementation with a single queue Q is still a single crawling process. Parallelization of the crawling process is discussed in Section 6.3.

- Crawlers should be respectful of servers and should not abuse the resources at the target site – otherwise administrators may decide to block access (Koster 1995). The Robots Exclusion Protocol (http://www.robotstxt.org/) allows webmasters to define which portions of a site are permissible for robots to fetch from. Under this protocol, certain directories or dynamically generated pages can be declared as off-limits for a crawler if they are listed in the file robots.txt located in the root directory. A well-behaved crawler should always avoid fetching excluded documents. Similarly, by inserting special a META tag in the HTML code, a webmaster may indicate that the contents of certain pages should not be indexed.

- Overloading sites or network links should also be avoided. For example, multiple threads should not fetch from the same server simultaneously or too often. Unfortunately, outlinks often point within the same site (this happens in particular for 'dot com' sites), leading to a relatively high degree of locality in a single queue. For example, Mercator (Heydon and Najork 2001) is a Web crawler that implements a sophisticated strategy for 'broadening' as much as possible the crawling fringe and increasing the elapsed time between two consecutive requests to the same server. It maintains a data structure formed by several FIFO queues, each containing only URLs pointing to a unique server, and by an index that associates a timestamp to each queue. Fetching threads must check in the index if enough time has passed since the last access to the same queue, and once they download a document they insert a timestamp in the corresponding queue.

- URLs that have already been fetched or discovered must not be inserted again into the crawling queue Q. The test in line 9, however, is not straightforward to implement. In particular, since Q and D are stored on disk, special care is needed in order to avoid any overheads due to external memory management. Heydon and Najork (2001) propose to use two caches. The first cache contains recently added URLs, exploiting locality during the crawl. The second cache contains popular URLs, as determined during the crawl. Heydon and Najork

(2001) report high hit rate for the second cache and this can be explained by the scale-free distribution for the number of incoming links of a given Web page (see Chapter 3 for details on these graphical properties of the Web).

- The real world is more complex than a mathematical model of the Web and in practice several precautions must be enforced in a real crawler to avoid problems like aliases and 'traps' (Heydon and Najork 1999). Aliases occur if the same document can be addressed by many distinct URLs, for example, if a site is registered under multiple domain names. It is possible to cope with aliases by using canonicalization, i.e. issuing a DNS request in order to get the canonical name of the host (see Section 2.3.2). A related problem of multiple URLs pointing to the same document occurs when the server embeds session IDs into the URL. Traps can be generated by malicious common gateway interface (CGI) scripts that keep generating 'on the fly' fake documents pointing to other fake documents.

- DNS lookup is necessary to map hostnames in URLs to IP addresses before downloading, or to canonicalize URLs before inserting them in the discovered queue. A DNS lookup can often take considerable time due to slow DNS server responses. Efficient crawlers may need to rewrite DNS resolution in a multithreaded fashion (Heydon and Najork 1999).

We should also bear in mind another important difference between graph visiting and crawling. A graph is normally thought of as a static object during the execution of a crawling algorithm. However, in reality, the Web is dynamic and changes continuously in terms of both content and topology. Fetching a billion documents can take weeks. Crawling is thus like taking a picture of a living scene with some objects remaining fixed and others moving at a very high speed. At the end of the process, D and E will unavoidably contain stale information. The problem can be somewhat addressed by more frequently downloading sites that are expected to change more often (e.g. portals, press, etc.) and less frequent downloading of more static sites. Accounting for temporal effects during a crawl is a relatively recent research topic – in Section 6.4 we will discuss specific crawling approaches that address Web dynamics.

2.6 Exercises

Exercise 2.1. Explain the differences amongst URI, URL, and URN using Venn diagrams and provide examples for each intersection amongst these sets.

Exercise 2.2. Write a program that extracts all the URIs from a given HTML document.

Exercise 2.3. Write a program to automatically query a given search engine. The program should accept a query string as input and build up a list of URIs extracted from the HTML pages returned by the search engine.

Exercise 2.4. Improve the program of Figure 2.6 and perform some basic sampling experiments. In particular:

- write a multithreaded version so that there will be no problems waiting for a longer timeout in the case of connection failure; and

- correct the random generation of IP addresses to exclude special values that are associated for example with default routers, or unassigned portions of the address space (see O'Neill *et al.* (1997) for details).

After completing your program collect some statistics from the Web, for example, the distribution of usage of different HTTP server software.

3

Web Graphs

3.1 Internet and Web Graphs

The Internet can be viewed as a graph in many different ways. For instance, in one representation, one could focus on the physical layer, and use a graph where nodes correspond to computers, routers, and so forth, at fixed locations, and edges correspond to direct physical connections. This graph, however, is somewhat limited. It misses the fast growing wireless boundary of the Internet and, more fundamentally, it does not really capture the graphical connections associated with the information available through the Internet. Thus, in the rest of this chapter, we focus on graphical representations where the nodes represent Internet pages and the edges are associated with hyperlinks that connect the information contained in different pages, the Web graph. Remarkably, as we shall see later in the chapter, the Web graph shares several interesting properties with the Internet hardware graph, as well as with many other networks.

When focusing on the graph of hyperlinks, there are a number of additional factors to be considered. First, the edges can be considered as directed, from the source page to the target page, or undirected. For many of the results discussed in this chapter, this distinction turns out not to be essential. Second, the graph itself is dynamic and undergoing constant changes: nodes and connections are being added almost every minute and, less frequently, old nodes or connections are being deleted. The content of existing nodes, i.e. Web pages, is also subject to change, although these changes may not alter the underlying graphical structure, unless new hyperlinks are created. One consequence of these dynamical aspects is that it is virtually impossible to know exactly what all the nodes and connections of the graph are at time t. Thus, as discussed in Chapter 2, the exact size of the graph is not really a useful or even well-defined quantity. Third, although most of the time we will focus on the primary connected component of the graph (at least in the undirected version), additional small disconnected components exist on the fringes. Finally, there are pages and hyperlinks that are created dynamically on the fly, for instance after entering a query in a search engine. These have typically a short life span and therefore will be excluded from the analysis.

Modeling the Internet and the Web   P. Baldi, P. Frasconi and P. Smyth
© 2003 P. Baldi, P. Frasconi and P. Smyth   ISBN: 0-470-84906-1

In short, we are dealing with a fuzzy graph that is constantly evolving on several different time scales. For simplicity the main analysis focuses on a large and slowly evolving subgraph of this graph. The dynamic aspect of course suggests other interesting questions, related to the temporal evolution of relevant variables. For example, how is the total number of nodes and links growing with time and what is the life expectancy of a typical page? Some of these temporal issues will also be addressed in this chapter and again later in Chapter 6.

There are many reasons, from mathematical, to social, to economic to study the Web graph. It is an example of a large, dynamic, and distributed graph where we have the privilege of being able to make a large number of measurements. In addition, as we are about to see, the Web graph shares many properties with several other complex graphs found in a variety of systems, ranging from social organizations to biological systems. The graph may reflect psychological and sociological properties of how humans organize information and their societies. Perhaps on a more pragmatic note, the structure of the graph may also provide insights regarding the robustness and vulnerability of the Internet and help us develop more efficient search and navigation algorithms. Indeed, one way to locate almost any kind of information today is to navigate to the appropriate Web page by following a sequence of hyperlinks, i.e. a path in the graph. Efficient navigation algorithms ought to be able to find short paths, at least most of the time. And finally, the behavior of users as they traverse the Web graph is also of great interest and is the focus of Chapter 7.

Having set our focus on the graph of relatively stable Web pages and their hyperlinks, several different questions can be asked regarding the size and connectivity of the graph, the number of connected components, the distribution of pages per site, the distribution of incoming and outgoing connections per site, and the average and maximal length of the shortest path between any two vertices (diameter).

Several empirical studies have been conducted to study the properties of the Web graph. These are based on random sampling methods or, more often, on some form of crawling applied to subgraphs of various sizes, ranging from a few thousand to a few hundred million pages. To a first-order approximation, these studies have revealed consistent emerging properties of the Web graph, observable at different scales. One particularly striking property is the fact that connectivity follows a power-law distribution. Remarkably, these results have had a sizable impact in graph theory, with the emergence of new problems and new classes of random graphs that are the focus of active mathematical investigation (see Bollobás and Riordan (2003) and additional references in the notes at the end of this chapter).

One preliminary but essential observation that holds for all the graphs to be considered in this chapter is that they are *sparse*, i.e. have a number of edges that is small ($|E| = O(n)$) or at least $o(n^2)$ compared to the number $O(n^2)$ of edges in a complete or dense graph. For the Web, this is intuitively obvious (at least in the foreseeable future), because the average number of hyperlinks per page can be expected to be roughly a constant, in part related to human information processing abilities. This sparseness is already a departure from the uniform random graph model where the probability of an edge p is a constant and the total number of edges $|E| \approx pn^2/2$.

3.1.1 Power-law size

As discussed in Chapter 2, it is difficult to estimate the exact number of documents on the Web at any given time and this number continues to grow at a fast pace. Simple 'back-of-the-envelope' estimates as well as more sophisticated ones (Lawrence and Giles 1999) suggest a size of well over a billion nodes, and several billion edges. At the time of this writing, major search engines such as Google claim an index size of roughly three billion URLs, confirming that the actual order of magnitude of the Web is in the vicinity of 10^{10}.

In Huberman and Adamic (1999), the distribution of site sizes measured by the number of pages, as determined by crawls of common search engines including several hundred thousand pages, is reported to follow a power-law distribution. This power-law distribution is observed over several orders of magnitude with an exponent γ in the range 1.6–1.9. This power law allows one, for instance, to estimate the expected number of sites of any size, even if that size has not been observed in a crawling experiment. This is achieved by using the scaling property of power laws seen in Chapter 1 by writing $P(m) = P(n) \times (m/n)^{-\gamma}$.

3.1.2 Power-law connectivity

Several studies (Barabási and Albert 1999; Broder *et al.* 2000; Kleinberg *et al.* 1999) have consistently reported that the distribution of the number of connections per node follows a power-law distribution, both in the directed and undirected cases. These power-law results, derived from crawls of particular organizations or various commercial crawls of the Web, are observed over several orders of magnitude and at different scales. Thus, the general 'feel' of the Web graph is the same for very different subgraphs, such as those spanned by a university, a country, or the crawl of a commercial search engine.

For instance, a study conducted by crawling the domain of the University of Notre Dame, Indiana (Albert *et al.* 1999; Barabási and Albert 1999) reported exponents of $\gamma = 2.45$ and $\gamma = 2.1$ for the outdegree and indegree distributions, respectively (Table 3.1). At the time of the study, the graph associated with the *.nd.edu domain contained 325 729 documents and 1 469 680 links. A larger similar study is described in Broder *et al.* (2000).

A power-law distribution for the connectivity is very different from the distribution found in traditional uniform random graphs (Bollobás 1985; Erdos and Rényi 1959, 1960), or in other related models discussed below (Watts and Strogatz 1998). In a random uniform graph with random independent edges of fixed probability p, with p small, the number of edges with k neighbors has a Poisson distribution with mean

$$\lambda = n \binom{n-1}{k} p^k (1-p)^{n-1-k} \tag{3.1}$$

and decays exponentially fast to zero as k increases toward its maximum value $n - 1$. Thus the power-law distribution is a remarkable example of emerging order in a large graph created by many agents, each of whom is completely free to create documents and hyperlinks.

The emergence of a power-law distribution is by itself intriguing enough to require an explanation. But in addition, the need for an explanation is exacerbated by the fact that a similar distribution has been observed for many other networks: business networks, social networks, transportation networks, telecommunication networks, biological networks (both molecular and neural) and so forth (Table 3.1). The following list of examples is not exhaustive.

- The Internet at the router and inter-domain level has a connectivity distribution that falls off as a power-law distribution with $\gamma \approx 2.48$ (Faloutsos *et al.* 1999). Power-law connectivity has also been reported at the level of peer-to-peer networks (Ripeanu *et al.* 2002).

- The call graphs associated with the calls handled by some subset of telephony carriers over a certain time period (Abello *et al.* 1998).

- The power grid of the western United States (Albert *et al.* 1999; Phadke and Thorp 1988) where, for instance, nodes represent generators, transformers, and substations and edges represent high-voltage transmission lines between them.

- The citation network where nodes represent papers and links are associated with citations (Redner 1998). A similar graph, smaller but famous in the mathematics community, is the graph of collaborators of Paul Erdös (see http:// www.acs.oakland.edu/~grossman/erdoshp.html).

- The collaboration graph of actors (http://us.imdb.com), where nodes correspond to actors and links to pairs of actors that have costarred in a film (Barabási and Albert 1999).

- The networks associated with metabolic pathways (Jeong *et al.* 2000) where the probability that a given substrate participates in k reactions decays as $k^{-\gamma}$ in a representative variety of organisms. These reactions have directions and both the indegrees and outdegrees follow a power-law distribution with similar exponents in general. In *Escherichia coli*, for instance, the exponent is $\gamma = 2.2$ for both the indegrees and the outdegrees. In contrast with what happens for the other networks, Jeong *et al.* (2000) report that in metabolic networks the diameter, measured by the average distance between substrates, does not seem to grow logarithmically with the number of molecular species, but rather appears to be constant for all organisms and independent of the number of molecular species. (They also observe that the ranking of the most connected substrates is essentially the same across all organisms). A constant diameter may confer increased flexibility during evolution.

- The networks formed by interacting genes and proteins as described in Maslov and Sneppen (2002). These authors also report the existence of a richer level of structure in the corresponding graphs. In particular, direct

Table 3.1 Power-law exponent, average degree, and average diameter (e.g. average minimal distance between pairs of vertices). Data partly from Barabási and Albert (1999) and Jeong *et al.* (2000). Ex., entries left as an exercise; NA, not applicable.

	Average Exponent γ	Average degree	diameter
actors	2.3	28.78	Ex.
power grid	4	2.67	Ex.
citation	3	Ex.	Ex.
protein interactions	2.5	Ex.	Ex.
E. coli metabolic indegree	2.2	3.7	3.2
E. coli metabolic outdegree	2.2	3.7	3.2
Internet routers	2.48	Ex.	Ex.
WWW (undirected)	Ex.	5–7	19
WWW (indegree)	2.1	5.46	Ex.
WWW (outdegree)	2.45	Ex.	Ex.
theoretical model (pref. attach.)	2.9	Ex.	Ex.
WWW (pages per site)	1.8	NA	NA
language	—	NA	NA

connections between proteins with high degrees are suppressed, whereas direct connections between highly connected and low-connected pairs of proteins are favored. This additional pattern of connectivity may reduce the likelihood of cross talk between different functional modules and confer additional robustness to the network by localizing the effect of perturbations.

- The graph of nervous connections in the model organism *Caenorhabditis elegans* (Achacoso and Yamamoto 1992). *C. elegans* is a little transparent worm that has a hardwired nervous system with a fixed number of neurons (vertices) and a genetically wired set of connections (edges) associated with synapses or gap junctions between the corresponding neurons. However, because the entire network contains only 302 neuronal nodes (and about 5000 chemical synapses), the statistical significance of this result is somewhat weaker.

Reports of power-law distributions have on occasion been controversial, as in the case of food webs (McCann *et al.* 1998; Paine 1992; Pimm *et al.* 1991), which are another example of networks that are nonrandom but also nonregular. For biological systems in particular, large-scale information about graphs of interacting elements is not easy to derive and is only beginning to become available with high-throughput technologies such as DNA and protein arrays. As more data become available, better assessments of power-law connectivity, its origins, its fluctuations, and how universal it is should become possible.

3.1.3   Small-world networks

A second empirical observation reported for the Web graph, which is also shared by
many of the networks above, is that the diameter of the Web graph is small relative to
the size of the overall system. This property is also often described in terms of 'small
world' or 'six degrees of separation', originating in the study of sociological webs of
relationships by Milgram in the late 1960s and the idea that any two human beings on
the planet are typically connected by a short chain of acquaintance relationships (Korte
and Milgram 1978; Milgram 1967; Travers and Milgram 1969). In more mathematical
terms, small-world networks are characterized by a diameter that is exponentially
smaller than the size, i.e. where the diameter is bounded by a polynomial in $\log n$.

Unlike the distribution of connectivity, an estimate of the diameter in terms of the
largest or average length of the shortest path between any two Web pages cannot
be derived by exhaustive analysis of n vertices. Finding the shortest path between a
source vertex and any other vertex in a graph takes $O(n^2)$ steps in general, which
can be reduced to $O(n \log n)$ in the case of a sparse graph like the Web, essentially
using dynamic programming (see Cormen *et al.* 2001 for more details). Even in a
sparse graph like the Web, this can be computationally demanding for large values of
n. Thus, diameter properties must be inferred from sampling, the average diameter
being easier to estimate than the maximal diameter.

By simulating a random graph with the same power-law connectivity as the one
found on the Web (Albert *et al.* 1999), it was found that the average distance between
two vertices is given by $d = 0.35 + 2.06 \log_{10} n$. Thus the Web is clearly a small-
world network. For $n = 10^9$, $d = E[l] = 18.89$, so that two randomly chosen documents are separated on average by 19 clicks (implicitly this implies that directed
paths are being considered, since current browsers, unlike search engines, can only
follow links in one direction). These authors also report that d has a roughly Gaussian
distribution. The logarithmic scaling of the diameter or the average distance between
documents on the Web is important to ensure that the future growth of information
on the Web remains manageable. Under these assumptions, a ten-fold increase in
overall Web size increases the average number of clicks along a minimal path by
only two. It should be noted, however, that even an increase of only two clicks on
average would probably be perceived as extremely annoying and better solutions are
needed – indeed this is in part the role of search engines, which are studied in the
coming chapters.

As a more general but important side issue, the minimal distance between two
randomly selected nodes is interesting from a theoretical standpoint, but does not
characterize actual browsing for a variety of reasons. First, humans do not necessarily
take the shortest path available in the graph because, for instance:

(a) they may not know it, nor be able to effectively approximate it;

(b) they may get distracted by something on the way;

(c) they bookmark sites, so that future 'traversals' may ignore the graph entirely.

Second, humans do not traverse links from one randomly selected site to another. Exploratory browsing aside, they go to a given page in response to an informational need, and generally start from a page that they believe will lead them to their target quickly – after all, this is what search engines are for.

One additional issue related to the power-law distribution, small network, and sampling size, is the behavior of the degree distribution for very large values of k and the difficulty in assessing the behavior of the tail of the distribution with finite data. Amaral *et al.* (2000) have analyzed several naturally occurring networks and reported the existence of three different classes of small-world networks:

(a) scale-free networks characterized by a power-law distribution of connectivity over the entire range of values;

(b) broad-scale networks, where the power-law distribution applies over a broad range but is followed by a sharp cutoff; and

(c) single scale networks with a connectivity distribution that decays exponentially, as in the case of standard random graph models.

The authors argue that in single and broad-scale networks there are in general additional constraints that limit the addition of new links, related for instance to the aging of vertices, their limited capacity, or the cost of adding links. It is not clear, however, that a given finite network can always be classified in a clean fashion into one of these three categories. Furthermore, it is also not clear whether the very large-scale tail of the degree distribution of the Web graph exhibits a cutoff or not, and if so whether it is due to additional operating constraints.

3.1.4 Power law of PageRank

When searching information on the Web, it is important to be able to assess the importance of a page relative to a query and to rank pages accordingly. A crude way of measuring the importance of a page is to look at its indegree, i.e. the number of pages that point to it. This index is not very reliable, however, because it is entirely local and very easy to spam. PageRank may be a better and more global index, originally used in the Google search engine (Brin and Page 1998). It will be described here informally and studied in more detail in Chapter 5. Informally, the PageRank of a page can be understood as the proportion of time a random surfer would spend on that page at steady state. The idea is to imagine a random first-order Markov surfer who, at each time step, travels from one Web page to the next by randomly and uniformly selecting an outgoing hyperlink on the current page. Thus, in this view, the Web graph is the transition graph of a Markov chain and all outgoing edges from a given node are assigned the same probability. Because this Markov chain could have isolated or absorbing components, a general parameter ϵ can be added so that, in addition, at each time step there is a probability $1 - \epsilon$ of moving to any other page on the Web. It is well known and easy to see that the steady-state distribution of such a Markov chain, representing the probability of being in any state, is obtained by looking at the eigenvector and eigenvalue of the transition matrix. Specifically, the page rank $r(v)$ of page v is the steady-state distribution obtained by solving the system of n linear equations given by

$$r(v) = \frac{1 - \epsilon}{n} + \epsilon \sum_{u \in \text{pa}[v]} \frac{r(v)}{|\text{ch}[u]|}, \tag{3.2}$$

where pa[v] is the set of parent nodes, i.e. of pages that point to page v, and $|\text{ch}[u]|$ denotes the outdegree, i.e. the size of the set of children nodes.

Analysis of the distribution of PageRank values in Pandurangan *et al.* (2002) indicates that PageRank also follows a power-law distribution with the same exponent (namely 2.1) as the indegree distribution and, as in the case of degree distributions, this distribution seems to be relatively insensitive to the particular snapshot of the Web used for the measurements.

3.1.5   The bow-tie structure

A more careful and comprehensive analysis of the topology of the Web must take into account the orientation of the edges and existence of disconnected components. The small-world property applies to certain components and to the 'core' of the Web graph but is not true for the entire Web graph. There are significant portions of the Web that cannot be reached from other significant portions, and there is a significant portion of pairs of nodes that can only be connected through fairly long paths.

A large-scale study using two Altavista crawls, each with over 200 million nodes and 1.5 billion links (Broder *et al.* 2000), reported a macroscopic bow-tie structure (Figure 3.1) in which over 90% of the nodes belong to a single connected component if the hyperlinks are treated as undirected edges. Incidentally, the study also confirmed the power-law distribution of connectivity both for page indegrees ($\gamma = 2.1$) and outdegrees ($\gamma = 2.72$).

The bow-tie structure contains four different components. The first component is the core forming a giant strongly connected component (SCC), whose pages can reach each other via directed paths. This component contains most of the prominent sites. The second component (IN) consists of upstream pages that can reach the core via a directed path of hyperlinks, but cannot be reached from the core in a similar way. The third downstream component (OUT) consists of pages that can be reached from the core via a directed path of hyperlinks, but cannot reach the core in a similar way. Finally, the fourth component is more heterogeneous and contains smaller disconnected components and 'tendrils' consisting of pages that can neither reach or be reached from the core. However, they can be reached via a directed path originating from pages in IN, or pages that can project into OUT via a directed path. Finally, using Broder *et al.*'s terminology, 'tubes' are directed paths involving tendrils from a page in IN to a page in OUT. In the reported experiments, each of the four components has roughly the same size and contains on the order of 50 million

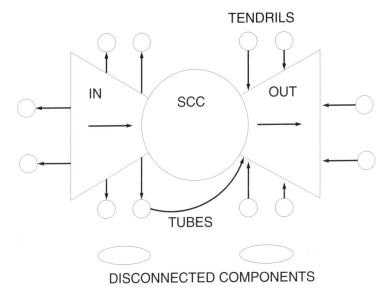

Figure 3.1 Internet bow-tie mode, adapted from Broder *et al.* (2000), with four broad regions of roughly equal size. A strongly connected component (SCC) within which each node can reach any other node via a directed path. An IN component of nodes that can reach nodes in SCC via a directed path but cannot be reached by nodes in SCC via a directed path. A similar OUT component downstream of the SCC component. The fourth component is more heterogeneous and contains (a) small disconnected components; and (b) tendrils associated with nodes that are not in SCC and can be reached by a directed path originating in IN or terminating in OUT. Tubes connect tendrils emanating from IN to tendrils projecting into OUT.

pages, although these proportions could vary in time and would need to be revisited periodically.

Because this study contains disconnected components, it is clear that both the maximal and average diameter are infinite. Thus, the author use the maximal minimal path length and the average path length restricted to pairs of points that can be connected to each other. With this caveat, they report in their study a maximal minimal diameter of at least 28 for the central core, and over 500 for the entire graph. For any randomly selected pair of points, the probability that a directed path exists from the first to the second is 0.24 and if a directed path exists, its average length is about 16, reasonably close to the estimate of 19 obtained in the previous section. The average length of an undirected path, when it exists, is close to 7. Thus in particular the core has a small-world structure: the shortest directed path from any page in the core to any other page in the core involves 16–20 links on average.

The study in Broder *et al.* (2000) also reports a power-law distribution in the sizes of the connected components with undirected edges, with an exponent $\gamma = 2.5$ over roughly five orders of magnitude. Bow-tie structure in subgraphs of the Webgraph associated, for instance, with a particular topic are described in Dill *et al.* (2001).

3.2   Generative Models for the Web Graph and Other Networks

There have been several proposals for stochastic models that can explain, or at least partly reproduce, the emergent properties observed for the distribution of site sizes and the connectivity of Web-like graphs described in the previous sections. In particular, there have been a number of such proposals for the power-law distributions and small-world network properties. If the processes that drive the growth of the WWW are highly decentralized, then the power-law distribution, for instance, must emerge from an aggregation of local behaviors. But what is the right model for this aggregation? In this section, we review some of the models that have been proposed and provide additional pointers at the end of the chapter. In the case of random Web-like graph models, these new models strongly differ from the classic random graph models (Bollobás 1985; Erdos and Rényi 1959, 1960), where the number n of vertices is fixed and each possible edge occurs independently of all other edges with a fixed probability p. In fact, it is fair to say that the study of the graphical properties of the Internet and the Web have created new problems and new areas of active investigation in the mathematical theory of random graphs.

3.2.1   Web page growth

Several empirical estimates of the size of the Web, the number of IP addresses, the number of servers, the average size of a page, and so forth, have been published, for instance in Lawrence and Giles (1999). Huberman and Adamic (1999) have observed a power-law distribution of site sizes and proposed a generative model to account for this distribution. The first component of this model is that sites have short-term (for instance, daily) size fluctuations up or down that are proportional to the size of the site. This is not an unreasonable assumption: a site with 100 000 pages may gain or loose a few hundred pages in one day, whereas the same effect would be rather surprising for a site with only 100 pages. A second ingredient is to assume that there is an overall growth rate α so that the size $S(t)$ satisfies

$$S(t + 1) = \alpha(1 + \eta_t \beta)S(t),\tag{3.3}$$

where η_t is the realization of a ± 1 Bernouilli random variable at time t with probability 0.5, and β is the absolute rate of the daily fluctuations. Thus, after T steps,

$$S(T) = \alpha^T S(0) \prod_{i=0}^{T-1}(1 + \eta_t \beta),\tag{3.4}$$

so that

$$\log S(T) = T \log \alpha + \log S(0) + \sum_{i=0}^{T-1} \log(1 + \eta_t \beta).\tag{3.5}$$

The last term can also be written as

$$l \log(1 + \beta) + (T - l) \log(1 - \beta),$$

where l is the number of positive fluctuations. Assuming that the variables η_t are independent, by the central limit theorem it is clear that for large values of T the variable $\log S(T)$ is normally distributed. Alternatively, $\log S(T)$ can be associated with a binomial distribution counting the number of times η_t is equal to $+1$ and the binomial distribution can be approximated by a Gaussian. In any case, it follows that $S(T)$ has a log-normal distribution which, as seen in Chapter 1, is related to but different from the observed power-law distribution. However, Huberman and Adamic (1999) report that, if this model is modified to include a wide distribution of growth rates across different sites and/or the fact that sites have different ages (with many more young sites than old ones), then simulations show that the distribution of sizes across sites obeys a power-law distribution (see Mitzenmacher (2002) for more general background about the relationship between the log-normal and power-law distributions).

It must be pointed out, however, that the simple Fermi model of Chapter 1 provides a somewhat cleaner and analytically tractable model that seems to capture the power-law distribution of website sizes, at least to a first degree of approximation. To see this, instead on focusing on daily fluctuations, which are more of a second-order phenomenon, it is sufficient to consider that websites are being continuously created, that they grow at a constant rate α during a growth period, after which their size remains approximately constant, and that the periods of growth follow an exponential distribution. Assuming, consistently with observations, a power-law exponent $\gamma = 1.08$ for the size, this would give a relationship $\lambda = 0.8\alpha$ between the rate λ of the exponential distribution and the growth rate α.

3.2.2 Lattice perturbation models: between order and disorder

Watts and Strogatz (1998) have argued that the small networks found in biological, technological, social, and man-made networks are somewhere between completely organized (regular) networks, and completely disorganized networks. They are highly clustered at the local level and have short characteristic path lengths (Table 3.2). Hence, they have suggested a model based on the idea of progressively increasing the level of disorder in a regular system.

More specifically, they start from a regular arrangements of points, such as a lattice with constant degree k and neighborhoods that are entirely local. Then they progressively visit all the vertices and edges in the graph and randomly rewire each of the local edges into a long-ranged edge with probability p, avoiding edge duplication. Obviously, $p = 0$ corresponds to complete order and, as p is increased, the graph becomes increasingly disordered until $p = 1$, at which stage the graph is completely random. Note, however, that the procedure does not alter the total number of vertices or edges in the graph, nor the average connectivity.

Two quantities can be used to monitor the evolution of the undirected graph structure during the rewiring process. First, the average diameter, $d = d(p)$, corresponding to the average distance (number of edges) between any two vertices, which is a global property of the graph. Second, a more local property measuring the average density of local connections, or cliquishness, defined as follows. If a vertex v has k_v neighbors, there are at most $k_v(k_v - 1)/2$ edges between the corresponding nodes. If c_v is the corresponding fraction of allowable edges that actually occur, then the global cliquishness is measured by the average $c = c(p) = \sum_v c_v/|V|$, where $|V|$ is the total number of vertices. In a social network, d is the average number of friendships in the shortest path between two people, and c_v reflects the degree to which friends of v are friends of each other.

To ensure that the graph is both sparse but connected, the value k satisfies

$$n \gg k \gg \log n \gg 1,$$

where $k \gg \log n$ ensures that a corresponding random uniform graph remain connected (Bollobás 1985). In this regime, Watts and Strogatz (1998) find that $d \approx n/2k \gg 1$ and $c \approx 3/4$ as $p \to 0$, while $d \approx d_{\text{random}} \approx \log n/\log k$ and $c \approx c_{\text{random}} \approx k/n \ll 1$ as $p \to 1$. Thus, the regular lattice with $p = 0$ is a highly clustered network where the average diameter d grows linearly with n, whereas the random uniform network $p = 1$ is poorly clustered, with d growing logarithmically with n. These extreme cases may lead one to conjecture that large c is associated with large d, and small c with small d, but this is not the case.

Through simulations, in particular for the case of a ring lattice, Watts and Strogatz find that for small values of p the graph is a small-world network, with a high cliquishness like a regular lattice and a small characteristic path length like a traditional random graph. As p is increased away from zero, there is a rapid drop in $d = d(p)$ associated with the small-world phenomena, during which $c = c(p)$ remains almost constant and equal to the value assumed for the regular lattice over a wide range of p. Thus, in this model, the transition to a small-world topology appears to be almost undetectable at the local level. For small values of p, each new long-range connection has a highly nonlinear effect on the average diameter d. In contrast, removing an edge from a clustered neighborhood has at most a small linear effect on the cliquishness c.

In the case of a one-dimensional ring lattice, a 'mean field solution' for the average path length and for the distribution of path lengths in the model is given in Newman *et al.* (2000). The basic idea behind the mean field approximation is to represent the distribution of relevant variables over many realizations by their average values. The authors apply the approximation to the continuous case first and use the fact that when the density of shortcuts is low, the discrete and continuous models are equivalent (see also Barthélémy and Amaral (1999) for an analysis of the transition from regular local lattice to small-world behavior and Amaral *et al.* (2000) for a classification of small-world networks).

Even as a very vague model of the Web graph, however, the lattice perturbation model has several limitations. First, there is no clear concept of an underlying lattice and the notion of short and long links on the Web is not clearly defined. While short

Table 3.2 Small-world networks according to Watts and Strogatz (1998). n, number of vertices; k, average degree; d, average distance or diameter; c, cliquishness. Random graphs have the same number of vertices and same average connectivity. The difference in average connectivity in the actor networks with respect to Table 3.1 results apparently from the inclusion of TV series in addition to films (A.-L. Barabási 2002, personal communication).

	n	k	d	d_{random}	c	c_{random}
Actors	225 226	61	3.65	2.99	0.79	0.000 27
Power grid	4 941	2.67	18.7	12.4	0.080	0.005
C. elegans	282	14	2.65	2.25	0.28	0.05

links could correspond to links within organizations and some correlation between link density and geographical distance may be expected, the whole point of the Web is to make a link between pages that are geographically distant as easy to create as a link between pages that are geographically close. More importantly perhaps, it can be shown that the edge rewiring procedure does *not* yield a graph with power-law distributed connectivity (see also Barthélémy and Amaral 1999). The degree distribution remains bounded and strongly concentrated around its mean value. Thus, although the lattice rewiring procedure yields small-world networks, it does not yield a good model for the connectivity of the Web graph. Other models are needed to try to account for the scale-free connectivity distribution.

3.2.3 Preferential attachment models, or the rich get richer

There are two aspects of many real networks that are not addressed by either the traditional random graph models, or by the models in Watts and Strogatz (1998) described above. First, in the traditional random graph approach or in the disordered lattice approach, the number of vertices is fixed in advance rather than growing in time. Second, in many real networks, edges are not created independently at random, but rather seem to follow some preferential attachment rule. The probability that a new actor is cast with a well-established actor is higher than the probability of them being cast with a less well-known actor. Likewise, a new Web page is more likely to be linked to a well-known page, which is already well connected.

In Barabási and Albert (1999) a rich-get-richer model is proposed that incorporates these features. The model starts with a set M_0 of vertices, at time zero. At each time step, a new node is created together with $m \leqslant M_0$ links. These new links connect the new node to m randomly selected nodes from the set of already existing nodes. Preferential attachment is modeled by assuming that the probability that a newly created vertex v is connected to an existing vertex w is proportional to the degree $k_w(t)$ of w, so that the corresponding probability of attachment is given by $k_w / \sum_r k_r$. This is the simplest formulation but other forms of preferential attachment are of course possible. After t time steps, the network has $M_0 + t$ vertices and mt edges. Simulations show that such a model evolves into a scale-free network, where the probability that a vertex has connectivity k follows a power-law distribution with exponent $\gamma = 2.9$ with $M_0 = m = 5$. Simulations also show that after a relatively short transient phase, this distribution becomes stationary and does not depend on the network size. The stationary power-law distribution is not observed if there is preferential attachment and the number of vertices is kept fixed or, vice versa, if the number of nodes grows but attachment is random and uniform (see also Barabási *et al.* 1999). In the case of uniform attachment for instance, the degrees are geometrically distributed (see Barabási *et al.* (1999) for an intuitive argument and Bollobás *et al.* (2001) for a rigorous proof).

This behavior can be understood using a simple 'mean field' treatment assuming continuous connectivity (Barabási *et al.* 1999). The basic principle behind a mean field analysis is to replace variable quantities by their averages. Consider vertex v after t time steps with a degree $k_v(t)$ when the total number of edges is mt. When a new vertex is added, the probability that it is joined to v is $mk_v(t)$ divided by the sum of all the degrees, which is twice the total number of edges. By taking expectations and assuming that multiple edges between nodes are allowed, we have from the preferential attachment rule that

$$k_v(t+1) - k_v(t) = m\frac{k_v(t)}{2mt} = \frac{k_v}{2t}. \tag{3.6}$$

Moving to a continuous formalism and replacing difference equations by differential equations, the rate at which a vertex v acquires edges is $\partial k_v/\partial t = k_v/2t$. If t_v is the time at which vertex v is added to the network, we have the initial condition $k_v(t_v) = m$ giving $k_v(t) = m(t/t_v)^{0.5}$. Thus, the connectivity of older vertices with a smaller t_v grows faster than the connectivity of younger vertices. This can also be used to get an estimate of the exponent γ. Indeed,

$$P(k_v(t) < k) = P\left(t_v > \frac{mt^2}{k^2}\right) = 1 - P\left(t_v \leqslant \frac{mt^2}{k^2}\right) = 1 - \frac{m^2 t}{k^2(t + M_0)}. \tag{3.7}$$

The last equality assumes that vertices are added at equal time intervals to the system, yielding the density $P(t_v) = 1/(M_0 + t)$. The probability density $P(k)$ satisfies $P(k) = \partial[k_v(t) < k]/\partial k$, which leads to the stationary solution $P(k) = 2m^2/k^3$ corresponding to $\gamma = 3$. The authors also report that nonlinear forms of preferential attachment do not lead to stationary power-law distributions.

A mathematically rigorous treatment of this model can be found in Bollobás *et al.* (2001), who note that the description of the model is rather vague in two respects. First, when there are no edges in the network, the preferential attachment distribution is not defined and if $k_v = 0$ it remains equal to zero in the future. This can easily be remedied by using a uniform distribution on the first M_0 vertices, until each one of them has degree at least one. More importantly, the question of how to choose the m earlier vertices of attachment at each time step is not precisely defined. Bollobás *et al.* (2001) propose a more precise version, where vertices are chosen one at a time and both multiple edges and loops are allowed. As it turns out, multiple edges and loops remain rare and therefore do not really impact the overall behavior, or the value of the exponent. Using this precise model, Bollobás *et al.* (2001) prove that $\gamma = 3$, first for the case of $m = 1$ and then for the general case for degrees up to $O(\sqrt{n})$, by deriving the general case from the case of $m = 1$.

It has been shown, via simulations, that this model of preferential attachment leads to small-world networks, i.e. graphs with small diameter, with an asymptotic diameter of roughly $\log n$, where n is the number of growth steps (or, equivalently, the number of vertices). Bollobás and Riordan (2003) have shown rigorously that $\log n$ is the right asymptotic value when $m = 1$ and that for $m \geqslant 2$ the correct asymptotic bound is in fact $\log n / \log \log n$. It is worth observing that for $m = 1$ the graphs are essentially trees so that the bound in this case, as well as other results, can be derived from the theory of random recursive trees (see Mahmoud and Smythe 1995; Pittel 1994). (A recursive tree is a tree on vertices numbered $\{1, 2, \ldots, n\}$, where each vertex other than the first is attached to an earlier vertex. In a random recursive tree this attachment occurs at random with a particular distribution such as uniform, preferential attachment, and so forth.)

Of interest is the observation that a diameter growing like $\log n$ is in fact very common in a variety of random graphs, including random regular graphs (Bollobás and de la Vega 1982). Thus, from a random graph standpoint, the small-world properties of the Internet are perhaps not that surprising. In fact, rather than our naive six-degree-of-separation surprise, the question becomes rather why the diameter would be so large! The smaller diameter of $\log n / \log \log n$ obtained when $m > 1$ is a step in that direction.

The model above only yields a characteristic exponent value of $\gamma = 3$ and produces only undirected edges. The graphs it produces are also too structured to be good models of the Internet: for instance when $m = 1$ the graph consists of M_0 trees, since there are M_0 initial components and $m = 1$ ensures that each component remains a tree. It is clear, however, that the model can be modified to produce directed edges, accommodate other exponents, and to produce more complex topologies.

To address the edge orientation problem, the orientation of each new edge could be chosen at random with a probability of 0.5 and the preferential attachment rule could take into consideration both indegrees and outdegrees. The fixed exponent problem can be obviated, for instance, with a richer mixture model where not only new nodes, but also new links are added (without adding nodes), and links are also rewired (Albert and Barabási 2000) (see also Cooper and Frieze (2001) for a similar model and an asymptotic mathematical analysis based on linear difference equations). In one implementation, the model above is extended by incorporating at each step the possibility of adding m new edges (without adding nodes) or rewiring m existing edges guided, in both cases, by the preferential attachment distribution. The three basic processes – adding nodes, rewiring, adding links – are assigned mixture probabilities p, q and $1 - p - q$. An extension of the mean field treatment given above shows that depending on the relative probability of each one of the three basic processes and m (the number of new nodes or edges at each step), the system can lead to either an exponential regime or a power-law regime, with different power-law exponents γ. Clearly these variations also break the tree-like structure of the simple model.

The preferential attachment model is quite different from the standard random graph models where the size is fixed. Even without the preferential attachment mechanism, when graphs grow older, nodes tend to have higher connectivity simply due to their age and this fact alone tends to remove some of the randomness (Callaway *et al.* 2001). The simplicity of the preferential attachment model is its main virtue but also its main weakness and it can only be viewed at best as a first-order approximation. While the model does reproduce the scale-free and small-world properties associated with the Web graph, it is clear that more realistic models need to take into account other higher order effects, including deletion of pages or links, differences in attachment rates, weak correlations to a variety of variables such as Euclidean distance, and so forth. Deviation from power-law scaling has been observed, especially for nodes with low connectivity. In addition, the deviations seem to vary for different categories of pages (Pennock *et al.* 2002). For example, the distribution of hyperlinks to university home pages diverges strongly from a power-law distribution and seems to follow a far more uniform distribution. Additional work is needed to better understand the details of the distribution and its fluctuations, and to create more precise generative models that can capture such fluctuations and other higher order effects.

3.2.4 Copy models

Kleinberg *et al.* (1999) have proposed a different 'copy' model, where vertices are 'born' one at a time but, instead of preferential attachment, each new vertex chooses an existing vertex and copies a random subset of its links (see also Gilbert 1997) or, alternatively, chooses new neighbors uniformly on its own. This copy process is in part suggested by the fact that often when a new page is created it is inspired by another page and retains a subset of its links. Naturally, processes for deletion of nodes and edges can also be added to the model.

To be more precise, in the simplest version of these models, at each time step a node is added. With probability p a new edge is created between the new node and one of the pre-existing nodes chosen uniformly. With probability $1 - p$, one of the pre-existing nodes and one of its pre-existing edges is chosen uniformly and the edge is copied (in the sense that a new link is created from the new page to the same page pointed to by the pre-existing edge). This can again be viewed as a mixture of two different elementary processes.

Simulations in Kumar *et al.* (1999a, 2000) report that this model also leads to power-law degree distribution and also to a high number of dense bipartite subgraphs found in the Web graph (see below). The copying process, however, introduces complex dependencies between the variables and analytical treatments of these copy models have not been developed so far. It is also clear that in its most simple form the copy model yields a tree with n vertices and $n - 1$ edges which is too crude to provide an approximation of the Web graph. Finally, while copy models were originally introduced as an alternative to the simple rich-get-richer model, they can in fact be viewed as a generalization. In the simple version above, with probability p an edge is created to an existing node uniformly, and with probability $1 - p$ an edge is created to an existing node with a probability that in the directed case strongly depends on its indegree. If an existing page or node v has indegree $|pa[v]|$ and each of its parents $i = 1, \ldots, |pa[v]|$ has outdegree $|ch[i]|$, then the probability of connecting to v through the copy mechanism is given by

$$\sum_{i=1}^{|pa[v]|} \frac{1}{n|ch[i]|}.$$

This is because once the copy mechanism has been selected, there is a $1/n$ chance of choosing a given parent i of v, and a $1/|ch[i]|$ chance of choosing the edge running from i to v. Thus, everything else being equal, the copying process favors creating a link to nodes that already have a large indegree and therefore is a form of preferential attachment based on vertex degrees (to be exact, the most favored attachment nodes are those with very high indegree and whose parent pages have low outdegree).

3.2.5   PageRank models

It should be clear by now that alternative or more general growth models can be created by using a mixture of elementary mechanisms, and by using some form of preferential attachment based on any kind of node or page metric. In particular, we have seen that the relevance of a page can be assessed by its PageRank and that in some ways this may be a good alternative to using the indegree of that page. As a result, it is possible to develop models of preferential attachment based on PageRank (or other measures) rather than connectivity. The model proposed in Pandurangan *et al.* (2002) is an example of a mixture model where a new node is uniformly connected with probability p to an existing node, and with probability $1 - p$ to an existing node in proportion to its PageRank. An obvious mixture generalization is to have a probability p for selecting nodes uniformly, a probability q ($p + q \leqslant 1$) of selecting nodes proportionally to their Page Rank, and a residual probability $1 - p - q$ of selecting nodes proportionally to their degree.

Experiments and mean field analysis are reported in Pandurangan *et al.* (2002) for the 'brown.edu' domain (roughly 100 000 pages and 700 000 hyperlinks), the smaller 'cs.brown.edu' domain (roughly 25 000 pages and 175 000 hyperlinks), and the larger WT10g corpus (for details see http://www.ted.cmis.csiro.au/TRECWeb/wt10ginfo.ps.gz), containing roughly 1.69 million documents. The authors confirm the power-law distribution of indegrees, outdegrees, degrees, and PageRanks across all three scales. Mean field analysis and simulations suggest that preferential attachment guided by PageRank alone is not sufficient to reproduce all power-law behaviors. The power-law behavior of the indegree, outdegree, and Page-Rank distributions can be fitted by a preferential attachment model based on degree alone. In addition, the three distributions can also be fitted by a uniform simulated mixture model, where the probability of attachment based on degree is equal to the probability of attachment based on PageRank (i.e. $p \approx q \approx 1 - p - q \approx 0.33$).

In time, additional components could be added to the mixture to reflect attachment biases that are driven by other considerations than the local (degree) or global (Page-Rank) popularity of a node such as, for instance, the thematic aspects associated with local communities and locally densely interconnected subgraph described below.

3.3 Applications

The graphical properties of the Web and other networks are worth investigating for their own sake. In addition, these properties may also shed light on some of the functional properties of these networks, such as robustness. In the case of the Web, they may also lead to better algorithms for exploiting the resources offered by the underlying network. Systems with small-world properties may have enhanced signal-propagation speed and synchronizability. In particular, on the Web, an intelligent agent that can interpret the links and follow the most relevant ones should be able to find information rapidly. An intelligent agent ought to be able also to exploit the correlations that exist between content and connectivity. Furthermore, thematic communities of various sorts ought to be detectable by patterns of denser connectivity that stand above the general background of hyperlinks.

3.3.1   Distributed search algorithms

The existence of short paths between two vertices in the Web and many other small-world networks arising in nature and technology is significant but raises the questions of whether such short paths, or an approximation to them, can be found efficiently using local knowledge. Because the average path length is typically bounded by a polynomial in $\log n$, mathematically it would be very satisfying if local algorithms could be developed that can find a short path between two vertices in the network in average time also bounded by a polynomial of $\log n$. This is in sharp contrast with the polynomial bound for finding the shortest path using global knowledge of the graph and breadth-first/dynamic programming (Cormen *et al.* 2001).

The existence of navigation algorithms with average time bounded by a polynomial in $\log n$ seems plausible for at least two reasons:

(a) we do not really need to find the shortest path, just a reasonably short path; and

(b) unlike the case of breadth first, which requires examining essentially all the possible paths, the local knowledge or latent structural clues available in the network may provide a means of restricting the search only to a small subset of most promising paths.

To some extent, the feasibility of routing with local information in certain complex networks is corroborated by experiments. In the original experiment described by Milgram (1967), several people in a small town in Nebraska were given the task of sending a letter to a target in Massachusetts, with the instruction that the letter be forwarded at each step by the sender to someone they knew by their first name. The experiment revealed that the typical letter path had a length of six, hence the term 'six degrees of separation'. The success of this experiment suggests that it may possible for intelligent agents to find short paths in complex networks using only local address information. Furthermore, the experiment also provides some clues about possible algorithms. 'The geographic movement of the [message] from Nebraska to Massachusetts is striking. There is a progressive closing in on the target areas as each new person is added to the chain' (Milgram 1967; see also Hunter and Shotland 1974; Killworth and Bernard 1978; White 1970).

This observation, as well as casual observation of how we seek information over the Web, suggests a trivial greedy algorithm, where at each step the agent locally chooses whichever link is available to move 'as close as possible' to the target. The problem lies in the definition of closeness and the kind of information available regarding closeness at each node. Clearly, in the extreme case, where distances are measured in terms of path lengths in the underlying graph and where the distance to the target node is available at each node (but not necessarily the path), an agent can find the shortest path requiring a number of steps equal to the distance (hence a polynomial in $\log n$ steps on average in a small-world network) using the obvious greedy algorithm. But agents navigating the Web using local information would not have precise access to graph distances. Thus, more sophisticated or mathematically interesting models must rely on a weaker form of local knowledge and weaker estimates or approximations to the distance to the target. At one other simple extreme, consider that, when the agent examines all the links emanating from one node, instead of having access to the precise information on how much closer or further away the link would bring him to the target node, the only local information he is given is noisy binary information on whether each link brings him closer to or further away from the target.

In Kleinberg (2000a,b, 2001) local greedy navigation algorithms are discussed for lattices that have been perturbed, along the lines of the model in Watts and Strogatz (1998), with directed connections. These are square $n = l \times l$ lattices in two dimensions or, more generally, $n = l^m$ cubic lattices in m dimensions. Each vertex is fully connected to all its neighbors within lattice distance l_0, with an additional q independent long-ranged connections to nodes at lattice distance r chosen randomly with probability proportional to $r^{-\alpha}$. When $\alpha = 0$, this corresponds to a uniform choice of long-ranged connections.

The greedy algorithm to navigate from u to v always chooses the connection that gets the closest to the target. In two dimensions ($m = 2$), the greedy algorithm has a rapid delivery time, measured by the expected number of steps, which is bounded by a function proportional to $(\log n)^2$ provided $\alpha = m = 2$. In particular, in the case of uniform long-range connections ($\alpha = 0$), no decentralized algorithm can achieve rapid delivery time. This uniform case corresponds to a 'world in which short chains exist but individuals, faced with a disorienting array of social contacts, are unable to find them' (Kleinberg 2000b). For $\alpha \neq m$, the delivery time is asymptotically much larger and greater than a polynomial in l. Similar results hold in m dimensions, with rapid (polynomial in $\log n$) delivery if and only if $\alpha = m$. Likewise, in m dimensions, the greedy algorithm has a rapid delivery time bounded by a polynomial in $\log n$ provided that $\alpha = m$. For $\alpha \neq m$, the delivery time is asymptotically much larger.

This result is interesting but should be viewed only as a first step. Its significance is undermined by the fact that the graphs being considered are regular instead of power-law distributed. Furthermore, the behavior does not seem to be robust and requires a very specific value of the exponent α. Finally, the routing algorithm requires knowledge of lattice distances between nodes and the global position of the target that may not be trivial to replicate in a realistic Web setting. Further extensions and analysis are the object of ongoing research in the context, for instance, of peer-to-peer systems, such as Gnutella (Adamic *et al.* 2001), or Freenet (Zhang and Iyengar 2002). A model for searchability in social networks that leverages the list of attributes identifying each individual is described in Watts *et al.* (2002).

From a different perspective, given the problem of navigating from u to v, an intelligent agent starting at u can be modeled as an agent that somehow knows whether any given edge brings it closer to its target v, with some error probability. This captures the idea that a human trying to find the 'date of birth of Albert Einstein' is more likely to follow a link to a page on 'Nobel prizes in physics' than a link to a page on 'Nobel prize in economics'. If, for the sake of the argument, in a small-world network the typical distance between u and v is $\log n$, then an agent that can detect good hyperlinks with an error smaller than $1/\log n$ will typically make no mistakes along a path of length $\log n$ and therefore will typically reach the target. However, even if the agent has an error rate of 0.5, then, by a simple diffusion (or random walk) argument, it will take on the order of $\log^2 n$ steps to reach the target, which is still polynomial in $\log n$. Thus, in this sense, the problem of building an intelligent agent, can be reduced to the problem of building an agent that can discern 'good' links with a 0.5 error rate or better.

3.3.2 Subgraph patterns and communities

Both the real life networks and their models studied in this chapter show that the property of a power-law degree distribution and small diameter do not entirely characterize all the statistical topological properties of a network, and that there are additional properties and structures that can be considered which may be more characteristic of certain specific networks. In particular, there are subgraphs of the Web graph that appear to occur more frequently than one would expect by chancel and which have been used both to discover content as well as 'communities' and to follow their dynamics (Chakrabarti *et al.* 1999a, 2002; Flake *et al.* 2000, 2002; Gibson *et al.* 1998; Kleinberg 1999; Kleinberg and Lawrence 2001; Kumar *et al.* 1999b). For example, clique-like structures (Kumar *et al.* 1999b) are overabundant in the Web graph: another departure from standard random graphs. In fact, particular subgraph motifs appear to form the building blocks and characteristic signatures of complex networks (Milo *et al.* 2002; Oltvai and Barabási 2002).

A statistically significant high density of hyperlinks among a set of Web pages is an indication that they may be topically related. We have seen that the basic background density of the Web graph corresponds to a sparse graph where the number of edges scales linearly with the number of vertices. Deviations from this background can be defined in many ways. In one possible definition, a community can be defined as any cluster of nodes such that the density of links between members of the community (in either direction) is higher than the density of links between members of the community and the rest of the network. Variations on this definition are possible to detect communities with varying levels of cohesiveness.

Another characteristic pattern in many communities is the hubs–authority bipartite graph pattern (Kleinberg and Lawrence 2001), where a hub is an index or resource page that points to many authority or reference pages in a correlated fashion. This is studied further in Chapter 5. Likewise, an authority page is a page that is pointed to by many hub pages. Other patterns discussed in the literature include the case where authorities are linked directly to other authorities (Brin and Page 1998).

Once a particular subgraph pattern is selected as a signature of a community, the Web graph can be searched to identify the occurrence of such signatures. In many cases, the general graphical search problem is NP-complete. However, polynomial time algorithms can be derived with some additional assumptions. In Flake *et al.* (2002), for instance, communities are associated with maximum flow algorithms.

Large-scale studies reported in the literature have identified hundreds of thousands of communities, including some unexpected ones (e.g. 'people concerned with oil spills off the coast of Japan'). Although analyses that are purely structural can discover interesting results, it is clear that it is possible to integrate structure and content analysis (Chakrabarti *et al.* 2002). For instance, once a community is found based on hyperlink structure, words that are common to the pages in the community can be identified and used to further define and reliably extract the community from the rest of the network. Likewise, textual analysis can be used to seed a community search.

Discovering communities and their evolution may be useful for a number of applications including:

(a) catching trends at early stages;

(b) facilitating the transmission of information (or misinformation) and bringing together individuals with common interests; and

(c) targeted marketing.

The importance of link analysis for search engines and information retrieval will be explored further in Chapter 5. It must be pointed out, however, that the utilization of link analysis in these applications opens the door for 'second-order' applications. For instance, search engines based on link analysis may tend to increase the popularity of pages that are already popular and it may be important to develop methods to attenuate this effect.

3.3.3   Robustness and vulnerability

There is a clear sense that many of the networks studied in this chapter are in some sense robust. In spite of frequent router problems, website problems, and temporary unavailability of many Web pages, we have not yet suffered a global Web outage. Likewise, mutation or even removal of a few enzymes in *E. coli* is unlikely to significantly disrupt the function of the entire system. Mathematical support for such intuition is provided by studying how the diameter or connectivity of the graph is affected by deleting nodes randomly. When a few nodes are deleted at random, the graph remains by and large connected and the small-world property remains intact. In other words, the systems are robust with respect to random errors or mutations. However, as pointed out in Albert *et al.* (2000), the systems considered in this chapter are vulnerable to targeted attacks that focus on the nodes with the highest connectivity. These targeted attacks can significantly disrupt the properties of the systems.

To be more specific, with the same number of nodes and connections, a scale-free graph has a smaller diameter in terms of average shortest path than a uniform random graph. In a random uniform graph (sometimes also called random exponential graph, when referring to the decay of the connectivity) all nodes are more or less equivalent. In simulations reported in Albert *et al.* (2000) using random exponential and scale-free graphs, when a small fraction f (up to a few percentage points) of nodes are removed at random, the diameter increases monotonically (linearly) in an exponential graph, whereas the diameter remains essentially constant in a scale-free graph with the same number of nodes and edges. Thus, in scale-free graphs, the ability of nodes to communicate is unaffected even by high rates of random failures. Intuitively, this is not surprising and is rooted in the properties of the degree distribution of each kind of graph. In an exponential network, all the nodes are essentially equivalent and have similar degrees. As a result, each node contributes equally to the connectivity and to the value of the diameter. In a scale-free graph with a skewed degree distribution, randomly selected nodes are likely to have small degrees and their removal does not disrupt the overall connectivity.

This situation, however, is somewhat reversed during an attack where specific nodes are targeted. An attack is simulated in Albert and Barabási (2000) by removing nodes in decreasing order of connectivity. In an exponential graph, there is essentially no difference between random deletions and attack. In contrast, the diameter of the scale-free graph increases rapidly in the scale-free network, roughly doubling when 5% of the nodes are removed. A higher degree of robustness of the undirected Web graph to attacks, however, has been reported in other studies (Broder *et al.* 2000).

Similar effects are also observed when the connected components of both kinds of graphs are studied, both under random error and targeted attack scenarios. In particular, the scale-free network suffers a substantial change in its topology under attack, in the sense that it undergoes fragmentation and breaks down into many different connected components when under attack, whereas it basically remains connected during random removal of a few nodes.

To be more precise, two statistics that can be used to monitor the connected components during node deletion are the size S of the largest connected component relative to the size of the system and the average size s of the connected components other than the largest one. In the case of an exponential graph, when f is small both $S \approx 1$ and $s \approx 1$ and only isolated nodes fall off the network. As f increases, s increases, until a phase transition occurs around $f = 0.28$, where the connectivity breaks down, the network fragments into many components, and s reaches a maximum value of 2. Beyond the transition point, s decreases progressively toward unity and S rapidly decreases toward zero. This behavior is observed both under random node deletion and under attack.

In contrast, in a scale-free network under random deletion of nodes as f is increased S decreases linearly and s remains close to unity. In case of attack, however, the network breaks down in a manner similar to but faster than the breakdown of the exponential network under random deletion: S drops rapidly to zero with a critical threshold of 0.18 instead of 0.28. s has a peak of two, reached at the critical point. When tested on a directed subgraph of the entire Web, containing 325 729 nodes and 1 469 680 links, this critical point is found to occur at $f = 0.067$ (Albert and Barabási 2000).

There are many other important aspects of Internet/Web networks' robustness, security, and vulnerability that cannot be addressed here for lack of space. We encourage the interested reader to explore the literature further and limit ourselves to providing two additional pointers. On the vulnerability side, it is worth mentioning that the architecture and protocols of the Internet can be exploited to carry parasitic computations in which servers unwittingly perform computations on behalf of a remote node (Barabási et al. 2001). On the robustness side, peer-to-peer networks such as Gnutella (Ripeanu et al. 2002) or Freenet (Clarke et al. 2000, 2002) are robust Internet network architectures that do not rely on centralized servers.

3.4   Notes and Additional Technical References

Besides growth models, there are different possible mathematical approaches to defining random power-law graphs. Some of these definitions generalize the original definitions of uniform random graphs given in Erdos and Rényi (1959, 1960). One possibility similar to $G(n, e)$ (where $e = |E|$) is to introduce a uniform distribution over all graphs that satisfy a particular power-law degree sequence. This is a special case of random graphs with a given degree sequence, for which there is an abundant literature. More precisely, the class $G(\alpha, \gamma)$ is defined by considering a uniform distribution over all graphs where the number y of nodes with degree x satisfies $y = e^{\alpha}/x^{\gamma}$. Alternatively one can use a definition similar to the definition of the class $G(n, p)$ for uniform random graphs, yielding an expected degree sequence rather than a fixed degree sequence (Chung and Lu 2002). As in the uniform case, the two models are essentially equivalent, at least in an asymptotic sense. Theorems on the connectivity of random power-law graphs are given in Aiello et al. (2001). In particular:

- when $\gamma < 1$, the graph is almost surely connected;

- when $1 < \gamma < 2$, there is a giant component of size $O(n)$ and all smaller components have size $O(1)$;

- when $2 < \gamma < \gamma_0 = 3.4785$, there is a giant component of size $O(n)$ and all smaller components have size $O(\log n)$ (the size is $O(\log n/\log\log n)$);

- when $\gamma_0 < \gamma$, almost surely there are no giant components.

Thus, in short, when γ is small there is a giant component and when $\gamma > \gamma_0$ there is no giant component, but a large number of small components. Additional results include the fact that for $\gamma > 4$, the number of components of a given size follows approximately a power-law distribution (see also Chung and Lu 2002). For comparison purposes, a different, simpler, and relevant class of random graphs with given degree sequence are random regular graph with fixed degree. For instance, the diameter of random directed graphs with constant outdegree is studied in Philips *et al.* (1990).

Because the networks studied in this chapter are sparse, another relevant class of random graphs is random sparse graphs, i.e. random graphs with $|E| = o(n^2)$ edges. This class can again be defined by, for instance, introducing a uniform distribution over all sparse graphs. The diameter of random sparse graphs is studied in Chung and Lu (2001). Finally, quasi-random graphs are studied in Chung *et al.* (1989) and sparse quasi-random graphs are studied in Chung and Graham (2002). Chung *et al.* (2001) studies the problem of constructing a weighted graph to optimally realize a set of pairwise distances between a subset of the vertices, with an application to what has been called Internet tomography (reconstruction of the whole from low-dimensional 'slices').

3.5   Exercises

Exercise 3.1. Provide at least two different 'back-of-the-envelope' estimates for the size (in terms of the number of vertices and the number of edges) for the graph associated with the documents and hyperlinks of the Web. Refine such estimates by sampling IP addresses (using the `ping` command) or Web pages by some kind of crawling. Discuss the limitations of these estimates and how they could be improved. For instance, the `ping` command samples only active machines that are authorized to reply to the `ping` query. Consider also the possibility of probing typical ports (e.g. 80, 8080).

Exercise 3.2. Crawl the entire website of your university or some other large organization and compute the statistics of the corresponding graph including the distribution of the connectivity and the diameter both in the directed and undirected cases. [Easier alternatives to the crawl consist in downloading a publicly available data set or using a program such as wget (`http://www.gnu.org/software/wget/wget.html`).]

Exercise 3.3. Show that, if the connectivity of a large graph with n vertices has a power law with exponent $\gamma \geqslant 2$, then the graph is sparse in the sense that it has $O(n)$ vertices.

Exercise 3.4. In this chapter, we have listed the power-law exponent observed for several phenomena. In general, the values of the exponent are relatively small, say less than 10. How general is this and why?

Exercise 3.5. Fill in some of the entries in Table 3.1.

Exercise 3.6. Explore possible relationships between the Fermi model of Chapter 1 and the models introduced in this chapter to account for the power-law distribution of the size of the sites, or the power-law distribution of the connectivity.

Exercise 3.7. Simulate directly random directed graphs where the indegrees and outdegrees of the nodes follow a power-law distribution with the exponents given in Table 3.1. Estimate the average and maximal shortest distance between vertices, and how it scales with the number n of vertices. Study the distribution of this shortest distance across all possible pairs of vertices.

Exercise 3.8. Run simulations of lattice perturbation models, along the lines of those found in Watts and Strogatz (1998), on ring, square, and cubic lattices. Progressively increase the amount of random long-ranged connections in these graphs and monitor relevant quantities such as degree distribution, cliquishness, and average/maximal distance between vertices.

Exercise 3.9. Simulate the 'rich get richer' model and monitor relevant quantities such as degree distribution, cliquishness, and average/maximal distance between vertices. Examine modifications of the basic model incorporating any of the following: edge orientation, edge rewiring, edge creation and monitor relevant quantities. Try to analyze these modifications using the 'mean field' approach associated with Equations (3.6) and (3.7). Examine a growth model with uniform, rather than preferential, attachment. Show through simulations or a mean field type of analysis that the distribution of degrees is geometric rather than power law.

Exercise 3.10. Simulate a version of the copy model and model relevant quantities such as degree distribution, cliquishness, and average/maximal distance between vertices.

Exercise 3.11. First simulate a random network with power-law connectivity and small network properties using, for instance, the preferential attachment model of Barabási and Albert (1999). Then consider an agent that can navigate the network and is faced with the typical task of finding a short path between two vertices u and v using only local information. Suppose each time the agent considers an edge, he has binary information on whether this edge brings it closer or not to the target, and that this information comes with some error rate p. Design a navigation strategy for the agent and estimate its time complexity by simulations and by a 'back-of-the-envelope' calculation. Simulate this strategy on several of the networks discussed in this chapter.

Study how the typical travel time varies with p. Study the robustness of this strategy with respect to fluctuations in p. Modify the strategy to include a bias toward nodes with high degree, or high PageRank and study its time complexity. How would you model the effect of a search engine in this framework?

Exercise 3.12. Run simulations to study the effect of random and selective node deletion on the diameter and fragmentation of scale-free and exponential graphs.

4

Text Analysis

Having focused in earlier chapters on the general structure of the Web, in this chapter we will discuss in some detail techniques for analyzing the textual content of individual Web pages. The techniques presented here have been developed within the fields of *information retrieval* (IR) and *machine learning* and include indexing, scoring, and categorization of textual documents.

The focus of IR is that of accessing as efficiently as possible and as accurately as possible a small subset of documents that is maximally related to some user interest. User interest can be expressed for example by a query specified by the user. Retrieval includes two separate subproblems: indexing the collection of documents in order to improve the computational efficiency of access, and ranking documents according to some importance criterion in order to improve accuracy. *Categorization* or classification of documents is another useful technique, somewhat related to information retrieval, that consists of assigning a document to one or more predefined categories. A classifier can be used, for example, to distinguish between relevant and irrelevant documents (where the relevance can be personalized for a particular user or group of users), or to help in the semiautomatic construction of large Web-based knowledge bases or hierarchical directories of topics like the Open Directory (http://dmoz.org/).

A vast portion of the Web consists of text documents – thus, methods for automatically analyzing text have great importance in the context of the Web. Of course, retrieval and classification methods for text, such as those reviewed in this chapter can be specialized or modified for other types of Web documents such as images, audio or video (see, for example, Del Bimbo 1999), but our focus in this chapter will be on text.

4.1 Indexing

4.1.1 Basic concepts

In order to retrieve text documents efficiently it is necessary to enrich the collection with specialized data structures that facilitate access to documents in response to

Modeling the Internet and the Web   P. Baldi, P. Frasconi and P. Smyth
© 2003 P. Baldi, P. Frasconi and P. Smyth   ISBN: 0-470-84906-1

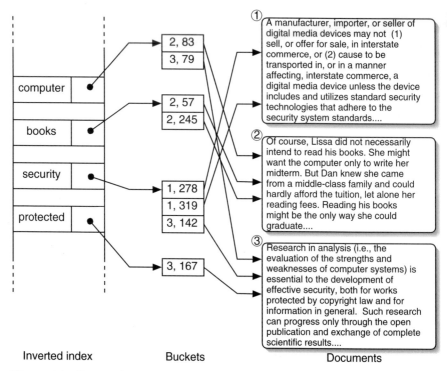

Figure 4.1 Structure of a typical inverted index for information retrieval. Each entry in the occurrence lists (buckets) is a pair of indices that identify the document and the offset within the document where the term associated to the bucket appears. Document (1) is an excerpt from Sen. Fritz Hollings's Consumer Broadband and Digital Television Promotion Act, published at http://www.politechbot.com/docs/cbdtpa/hollings.s2048.032102.html. Document (2) is an excerpt from Stallman (1997) and (3) is an excerpt from the declaration submitted by ACM in the District Court of New Jersey on the case of Edward Felten versus the Recording Industry Association of America (http://www.acm.org/usacm/copyright/felten_declaration.html).

user queries. A substring search, even when implemented using sophisticated algorithms like suffix trees or suffix arrays (Manber and Myers 1990), is not adequate for searching very large text collections. Many different methods of text retrieval have been proposed in the literature, including early attempts such as clustering (Salton 1971) and the use of signature files (Faloutsos and Christodoulakis 1984). In practice, *inversion* (Berry and Browne 1999; Witten *et al.* 1999) is the only effective technique for dealing with very large sets of documents. The method relies on the construction of a data structure, called an *inverted index*, which associates lexical items to their occurrences in the collection of documents. Lexical items in text retrieval are called *terms* and may consist of words as well as expressions. The set of terms of interest is called the *vocabulary*, denoted V. In its simplest form, an inverted index is a dictionary where each key is a term $\omega \in V$ and the associated value $b(\omega)$ is a pointer to an additional intermediate data structure, called a *bucket* or posting list. The bucket associated with a certain term ω is essentially a list of pointers marking all the occurrences of ω in the text collection.

The general structure is illustrated in Figure 4.1. In the simplest case, the entries in each bucket can simply consist of the document identifier (DID), the ordinal number of the document within the collection. However, in many cases it will be more useful to have a separate entry for each occurrence of the term, where each entry consists of the DID and the offset (in characters) of the term's occurrence within the document. The latter approach has two advantages. First, recording all the occurrences of the term allows us to present a user with a short context (i.e. a fragment of surrounding text) in which the term occurs in the retrieved document. Second, the occurrence information enables vicinity queries such as retrieving all documents where two assigned terms are positionally close in the text.

The size of the inverted index is $\Omega(|V|)$ and this data structure can typically be stored in main memory.

It can be implemented using a hash table so that the expected access time is independent of the size of the vocabulary. Buckets and the document repository must typically be stored on disk. This poses a challenge during the construction phase of the inverted index. If the buckets can be stored in the main memory, the construction algorithm is trivial. It simply consists of parsing documents, extracting terms, inserting the term in the inverted index if not present, and finally inserting the occurrence in the bucket. However, if the buckets are on disk, this approach will generate a tremendous amount of disk accesses making the algorithm impractical, since accessing a random byte of information on disk can take up to 10^5 times longer than accessing a random byte in main memory. Thus, for very large collections of text it is necessary to resort to specialized secondary memory algorithms (Witten *et al.* 1999).

Searching for a single term ω in an indexed collection of documents is straightforward. First, we access the inverted index to obtain $b(\omega)$. Then we simply scan the bucket pointed to by $b(\omega)$ to obtain the list of occurrences. The case of Boolean queries with multiple terms is only slightly more complicated. Suppose the query contains literals associated with the presence or absence of k terms. The previous procedure can be repeated separately for each term, yielding k lists of occurrences. These lists are then combined by elementary set operations that correspond to the Boolean operators in the query: intersection for AND, union for OR, and complement for NOT (Harman *et al.* 1992).

4.1.2 Compression techniques

The memory required for each pointer in the buckets can be reduced by sorting the occurrences of each term by its DID. In this way, rather than storing each DID, it suffices to store the sequence of differences between successive DIDs as a list of *gaps*. For example, the sequence of DIDs

$$(14, 22, 38, 42, 66, 122, 131, 226, 312, 331)$$

may be stored as a sequence of gaps

$$(14, 8, 16, 4, 24, 56, 9, 95, 86, 19).$$

The advantage is that frequent terms will produce many small gaps, and small integers can be encoded by short variable-length codewords. In particular, γ encoding (Elias 1975) represents a gap g as the pair of bit strings (u_g, b_g), where u_g is the unary code for the number $\lfloor \log_2 g \rfloor + 1$ (i.e. $\lfloor \log_2 g \rfloor$ ones followed by a zero) and b_g is the standard binary code for the number $g - 2^{\lfloor \log_2 g \rfloor}$. For example, the integer '7' would be represented as $(110, 11)$ and '9' would be represented as $(1110, 001)$. In contrast, for a collection of n documents, storing an integer word for a DID would require $\lceil \log_2 n \rceil$ bits. Because terms are distributed according to the Zipf distribution (see Chapter 1), on average this represents a significant memory saving (see Exercise 4.1 for details). Moffat and Zobel (1996) discuss this and other index compression techniques, together with fast algorithms for query evaluation and ranking.

In the case of large collections of Web pages, efficient indexing may be challenging and traditional approaches may quickly became inadequate. The design of an efficient distributed index for Web documents is discussed in Melnik *et al.* (2001).

4.2 Lexical Processing

4.2.1 Tokenization

Tokenization is the extraction of plain words and terms from a document, stripping out administrative metadata and structural or formatting elements, e.g. removing HTML tags from the HTML source file for a Web page. This operation needs to be performed prior to indexing or before converting documents to vector representations that are used for retrieval or categorization (see Section 4.3.1). Tokenization appears to be a straightforward problem but in many practical situations the task can actually be quite challenging.

The case of HTML documents is relatively easy. The simplest approach consists of reducing the document to an *unstructured* representation, i.e. a plain sequence of words with no particular relationship among them other than serial order. This can be achieved by only retaining the text enclosed between <html> and </html> (see Section 2.1), removing tags, and perhaps converting strings that encode international characters to a standard representation (e.g. Unicode). The resulting unstructured representation allows simple queries related to the presence of terms and to their positional vicinity. More generally, the additional information embedded in HTML can be exploited to map documents to *semi-structured* representations. In this case, the representation is sensitive to the presence of terms in different elements of the document and allows more sophisticated queries like 'find documents containing *population* in a table header and *famine* in the title'. Extracting semi-structured representations from HTML documents is in principle very easy since tags provide all the necessary structural information. However, the reader should be aware that, although HTML has a clearly defined formal grammar, real world browsers do not strictly enforce syntax correctness and, as a result, most Web pages fail to rigorously comply to the HTML syntax.[1] Hence, an HTML parser must tolerate errors and include recovery mechanisms to be of any practical usefulness. A public domain parser is distributed with LIBWWW, the W3C Protocol Library `http://www.w3.org/Library/`. An HTML parser written in Perl (with methods for text extraction) is available at `http://search.cpan.org/dist/HTML-Parser/`.

Obviously, after plain text is extracted, punctuation and other special characters need to be stripped off. In addition, the character case may be *folded* (e.g. to all lowercase characters) to reduce the number of index terms.

Besides HTML, textual documents in the Web come in a large variety of formats. Some formats are proprietary and undisclosed and extracting text from such file types is severely limited by the fact that the associated formats have not been disclosed. Other formats are publicly known, such as PostScript[2] or Portable Document Format (PDF).

Tokenization of PostScript or PDF files can be difficult to handle because these are not data formats but algorithmic descriptions of how the document should be rendered. In particular, PostScript is an interpreted programming language and text rendering is controlled by a set of *show* commands. Arguments to show commands are text strings and two-dimensional coordinates. However, these strings are not necessarily entire words. For example, in order to perform typographical operations such as kerning, ligatures, or hyphenation, words are typically split into fragments and separate show commands are issued for each of the fragments. Show commands do not need to appear in reading order, so it is necessary to track the two-dimensional position of each 'shown' string and use information about the font in order to correctly reconstruct word boundaries. PostScript is almost always generated automatically by other programs, such as typesetting systems and printer drivers, which further complicates matters because different generators follow different conventions and approaches. In fact perfect conversion is not always possible. As an example of efforts in this area, the reader can consult Neville-Manning and Reed (1996) for details on PreScript, a PostScript-to-plain-text converter developed within the New Zealand Digital Library project (Witten *et al.* 1996). Another converter is Pstotext, developed within the Virtual Paper project (`http://research.compaq.com/SRC/virtualpaper/home.html`).

PDF is a binary format that is based on the same core imaging model as PostScript but can contain additional pieces of information, including descriptive and administrative metadata, as well as structural elements, hyperlinks, and even sound or video. In terms of delivered contents, PDF files are therefore much closer in structure to Web pages than PostScript files are. PDF files can (and frequently do, in the case of digital libraries) embed raster images of scanned textual documents. In order to extract text

---

[1] See `http://validator.w3.org/` for a public HTML validation service.

[2] PostScript is a registered trademark of Adobe Systems Incorporated.

from raster images, it is necessary to resort to optical character recognition (OCR) systems (Mori *et al.* 1992).

4.2.2   Text conflation and vocabulary reduction

Often it is desirable to reduce all the morphological variants of a given word to a single index term. For example, a document of interest might contain several occurrences of words like *fish*, *fishes*, *fisher*, and *fishers* but would not be retrieved by a query with the keyword *fishing* if the term 'fishing' never occurs in the text. Stemming consists of reducing words to their root form (such as *fish* in our example), which becomes the actual index term. Besides its effect on retrieval, stemming can be also useful to reduce the size of the index.

One well known stemming algorithm for English was developed by Porter (1980) as a simplification and systematization of previous work by Lovins (1968). It relies on a preconstructed suffix list with associated rules. An example of a rewriting rule is 'if the suffix of the word is IZATION and the prefix contains at least one vowel followed by a consonant, then replace the suffix with IZE'. This would transform, for example, the word BINARIZATION into the word BINARIZE. Porter's stemmer applies rules in five consecutive steps. Source code in several languages and a detailed description of the algorithm are available at http://www.tartarus.org/~martin/PorterStemmer/.

Another technique that is commonly used for controlling the vocabulary is the removal of stop words, i.e. common words such as articles, prepositions, and adverbs that are not informative about the semantic content of a document (Fox 1992). Since stop words are very common, removing them from the vocabulary can also significantly help in reducing the size of the index. In practice, stopwords may account for a large percentage of text, up to 20–30%. Naturally, removal of stop words also improves computational efficiency during retrieval.

4.3   Content-Based Ranking

A Boolean query to a Web search engine may return several thousands of matching documents, but a typical user will only be able to examine a small fraction of these. Ranking matching documents according to their relevance to the user is therefore a fundamental problem. In this section we review some classic approaches that do not exploit the hyperlinked structure of the Web. Ranking algorithms that take Web topology into account will be discussed in Chapter 5.

4.3.1   The vector-space model

Text documents can be conveniently represented in a high-dimensional vector space where terms are associated with vector components. More precisely, a text document

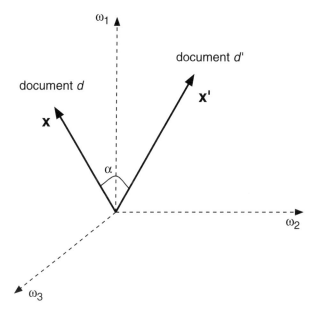

Figure 4.2   Cosine measure of document similarity.

d can be represented as a sequence of terms, $d = (\omega(1), \omega(2), \ldots, \omega(|d|))$, where $|d|$ is the length of the document and $\omega(t) \in V$. A vector-space representation of d is then defined as a real vector $x \in \mathbb{R}^{|V|}$, where each component x_j is a statistic related to the occurrence of the jth vocabulary entry in the document. The simplest vector-based representation is Boolean, i.e. $x_j \in \{0, 1\}$ indicates the presence or the absence of term ω_j in the document being represented.

Vector-based representations are sometimes referred to as a 'bag of words', emphasizing that document vectors are invariant with respect to term permutations, since the original word order $\omega(1), \ldots, \omega(|v|)$ is clearly lost. Representations of this kind are appealing for their simplicity. Moreover, although they are necessarily lossy from an information theoretic point of view, many text retrieval and categorization tasks can be performed quite well in practice using the vector-space model. Note that typically the total number of terms in a set of documents is much larger than the number of distinct terms in any single document, $|V| \gg |d|$, so that vector-space representations tend to be very *sparse*. This property can be advantageously exploited for both memory storage and algorithm design.

4.3.2   Document similarity

We can define similarity between two documents d and d' as a function $s(d, d') \in \mathbb{R}$. This function, among other things, allows us to rank documents with respect to a query (by measuring the similarity between each document and the query). A classic approach is based on the vector-space representation and the metric defined by the *cosine coefficient* (Salton and McGill 1983). This measure is simply the cosine of the angle formed by the vector-space representations of the two documents, x and x' (see Figure 4.2),

$$\cos(x, x') = \frac{x^\mathrm{T} x'}{\|x\| \cdot \|x'\|} = \frac{x^\mathrm{T} x'}{\sqrt{x^\mathrm{T} x} \cdot \sqrt{x'^\mathrm{T} x'}}, \tag{4.1}$$

where the superscript 'T' denotes the 'transpose' operator and $x^\mathrm{T} y$ indicates the dot product or inner product between two vectors $x, y \in \mathbb{R}^m$, defined as

$$x^\mathrm{T} y \doteq \sum_{i=1}^{m} x_i y_i. \tag{4.2}$$

Note that in the case of two sparse vectors x and y associated with two documents d and d', the above sum can be computed efficiently in time $\Omega(|d| + |d'|)$ (see Exercise 4.2).

Several refinements can be obtained by extending the Boolean vector model and introducing real-valued weights associated with terms in a document. A more informative weighting scheme consists of counting the actual number of occurrences of each term in the document. In this case $x_j \in \mathbb{N}$ counts term occurrences in the corresponding document (see Figure 4.3). x may be multiplied by the constant $1/|d|$ to obtain a vector of term frequencies (TF) within the document.

An important family of weighting schemes combines term frequencies (which are relative to each document) with an 'absolute' measure of term importance called inverse document frequency (IDF). IDF decreases as the number of documents in which the term occurs increases in a given collection. So terms that are globally rare receive a higher weight.

Formally, let $D = \{d_1, \ldots, d_n\}$ be a collection of documents and for each term ω_j let n_{ij} denote the number of occurrences of ω_j in d_i and n_j the number of documents that contain ω_j at least once. Then we define

$$\mathrm{TF}_{ij} \doteq \frac{n_{ij}}{|d_i|}, \tag{4.3}$$

$$\mathrm{IDF}_j \doteq \log \frac{n_j}{n}. \tag{4.4}$$

Here the logarithmic function is employed as a damping factor.

The TF–IDF weight (Salton *et al.* 1983) of ω_j in d_i can be computed as

$$x_{ij} = \mathrm{TF}_{ij} \cdot \mathrm{IDF}_j \tag{4.5}$$

or, alternatively, as

$$x_{ij} = \frac{\mathrm{TF}_{ij}}{\max_{\omega_k \in d_i} \mathrm{TF}_{ik}} \frac{\mathrm{IDF}_j}{\max_{\omega_k \in d_i} \mathrm{IDF}_k}. \tag{4.6}$$

The IDF weighing is commonly used as an effective heuristic. A theoretical justification has been recently proposed by Papineni (2001), who proved that IDF is the

According to news.com, Apple has warned one of its own dealers to stop handing out a patch to allow DVD burning with iDVD on non-Apple hardware

According	1	.69	0	0
Act	0	0	1	.69
Apple	1	0	1	0
Computer	0	0	1	.69
Digital	0	0	1	.69
DVD	1	0	1	0
DVDs	0	0	0	0
Millennium	1	.69	0	.69
a	1	.69	0	0
allow	1	.69	0	0
burning	1	0	1	0
com	1	.69	0	0
customers	0	0	1	.69
dealers	1	.69	0	0
drivers	0	0	1	.69
external	0	0	1	.69
handing	1	.69	0	0
hardware	1	.69	0	0
has	1	0	1	0
iDVD	1	.69	0	0
invoked	0	0	1	.69
its	0	0	0	0
news	1	.69	0	0
non	1	.69	0	0
of	1	.69	0	0
on	1	0	1	0
one	1	.69	0	0
out	1	.69	0	0
own	1	.69	0	0
patch	1	.69	0	0
prevent	0	0	1	.69
stop	1	.69	0	0
to	3	0	1	0
the	0	0	1	.69
warned	1	.69	0	0
with	1	.69	0	0

Apple Computer has invoked the Digital Millennium Copyright Act to prevent its customers from burning DVDs on external drives

Figure 4.3   Vector-space representations. For each of the two documents, the left vector counts the number of occurrences of each term, while the right vector is based on TF–IDF weights (Equation 4.5).

optimal weight of a term with respect to the minimization of a distance function that generalizes Kullback–Leibler divergence or relative entropy (see Appendix A).

4.3.3   Retrieval and evaluation measures

Let us consider a collection of n documents D. Each document is represented by an m-dimensional vector, where $m = |V|$ and V is the set of terms that occurred in the collection. Let $q \in \mathbb{R}^m$ denote the vector associated with a user query (terms that are present in the query but not in V will be stripped off). Each document is then assigned a score, relative to the query, by computing $s(x_i, q)$, $i = 1, \ldots, n$. The set R of *retrieved documents* that are presented to the user can be formed by collecting the top-ranking documents according to the similarity measure. The quality of the returned collection can be defined by comparing R to the set of documents $R^\star$ that is actually relevant to the query.[3]

Two common metrics for comparing R and $R^\star$ are *precision* and *recall*. Precision π is defined as the fraction of retrieved documents that are actually relevant. Recall that ρ is defined as the fraction of relevant documents that are retrieved by the system. More precisely,

$$\pi \doteq \frac{|R \cap R^\star|}{|R|}, \qquad \rho \doteq \frac{|R \cap R^\star|}{|R^\star|}.$$

Note that in this context the ratio between relevant and irrelevant documents is typically very small. For this reason, other common evaluation measures like accuracy or error rate (see Section 4.6.5), where the denominator consists of $|D|$, would be inadequate (it would suffice to retrieve nothing to get very high accuracy). Sometimes precision and recall are combined into a single number called F_β measure defined as

$$F_\beta \doteq \frac{(\beta^2 + 1)\pi\rho}{\beta^2 \pi + \rho}. \tag{4.7}$$

Note that the F_1 measure is the harmonic mean of precision and recall. If β tends to zero (∞) the F_β measure tends to precision (recall).

4.4 Probabilistic Retrieval

Researchers in IR have long posed the question of how to define an optimality criterion for ranking retrieved documents in response to a specified user interest. Robertson (1977) formulated an optimality principle called the probabilistic ranking principle (PRP), stating that

> if a reference retrieval system's response to each request is a ranking of the documents in order of decreasing probability of relevance to the user who submitted the request where the probabilities are estimated as accurately as possible on the basis of whatever data have been made available to the system for this purpose, the overall effectiveness of the system to its user will be the best that is obtainable on the basis of those data.

---

[3] Of course this is possible only on controlled collections, such as those prepared for the Text REtrieval Conference (TREC) (data, papers describing methodologies and description of evaluation measures and assessment criteria are available at the TREC website http://trec.nist.gov/).

The PRP resembles the optimal Bayes decision rule for classification, a concept that is well known for example in the pattern recognition community (Duda and Hart 1973). The Bayes optimal separation is obtained by ensuring that the posterior probability of the correct class (given the observed pattern) should be greater than the posterior probability of any other class. PRP can be mathematically stated by introducing a Boolean variable R (relevance) and by defining a model for the conditional probability $P(R \mid d, q)$ of relevance of a document d for a given user need (for example, expressed through a query q). Its justification follows from a decision-theoretic argument as follows. Let c denote the cost of retrieving a relevant document and $\bar{c}$ the cost of retrieving an irrelevant document, with $c < \bar{c}$. Then in order to minimize the overall cost, a document d should be retrieved next if

$$cP(R \mid d, q) + \bar{c}(1 - P(R \mid d, q)) \leqslant cP(R \mid d', q) + \bar{c}(1 - P(R \mid d', q)) \quad (4.8)$$

for every other document d' that has not yet been retrieved. But, since $c < \bar{c}$, the above condition is equivalent to

$$P(R \mid d, q) \geqslant P(R \mid d', q),$$

that is, documents should be retrieved in order of decreasing probability of relevance.

In order to design a model for the probability of relevance some simplifications are needed. The simplest possible approach is called binary independence retrieval (BIR) (Robertson and Spärck Jones 1976) as also used in the Bernoulli model for the Naive Bayes classifier, which we will discuss later, in Section 4.6.2. This model postulates a form of conditional independence amongst terms. Following Fuhr (1992), let us introduce the odds of relevance and apply Bayes' theorem:

$$O(R \mid d, q) = \frac{P(R = 1 \mid d, q)}{P(R = 0 \mid d, q)} = \frac{P(d \mid R = 1, q)}{P(d \mid R = 0, q)} \cdot \frac{P(R = 1 \mid q)}{P(R = 0 \mid q)}. \quad (4.9)$$

Note that the last fraction is the odds of R given q, a constant quantity across the collection of documents (it only depends on the query). The BIR assumption concerns the first fraction and was originally misidentified as a marginal form of term independence (Cooper 1991). We can state it as

$$\frac{P(d \mid R = 1, q)}{P(d \mid R = 0, q)} = \prod_{j=1}^{|V|} \frac{x_j P(\omega_j \mid R = 1, q) + (1 - x_j)(1 - P(\omega_j \mid R = 1, q))}{x_j P(\omega_j \mid R = 0, q) + (1 - x_j)(1 - P(\omega_j \mid R = 0, q))}.$$

$$(4.10)$$

The parameters to be estimated are therefore

$$\theta_j \doteq P(\omega_j \mid R = 1, q)$$

and $\eta_j \doteq P(\omega_j \mid R = 0, q)$. If we further assume that $\theta_j = \eta_j$ if ω_j does not appear in q we finally have

$$O(R \mid d, q) = O(R \mid q) \prod_{j:\omega_j \in q} x_j \frac{\theta_j}{\eta_j} \cdot (1 - x_j) \frac{1 - \theta_j}{1 - \eta_j}, \quad (4.11)$$

where the product only extends to indices j whose associated terms ω_j appear in the query. This can also be rewritten as

$$O(R \mid d, q) = O(R \mid q) \cdot \prod_{j \in q} \frac{1 - \theta_j}{1 - \eta_j} \cdot \prod_{\substack{j:\omega_j \in q, \\ x_j = 1}} \frac{\theta_j(1 - \eta_j)}{\eta_j(1 - \theta_j)}, \qquad (4.12)$$

where the last factor is the only part that depends on the document. The retrieval status value (RSV) of a document is thus computed by taking the logarithm of the last factor:

$$\text{RSV}(d) = \sum_{j:\omega_j \in d \cup q} \log \frac{\theta_j(1 - \eta_j)}{\eta_j(1 - \theta_j)}. \qquad (4.13)$$

Documents are eventually ranked in reverse order of RSV.

4.5   Latent Semantic Analysis

The retrieval approach based on vector-space similarities can reach satisfactory recall rates only if the terms in the query are actually present in the relevant documents. Users, on the other hand, often formulate queries choosing terms that express *concepts* related to the contents they want to retrieve. Natural language has a very rich expressive power and even at the lexical level, the large variability due to synonymy and polysemy causes serious problems to retrieval methods based on term matching. Synonymy means that the same concept can be expressed using different sets of terms (e.g. bandit, brigand, and thief). Synonymy negatively affects recall. Polysemy means that identical terms can be used in very different semantic contexts. For example, in English, the word 'bank' refers to both a repository where important material is saved (as in blood bank) and also to the slope beside a body of water (as in the bank of the Thames). Polysemy negatively affects precision. Overall, synonymy and polysemy lead to a complex relation between terms and concepts that cannot be captured through simple matching.

Thus, although a query may conceptually be very close to a given set of documents, its associated vector could be orthogonal or nearly orthogonal to those document vectors, simply because the authors of the document and the user have a different usage of language. Another more statistical way of thinking about this is to observe that the number of terms that are present in a document is a rather small fraction of the entire dictionary, reducing the likelihood that two documents use the same set of words to express the same concept.

Latent semantic indexing (LSI) is a statistical technique that attempts to estimate the hidden structure that generates terms given concepts. It uses a linear algebra technique known as singular value decomposition (SVD) to discover the most important associative patterns between words and concepts. LSI is a data-driven method, where a large collection of sentences or documents is employed to discover the statistically most significant co-occurrences of terms.

4.5.1 LSI and text documents

Let X denote a term–document matrix, defined as

$$X = [x_1 \cdots x_n]^{\mathrm{T}}. \tag{4.14}$$

Each row in the matrix is simply the vector-space representation of a document. In LSI we use occurrence counts as components. Each column contains the occurrences of a term in each document in the data set. The LSI technique consists of computing the SVD of X and setting to zero all the singular values except the largest K ones. In practical applications K is often set to values around between 100 and 1000. This can be seen as analogous to the PCA dimensionality reduction discussed in Section A.3: documents are mapped to a lower-dimensional latent semantic space induced by selecting directions of maximum covariance.

We will illustrate more in detail the LSI procedure through an example. Consider the collection of 10 documents in Table 4.1. The first five documents are related to the Linux operating system, and the remaining five to news in the area of genomes. For simplicity we focus only on frequent terms, in particular those occurring at least twice. We remove common words as 'the' and 'goes'. The resulting term–document matrix X^{T} is shown in the middle of Table 4.1 (we use the transpose of X for the sake of space). Performing SVD in this example yields a singular values matrix $\Sigma = \mathrm{diag}(2.57, 2.49, 1.99, 1.9, 1.68, 1.53, 0.94, 0.66, 0.36, 0.10)$. By setting $\hat{\Sigma} = \mathrm{diag}(2.57, 2.49, 0, 0, 0, 0, 0, 0, 0, 0)$, we obtain the reconstruction of X, $\hat{X} = U\hat{S}V^{\mathrm{T}}$, shown at the bottom of table 4.1.

As we can see, the reconstructed matrix $\hat{X}$ is not as sparse as the original matrix X. The new term weights account for word co-occurrences and appear to infer relations among words that pertain to synonymy, at least in a loose sense. For example, a query on the term 'Linux' would now assign a relatively high score to documents 2 and 3, although they do not contain the search keyword at all. We can explain this result arguing that SVD has inferred a link between 'Debian' or 'Gentoo' and 'Linux' through the co-occurrences of other words (they never occur together in our sample). Similarly, we can note in Table 4.1 that both 'DNA' and 'Dolly' have document vectors closer to the document vector of 'genome'.

When performing the numerical computation of the SVD of a very large document matrix, it is important to take advantage of sparsity. A variety of well-known numerical methods for computing the SVD of sparse matrices can be brought to bear (Berry 1992; Berry and Browne 1999; Letsche and Berry 1997).

4.5.2 Probabilistic LSA

The latent semantic indexing method described so far does not have a completely sound probabilistic interpretation. As we have discussed, the approximation of X by $\hat{X}$ obtained by setting to zero all but their first K singular values is optimal in the sense that $\|X - \hat{X}\|_2$ is minimized (amongst all projections in 'latent spaces' of dimension

Table 4.1 Example application of LSI. Top: a collection of documents. Terms used in the analysis are underlined. Center: the term–document matrix X^T. Bottom: the reconstructed term–document matrix $\hat{X}^T$ after projecting on a subspace of dimension $K = 2$.

d_1:	Indian government goes for open-source software
d_2:	Debian 3.0 Woody released
d_3:	Wine 2.0 released with fixes for Gentoo 1.4 and Debian 3.0
d_4:	gnuPOD released: iPod on Linux... with GPLed software
d_5:	Gentoo servers running an open-source mySQL database
d_6:	Dolly the sheep not totally identical clone
d_7:	DNA news: introduced low-cost human genome DNA chip
d_8:	Malaria-parasite genome database on the Web
d_9:	UK sets up genome bank to protect rare sheep breeds
d_{10}:	Dolly's DNA Damaged

	d_1	d_2	d_3	d_4	d_5	d_6	d_7	d_8	d_9	d_{10}
open-source	1	0	0	0	1	0	0	0	0	0
software	1	0	0	1	0	0	0	0	0	0
Linux	1	0	0	1	0	0	0	0	0	0
released	0	1	1	1	0	0	0	0	0	0
Debian	0	1	1	0	0	0	0	0	0	0
Gentoo	0	0	1	0	1	0	0	0	0	0
database	0	0	0	0	1	0	0	1	0	0
Dolly	0	0	0	0	0	1	0	0	0	1
sheep	0	0	0	0	0	1	0	0	1	0
genome	0	0	0	0	0	0	1	1	1	0
DNA	0	0	0	0	0	0	2	0	0	1

	d_1	d_2	d_3	d_4	d_5	d_6	d_7	d_8	d_9	d_{10}
open-source	0.34	0.28	0.38	0.42	0.24	0.00	0.04	0.07	0.02	0.01
software	0.44	0.37	0.50	0.55	0.31	−0.01	−0.03	0.06	0.00	−0.02
Linux	0.44	0.37	0.50	0.55	0.31	−0.01	−0.03	0.06	0.00	−0.02
released	0.63	0.53	0.72	0.79	0.45	−0.01	−0.05	0.09	−0.00	−0.04
Debian	0.39	0.33	0.44	0.48	0.28	−0.01	−0.03	0.06	0.00	−0.02
Gentoo	0.36	0.30	0.41	0.45	0.26	0.00	0.03	0.07	0.02	0.01
database	0.17	0.14	0.19	0.21	0.14	0.04	0.25	0.11	0.09	0.12
Dolly	−0.01	−0.01	−0.01	−0.02	0.03	0.08	0.45	0.13	0.14	0.21
sheep	−0.00	−0.00	−0.00	−0.01	0.03	0.06	0.34	0.10	0.11	0.16
genome	0.02	0.01	0.02	0.01	0.10	0.19	1.11	0.34	0.36	0.53
DNA	−0.03	−0.04	−0.04	−0.06	0.11	0.30	1.70	0.51	0.55	0.81

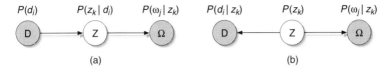

Figure 4.4 $\quad$ Bayesian networks describing the aspect model. Concepts (Z) are never observed in the data but their knowledge would render words (Ω) and documents (D) independent. The two networks are independence equivalent but use different parameterizations.

K). However, the L_2 matrix norm is algebraically well suited for Gaussian variables and is hard to justify its use in the case of the discrete space defined by vectors of counts of words.

We now discuss a probabilistic approach for describing a latent semantic space called the *aspect model* (Hofmann *et al.* 1999) or aggregate Markov model (Saul and Pereira 1997). Let an event in this model be the occurrence of a term ω in a document d. Let $z \in \{z_1, \ldots, z_K\}$ denote a latent (hidden) variable associated with each event according to the following generative scheme. First, select a document from a density $P(d)$. Second, select a latent concept z with probability $P(z \mid d)$. Third, choose a term ω to describe the concept linguistically, sampling from $P(\omega \mid z)$, i.e. assume that once the latent concept has been specified the document and the term are conditionally independent, as shown by the Bayesian network representation of Figure 4.4. The probability of each event (ω, d) is therefore

$$P(\omega, d) = P(d) \sum_z P(\omega \mid z) P(z \mid d). \tag{4.15}$$

The aspect model has a direct interpretation in terms of dimensionality reduction. Each document is uniquely determined by the mixing coordinates $P(z_k \mid d)$, $k = 1, \ldots, K$, that form a nonnegative vector belonging to a probabilistic latent semantic space. In other words, rather than being represented through terms, a document is represented through latent variables that in turn are responsible for generating terms. So if $K \ll |V|$ dimensionality reduction occurs.

The analogy with respect to LSI can be established by introducing the matrices U, V, and Σ whose elements are $u_{ik} = P(d_i \mid z_k)$, $v_{j,k} = P(\omega_j \mid z_k)$ and $\sigma_{kk} = P(z_k)$ for $i = 1, \ldots, n$, $j = 1, \ldots, m$, and $k = 1, \ldots, K$. In this way, all the $n \times m$ document–term joint probabilities can be collected in the matrix

$$P = U \Sigma V^{\mathrm{T}}. \tag{4.16}$$

This matrix can be directly compared to the SVD representation used in traditional LSI. Unlike $\hat{X}$ in LSI, P is a properly normalized probability distribution and its entries cannot be negative. Moreover, the coordinates of terms in the latent space can be interpreted as the distribution of terms, conditional on the hidden z_k values.

Hofmann (2001) shows how to fit the parameters of this model from data. In the simplest case, they are estimated by maximum likelihood using the EM algorithm (see Chapter 1) because of the latent variable Z. We assume the parameterization of

Figure 4.4b. During the E step, the probability that the occurrence of ω_j in d_i is due to latent concept z_k is computed as

$$P(z_k \mid d_i, \omega_j) \propto P(w_j \mid z_k)P(d_i \mid z_k)P(z_k). \quad (4.17)$$

This can be seen as a special case of belief propagation (see Chapter 1) in the network of Figure 4.4b. The M step simply consists of updating parameters as follows:

$$P(\omega_j \mid z_k) \propto \sum_{i=1}^{n} n_{ij} P(z_k \mid d_i, \omega_j), \quad (4.18)$$

$$P(z_k \mid d_i) \propto \sum_{j=1}^{|V|} n_{ij} P(z_k \mid d_i, \omega_j), \quad (4.19)$$

$$P(z_k) \propto \sum_{i=1}^{n} \sum_{j=1}^{|V|} n_{ij} P(z_k \mid d_i, \omega_j). \quad (4.20)$$

A possible refinement of the above EM algorithm consists of replacing Equation (4.17) by

$$P(z_k \mid d_i, \omega_j) \propto (P(\omega_j \mid z_k)P(d_i \mid z_k))^\beta P(z_k). \quad (4.21)$$

where β is a positive parameter that is initialized to unity and exponentially decreased in an outer loop that wraps around a full EM optimization (performed with constant β). This outer loop iteration continues as long as predictive performance on a validation set keeps improving. This refinement can help to prevent overfitting and has analogies with the annealing process in physics. The resulting algorithm is called *tempered* EM (TEM). Intuitively, smaller values of β in Equation (4.21) increase the entropy of the posterior distribution of Z performing a special form of regularization.

Once the model is trained, to define a similarity measure based on PLSA, we need to estimate which concepts are being expressed in the query q. As in LSI, documents and queries that are not in the original matrix need to be folded in. In the case of PLSA a point in the latent semantic space is a (posterior) distribution on Z, so folding in consists of calculating $P(z_k \mid q)$ for $k = 1, \ldots, K$. This is done by using the above TEM procedure while keeping $P(\omega_j \mid z_k)$ fixed.

The original aspect model described above is not without its limitations. The fact that there is a separate set of parameters for each document in the collection means both that there are a very large number of parameters to be estimated and that the model does not directly contain the concept of 'a new as yet unseen document' – this can lead to overfitting of the data. More recent work by Blei *et al.* (2002a) has begun to address some of these issues by placing the aspect model in a more Bayesian context.

4.6 Text Categorization

Text categorization consists of grouping textual documents into different fixed classes or categories. This may be useful, for example, to predict the topic of a Web page, or to decide whether it is relevant with respect to the interests of a given user. This problem is particularly well suited for a machine-learning solution. In the following we briefly review three of the most important and effective machine-learning algorithms that are often applied to text categorization: k nearest neighbors (k-NN), Naive Bayes, and support vector machines. Readers not familiar with general concepts in machine learning are urged to read Section 1.5 before continuing this section.

Several additional machine-learning methods not covered here have been applied to text categorization including decision trees (Lewis and Catlett 1994), neural networks (Wiener *et al.* 1995), and symbolic approaches such as inductive logic programming (Cohen 1995) and the induction of classification rules (Apté *et al.* 1994; Cohen 1996). A review of these and other methods can be found in Sebastiani (2002).

4.6.1 *k* nearest neighbors

k-NN is a memory based classifier that learns by simply storing all the training instances. During prediction, k-NN first measures the distances between a new point x and all the training instances, returning the set $N(x, D, k)$ of the k points that are closest to x. For example, if training instances are represented by real-valued vectors x, we could use Euclidean distance to measure the distance between x and all other points in the training data, i.e. $\|x - x_i\|_2, i = 1, \ldots, n$. After calculating the distances, the algorithm predicts a class label for x by a simple majority voting rule using the labels in the elements of $N(x, D, k)$, breaking ties arbitrarily. In spite of its apparent simplicity, k-NN is known to perform well in many domains. An early result by Cover and Hart (1967) shows that the asymptotic error rate of the 1-NN classifier (as the size of the training data set gets infinitely large) is always less than twice the optimal Bayes error (which is the lowest possible error rate achievable by *any* classifier in a particular feature space x).

In the case of text, majority voting can be replaced by a smoother metric where, for each class c, a scoring function

$$s(c \mid x) = \sum_{x' \in N_c(x, D, k)} \cos(x, x') \qquad (4.22)$$

is computed through vector-space similarities between the new documents and the subset of the k neighbors that belong to class c (Yang 1999), where $N_c(x, D, k)$ is the subset of $N(x, D, k)$ containing only points of class c. Despite the simplicity of the method, the performance of k-NN in text categorization is quite often satisfactory in practice (Joachims 1998; Lam and Ho 1998; Yang 1999; Yang and Liu 1999). Han *et al.* (2001) have proposed a variant of k-NN where the weights associated with features are learned iteratively – other statistically motivated techniques that extend the basic k-NN classifier are discussed in Hastie *et al.* (2001).

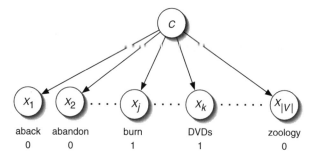

Figure 4.5 A Bayesian network for the Naive Bayes classifier under the Bernoulli document-based event model. The example document is the sentence 'burn DVDs and we will burn you.'

4.6.2 The Naive Bayes classifier

This classifier attempts to estimate the conditional probability of the class given the document, namely $P(c \mid d)$, for $c = 1, \ldots, S$. Using Bayes' theorem, we can write this probability as

$$P(c \mid d, \boldsymbol{\theta}) = \frac{P(d \mid c, \boldsymbol{\theta})P(c \mid \boldsymbol{\theta})}{P(d \mid \boldsymbol{\theta})} \propto P(d \mid c, \boldsymbol{\theta})P(c \mid \boldsymbol{\theta}), \qquad (4.23)$$

where $\boldsymbol{\theta}$ are the parameters of the model. In what follows we only include the dependence on $\boldsymbol{\theta}$ when necessary – otherwise it can be implicitly assumed to be present.

Note that, since the classes are assumed to be mutually exclusive, we do not really need to know $P(d)$, a term that can be thought of as a normalization factor guaranteeing that $\sum_c P(c \mid d) = 1$. The key idea behind the Naive Bayes classifier is the assumption that the terms in a document are conditionally independent given the class. This assumption is clearly false in many if not most practical situations, but it is often adequate to first-order for the bag-of-words representation, where word order in the document is not taken into account. Moreover, we should keep in mind that we are interested in a good approximation of $P(c \mid d)$, which does not necessarily mean that we need a perfect model of $P(d \mid c)$. In other words, we are interested in discrimination among classes, not in a high quality generative model of the document given the class (Ng and Jordan 2002). In practice, the classifier is known to work satisfactorily even when the conditional independence assumption is known not to hold (Domingos and Pazzani 1997).

There is a subtle issue concerning the interpretation of the document in terms of a probabilistic event (Lewis 1998; McCallum and Nigam 1998). If the document as a whole is considered to be an event, then it should be naturally described by a bag of words, and the words are the *attributes* of this event. In this case, each vocabulary term is associated with a Bernoulli attribute whose realization is unity if the term appears in the document, and zero otherwise. In addition to reducing documents to bags of words, the Naive Bayes model postulates that binary attributes are mutually independent given the class. This approach is substantially the same as the binary independence retrieval model that was popular for probabilistic retrieval in the 1970s (Robertson and Spärck Jones 1976) and has even more ancient origins (Maron 1961). The conditional independence assumption in this model can be depicted graphically using a Bayesian network such as that in Figure 4.5, suggesting that the class is the only cause of the appearance of each word in a document. Under this model, generating a document is like tossing $|V|$ independent coins and the occurrence of each word in the document is a Bernoulli event. Therefore we can rewrite the generative portion of Equation (4.23) as

$$P(d \mid c, \boldsymbol{\theta}) = \prod_{j=1}^{|V|} x_j P(\omega_j \mid c) + (1 - x_j)[1 - P(\omega_j \mid c)], \qquad (4.24)$$

where $x_j = 1$ [0] means that word ω_j does [does not] occur in d and $P(\omega_j \mid c)$ is the probability of observing word ω_j in documents of class c. Here $\boldsymbol{\theta}$ represents the set of probabilities (or parameters) $P(\omega_j \mid c)$, which is the probability of the binary event that word ω_j is 'turned on' within class c.

Alternatively, we may think of a document as a sequence of events $W_1, \ldots, W_{|d|}$. Each observed W_t has a vocabulary entry (from 1 to $|V|$) as an admissible realization. Note that the number of occurrences of each word, as well as the length of the document, are taken into account under this interpretation. In addition, since the document is a sequence, serial order among words should also be taken into account when modeling $P(W_1, \ldots, W_{|d|} \mid c)$. This could be done, for example, by using a Markov chain. A simplifying assumption, however, is that word occurrences are independent of their (relative) positions, given the class (Lewis 1992; Lewis and Gale 1994; Mitchell 1997). Equivalently, we assume that the bag-of-words representation retains all the relevant information for assessing the probability of a document whose class is known.

Under the word-based event model, generating a document is like throwing a die with $|V|$ faces $|d|$ times, and the occurrence of each word in the document is a multinomial event. Hence, the generative portion of the model is a multinomial distribution

$$P(d \mid \boldsymbol{\theta}) = G P(|d|) \prod_{j=1}^{|V|} P(\omega_j \mid c)^{n_j}, \qquad (4.25)$$

where n_j is the number of occurrences of word ω_j in d, and $P(\omega_j \mid c)$ is the probability that word ω_j occurs at any position $t \in [1, \ldots, |d|]$; because of the bag-of-words assumption this does not depend on t. Here the parameters $\boldsymbol{\theta}$ are the set of probabilities $P(\omega_j \mid c)$, where now $\sum_{j=1}^{|V|} P(\omega_j \mid c) = 1$. McCallum and Nigam (1998) found empirically that the multinomial model outperforms the Bernoulli model in several evaluation benchmarks. Note that as in the case of the document-based event model of Equation (4.24), the 'bag-of-words' assumption results in a factorization of $P(W_1, \ldots, W_{|d|} \mid c)$ explaining, perhaps, why 'Naive Bayes' is often used in the literature to refer to both event models.

The normalization factor is the multinomial coefficient

$$G = \frac{|d|!}{\prod_j n_j!}.$$

Neither $P(|d|)$ nor G are needed for classification, since $|d|$, the number of words or terms in a document, is assumed to be independent of the class. This last assumption can be removed and $P(|d| \mid c)$ explicitly modeled. Models of document length (e.g. based on Poisson distributions) have been used for example in the context of probabilistic retrieval (Robertson and Walker 1994). Note that in the case of the Bernoulli model there are $2^{|V|}$ possible different documents, while in the case of the multinomial model there is an infinite (but countable) number of different documents.

An additional model that may be developed and that lies somewhat in between the Bernoulli and the multinomial models consists of keeping the document-based event model but extending Bernoulli distributions to integer distributions, such as the Poisson (Lewis 1998). Finally, extensions of the basic Naive Bayes approach that allow limited dependencies among features have been proposed (Friedman and Goldszmidt 1996; Pazzani 1996). However, these models are characterized by a larger set of parameters and may overfit the data (Koller and Sahami 1997).

Learning a Naive Bayes classifier consists of estimating the parameters θ from the available data. We will assume that the training data set consists of a collection of labeled documents $\{(d_i, c_i), \ i = 1, \ldots, n\}$. In the Bernoulli model, the parameters θ include $\theta_{c,j} = P(\omega_j \mid c)$, $j = 1, \ldots, |V|$, $c = 1, \ldots, K$. These are estimated as normalized sufficient statistics

$$\hat{\theta}_{c,j} = \frac{1}{N_c} \sum_{i:c_i=c}^{n} x_{ij}, \tag{4.26}$$

where $N_c = |\{i : c_i = c\}|$ and $x_{ij} = 1$ if ω_j occurs in d_i. Additional parameters are the class prior probabilities $\theta_c = P(c)$, which are estimated as

$$\hat{\theta}_c = \frac{N_c}{n}. \tag{4.27}$$

Note that the estimates above correspond to ML estimates of the parameters. In Chapter 1 we discussed how we can also derive Bayesian parameter estimates that combine the observed data with prior knowledge. This can be achieved by using (for example) Dirichlet priors for the Bernoulli likelihood model. Bayesian estimates can be quite useful when estimating parameters from sparse data. For words in documents, we might have a weak noninformative prior that any word can in theory appear in documents from any class. This would avoid problems in making predictions on future documents when such a word shows up in some class c where it did not appear in the training data. A model based on ML parameter estimates would assign a probability of zero to that document, irrespective of how well the other words in the document matched class c. On the other hand, a model based on Bayesian estimates would assign that word-class combination a low nonzero probability and still allow the other words to play a role in determining the final class prediction for the document. We might also have more informative priors available, e.g. a Dirichlet prior on the distribution of words in each class that reflect word distributions as estimated from previous studies, or that reflect an expert's belief in what words are likely to appear in each class.

In the case of the multinomial model of Equation (4.25), the generative parameters are $\theta_{c,j} = P(\omega_j \mid c)$. Note that these parameters must satisfy $\sum_j \theta_{c,j} = 1$ for each class c. To estimate these parameters it is common practice to introduce Dirichlet priors (see Chapter 1 for details). The resulting estimation equations are derived as follows. In the case of the distributions of terms given the class, we introduce a Dirichlet prior with hyperparameters q_j and α, resulting in the estimation formula

$$\hat{\theta}_{c,j} = \frac{\alpha q_j + \sum_{i:c_i=c}^{n} n_{ij}}{\alpha + \sum_{l=1}^{|V|} \sum_{i:c_i=c} n_{il}},$$ (4.28)

where n_{ij} is the number of occurrences of ω_j in d_i. A simple noninformative prior assigns $q_j = 1/|V|$ and $\alpha = |V|$. Intuitively, this prior corresponds to the assumption that we have observed each word exactly once in one document of each class. This method (also known as Laplace smoothing) prevents the problem of estimating a null value for a parameter if a certain term ω_j never occurs in documents of a given class c in the training set (without the smoothing, the probability that a new document containing ω_j belong to c would be zero). Similarly, the estimation formula for the (unconditional) class probabilities is

$$\hat{\theta}_c = \frac{q_c'\alpha' + N_c}{\alpha' + n}.$$ (4.29)

Also in this case we could assume a noninformative Dirichlet hyperparameters $q_c' = 1/K$ and $\alpha' = K$.

4.6.3 Support vector classifiers

Support vector machines (SVMs) were introduced in Cortes and Vapnik (1995), extending earlier seminal work by Vapnik on statistical learning theory (Vapnik 1982). The basic underlying idea, often referred to as *structural risk minimization* is closely related to the theory of regularization but also to Bayesian approaches to learning (Evgeniou *et al.* 2000) and is essentially guided by the principle that the hypothesis that explains a finite set of examples should be searched in an appropriately 'small' hypothesis space.

SVMs are particularly well suited to deal with high-dimensional data such as vector-space representations of text documents. In their standard formulation, they deal with binary classification problems where the number of classes is restricted to two (see later for a discussion on multiclass SVM).

Consider a training set $D = \{(x_i, y_i), \ i = 1, \ldots, n\}$ with $x_i \in \mathbb{R}^m$, and where $y_i \in \{-1, 1\}$ is an integer that specifies whether x_i is a positive or a negative example. A linear discriminant classifier is then defined by introducing the *separating hyperplane*

$$\{x : f(x) = w^T x + w_0 = 0\}, \tag{4.30}$$

where $w \in \mathbb{R}^m$ and $w_0 \in \mathbb{R}$ are adjustable coefficients that play the role of model parameters. A binary classification function $h : \mathbb{R}^m \rightarrow \{0, 1\}$ can be obtained by taking the sign of $f(x)$, i.e.

$$h(x) = \begin{cases} 1, & \text{if } f(x) > 0, \\ 0, & \text{otherwise.} \end{cases} \tag{4.31}$$

Learning in this class of models consists of determining w and w_0 from data. The training examples are said to be *linearly separable* if there exists a hyperplane whose associated classification function is consistent with all the labels, i.e. if $y_i f(x_i) > 0$ for each $i = 1, \ldots, n$. Under this hypothesis, Rosenblatt (1958) proved that the following simple iterative algorithm terminates and returns a separating hyperplane:

PERCEPTRON(D)
1   **$w \leftarrow 0$**
2    $w_0 \leftarrow 0$
3   **repeat**
4        $e \leftarrow 0$
5   **for** $i \leftarrow 1, \ldots, n$
6   **do** $s \leftarrow \text{sgn}(y_i(w^T x_i + w_0))$
7   **if** $s < 0$
8   **then** $w \leftarrow w + y_i x_i$
9                 $w_0 \leftarrow w_0 + y_i$
10                $e \leftarrow e + 1$
11   **until** $e = 0$
12   **return** (w, w_0).

It can be shown that a sufficient condition for D to be linearly separable is that the number of training examples $n = |D|$ is less than or equal to $m + 1$ (see Exercise 4.4). This is particularly likely to be true in the case of text categorization, where the vocabulary typically includes several thousands of terms, and is often larger than the number of available training examples n (see Exercise 4.5 for details).

Unfortunately, learning with the Perceptron algorithm offers little defense against overfitting. A thorough explanation of this problem requires concepts from statistical learning theory that are beyond the scope of this book. To gain an intuition of why this is the case, consider the scenario in Figure 4.6. Here we suppose that positive and negative examples are generated by two Gaussian distributions (see Appendix A) with the same covariance matrix and that positive and negative points are generated with the same probability. In such a setting, the optimal (Bayes) decision boundary is the one that minimizes the posterior probability that a new point is misclassified and, as it turns out, this boundary is the hyperplane that is orthogonal to the segment

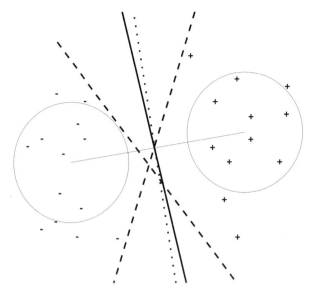

Figure 4.6 Alternative linear decision boundaries for a binary classification problem (see text for explanation).

connecting the centers of mass of the two distributions (dotted line in Figure 4.6). Clearly, a random hyperplane that just happens to separate training points (dashed line) can be substantially far away from the optimal separation boundary, leading to poor generalization to new data. The difficulty grows with the dimensionality of the input space m since for a fixed n the set of separating hyperplanes grows exponentially with m (a problem known as the *curse of dimensionality*). Remember that in the case of text categorization m may be significantly large (several thousands).

The statistical learning theory developed by Vapnik (1998) shows that we can define an *optimal separating hyperplane* (relative to the training set) having two important properties: it is unique for each linearly separable data set, and its associated risk of overfitting is smaller than for any other separating hyperplane. We define the *margin M* of the classifier to be the distance between the separating hyperplane and the closest training examples. The optimal separating hyperplane is then the one having maximum margin (see Figure 4.7). Going back to Figure 4.6, the theory suggests that the risk of overfitting for the maximum margin hyperplane (solid line) is smaller than for the dashed hyperplane. Indeed, in our example the maximum margin hyperplane is significantly closer to the Bayes optimal decision boundary.

In order to compute the maximum margin hyperplane, we begin by observing that the distance of a point x from the separating hyperplane is

$$\frac{1}{\|w\|}(w^{\mathrm{T}}x + w_0)$$

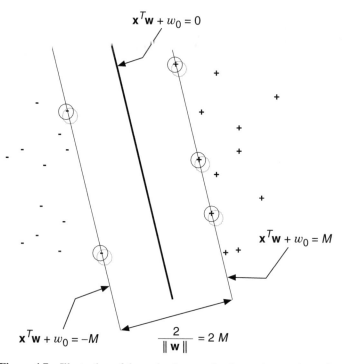

Figure 4.7 Illustration of the optimal separating hyperplane and margin.
Circled points are support vectors.

(see Figure 4.7). Thus, the optimal hyperplane can be obtained by solving the constrained optimization problem

$$\max_{\boldsymbol{w}, w_0} M \quad \text{subject to} \quad \frac{1}{\|\boldsymbol{w}\|} y_i(\boldsymbol{w}^{\mathrm{T}}\boldsymbol{x}_i + w_0) \geqslant M, \quad i = 1, \ldots, n, \qquad (4.32)$$

where the constraints require that each training point should lie in the correct semispace and at a distance not less than M from the separating hyperplane. Note that although $(\boldsymbol{w}, w_0)$ comprise $n + 1$ real numbers, there are only n degrees of freedom since multiplying $\boldsymbol{w}$ and w_0 by a scalar constant does not change the hyperplane. Thus we can arbitrarily set $\|\boldsymbol{w}\| = 1/M$ and rewrite the optimization problem (4.32) as

$$\min_{\boldsymbol{w}, w_0} \|\boldsymbol{w}\| \quad \text{subject to} \quad y_i(\boldsymbol{w}^{\mathrm{T}}\boldsymbol{x}_i + w_0) \geqslant 1, \quad i = 1, \ldots, n. \qquad (4.33)$$

The above problem can be transformed to its dual by first introducing the vector of Lagrangian multipliers $\boldsymbol{\alpha} \in \mathbb{R}^n$ and writing the Lagrangian function

$$\mathcal{L}(D) = -\tfrac{1}{2}\|\boldsymbol{w}\|^2 + \sum_{i=1}^{n} \alpha_i[y_i(\boldsymbol{w}^{\mathrm{T}}\boldsymbol{x} + w_0) - 1] \qquad (4.34)$$

and subsequently setting to zero the derivatives of (4.34) with respect to $\boldsymbol{w}$, w_0, obtaining

$$\max_{\boldsymbol{\alpha}} -\tfrac{1}{2}\boldsymbol{\alpha}^{\mathrm{T}}\Lambda\boldsymbol{\alpha} + \sum_{i=1}^{n} \alpha_i \quad \text{subject to } \alpha_i \geqslant 0, \quad i = 1, \ldots n, \qquad (4.35)$$

where Λ is an $n \times n$ matrix with $\lambda_{ij} = y_i y_j \boldsymbol{x}_i^{\mathrm{T}} \boldsymbol{x}_j$. This is a quadratic programming (QP) problem that can be solved, in principle, using standard optimization packages. For each training example i, the solution must satisfy the Karush–Kuhn–Tucker condition,

$$\alpha_i [y_i(\boldsymbol{w}^{\mathrm{T}}\boldsymbol{x} + w_0) - 1] = 0, \qquad (4.36)$$

and therefore either $\alpha_i = 0$ or $y_i(\boldsymbol{w}^{\mathrm{T}}\boldsymbol{x} + w_0) = 1$. In other words, if $\alpha_i > 0$ then the distance of point $\boldsymbol{x}_i$ from the separating hyperplane must be exactly M (see Figure 4.7).

Points with associated $\alpha_i > 0$ are called *support vectors*. The decision function $h(\boldsymbol{x})$ can be computed via Equation (4.30) or, equivalently, from the following dual form:

$$f(\boldsymbol{x}) = \sum_{i=1}^{n} y_i \alpha_i \boldsymbol{x}^{\mathrm{T}} \boldsymbol{x}_i. \qquad (4.37)$$

We remark that in the case of text categorization it may be very important to exploit the sparseness of the data vectors while computing dot products in Equations (4.35) and (4.37).

If the training data are not linearly separable, then this analysis can be generalized by introducing m nonnegative *slack variables* ξ_i and replacing the optimization problem in Equation (4.33) with

$$\min_{\boldsymbol{w}, w_0} \|\boldsymbol{w}\| + C \sum_{i=1}^{n} \xi_i \quad \text{subject to } \begin{cases} y_i(\boldsymbol{w}^{\mathrm{T}}\boldsymbol{x}_i + w_0) \geqslant 1 - \xi_i, & i = 1, \ldots, n, \\ \xi_i \geqslant 0, & i = 1, \ldots, n, \end{cases} \qquad (4.38)$$

where the constant C controls the cost associated with misclassifications.

This problem can also be dualized in the form

$$\max_{\boldsymbol{\alpha}} -\tfrac{1}{2}\boldsymbol{\alpha}^{\mathrm{T}}\Lambda\boldsymbol{\alpha} + \sum_{i=1}^{n} \alpha_i \quad \text{subject to } 0 \leqslant \alpha_i \leqslant C, \quad i = 1, \ldots n, \qquad (4.39)$$

yielding another QP problem. The classifier obtained in this way is commonly referred to as a support vector machine (SVM).

Note that solving the QP problem using standard optimization packages would take time $O(n^3)$ (assuming the number of support vector grows linearly with the number of training example). This time complexity is a practical drawback for SVMs. However, several approaches that can reduce complexity substantially have been proposed in the literature (Joachims 1999a; Platt 1999).

If the data are considerably nonlinearly separable then an SVM classifier will have a low accuracy, i.e. even the best linear hyperplane may be quite inferior in terms of prediction accuracy relative to a good nonlinear decision boundary. The methods developed in this section can be further extended to accommodate nonlinear separation by using kernel functions that map points $x \in \mathbb{R}^m$ into a higher-dimensional (possibly infinite-dimensional) space called the *feature space*, where data are linearly separable. Details on kernel methods can be found in Cristianini and Shawe-Taylor (2000); Schoelkopf and Smola (2002) and further details on the application of SVM to text categorization can be found in Joachims (2002).

The SVM presented here can only deal with binary classification problems. For more than two classes, a standard approach consists of *reducing* a multiclass problem into several binary sub-problems, coding each class with a binary string. The simplest (redundant) coding strategy is often referred to as 'one shot' or 'one vs all' and consists of defining as many dichotomies of the instance space as there are classes, where each class is considered as 'positive' in one and only one dichotomy. A more general reduction scheme is the information theoretic method based on error-correcting output codes, introduced by Dietterich and Bakiri (1995) and more recently extended in Allwein *et al.* (2000). Typically, these binary classifiers are trained independently but recent work (Crammer and Singer 2000; Guermeur *et al.* 2000) also considers the case where classifiers are trained simultaneously. The extraction of conditional probabilities from multiclass SVM is studied in Passerini *et al.* (2002). Note that the use of error-correcting codes is not limited to inherently binary classifiers like SVMs. For example, Ghani (2000) apply error-correcting codes in conjunction with the Naive Bayes classifier, reporting good results in text categorization tasks with many classes.

4.6.4　Feature selection

Even methods like SVMs that are especially well suited for dealing with high dimensional data (such as vectorial representations of text) can suffer if many terms are *irrelevant* for class discrimination. Feature selection is a dimensionality reduction technique that attempts to limit overfitting by identifying irrelevant components of data points and has been extensively studied in pattern recognition (Kittler 1986) and in machine learning (Kira and Rendell 1992; Koller and Sahami 1996; Langley 1994; Liu and Motoda 1998).

Methods essentially fall into one of two categories: filters and wrappers. Filters attempt to determine which features are relevant *before* learning actually takes place. Wrapper methods, on the other hand, are based on estimates of the generalization error computed by running a specific learning algorithm and searching for relevant features by minimizing the estimated error (Kohavi and John 1997). Although wrapper methods are in principle more powerful, in practice their usage is often hindered by the computational cost. Moreover, they can overfit the data if used in conjunction with classifiers having high capacity.

Information theoretic approaches are known to perform well for filter methods. In particular, information gain is a popular metric in several machine-learning contexts, including the induction of decision trees (Quinlan 1986). It measures the information about the class that is provided by the observation of each term. More precisely, let us denote by W_j an indicator variable such that $W_j = 1$ means that ω_j appears in a certain document. The information gain of W_j is then the mutual information $\mathcal{I}(C, W_j)$ between the class C and W_j (see Appendix A for a review of the main concepts in information theory). This is also the difference between the marginal entropy of the class $\mathcal{H}(C)$ and the conditional entropy $\mathcal{H}(C \mid W_j)$ of the class, given W_j:

$$G(W_j) = \mathcal{H}(C) - \mathcal{H}(C \mid W_j) = \sum_{c=1}^{K} \sum_{\omega_j=0}^{1} P(c, \omega_j) \log \frac{P(c, \omega_j)}{P(c)P(\omega_j)}. \qquad (4.40)$$

Note that if C and W_j are independent, the mutual information is zero. Filtering index terms simply consists of sorting terms by information gain and keeping only the k terms with highest gain. Information gain has been used extensively for text categorization (Craven *et al.* 2000; Joachims 1997; Lewis and Ringuette 1994; McCallum and Nigam 1998; Yang 1999) and has been generally found to improve classification performance compared to using all words.

One limitation of information gain is that relevance assessment is done separately for each attribute and the effect of co-occurrences is not taken into account. However, terms that individually bring little information about the class might bring significant information when considered together. In the framework proposed by Koller and Sahami (1996), whole sets of features are tested for relevance about the class. Let x denote a point in the complete feature space (i.e. the entire vocabulary V in the case of text documents) and x_G be the projection of x into a subset $G \subset V$. In order to evaluate the quality of G to represent the class, we measure the distance between $P(c \mid x)$ and $P(c \mid x_G)$ using the average relative entropy:

$$\Delta_G = \sum_{x} P(x)P(c \mid x) \log \frac{P(c \mid x)}{P(c \mid x_G)}. \qquad (4.41)$$

The optimal set of features should yield a small Δ_G. Clearly this setup is only theoretical, since Equation (4.41) is computationally intractable and the distributions involved are hard to estimate accurately. Koller and Sahami (1996) use the notion of a *Markov blanket* to reduce complexity. A set of features $M \subset V$ (with W_j not in M) is a Markov blanket for W_j if W_j is conditionally independent of all the features in $V \setminus (M \cup \{W_j\})$ given M (see also the theory of Bayesian networks in Chapter 1). In other words, once the features in M are known, W_j and the remaining features are independent. Thus the class C is also conditionally independent of W_j given M and, as a result, if G contains a Markov blanket for W_j then $\Delta_{G_j} = \Delta_G$, where $G_j = G \setminus \{W_j\}$. The feature selection algorithm can then proceed by removing those features for which a Markov blanket can be found. Koller and Sahami (1996) prove that a greedy approach where features are removed iteratively is correct in the sense that a feature deemed irrelevant and removed during a particular iteration cannot became relevant again after some other features are later removed. Moreover, since finding a Markov blanket may be computationally very hard and an 'exact' blanket may not exist, they suggest the following approximate algorithm:

$\textsc{ApproxMarkovBlanket}(D, V, k, n')$

1    $G \leftarrow V$
2   **repeat**
3   **for** $W_j \in G$
4   **do for** $W_i \in G \setminus \{W_j\}$
5   **do** $\rho_{ij} \leftarrow \dfrac{\text{cov}[W_i, W_j]}{\sqrt{\text{var}[W_i]\,\text{var}[W_i]}}$
6                        $M_j \leftarrow k$ features having highest ρ_{ij}
7                        $\boldsymbol{p}_j \leftarrow P(c \mid \boldsymbol{X}_{M_j} = \boldsymbol{x}_{M_j}, W_j = x_j)$
8                        $\boldsymbol{p}'_j \leftarrow P(c \mid \boldsymbol{X}_{M_j} = \boldsymbol{x}_{M_j})$
9                        $D(\boldsymbol{x}_{M_j}, x_j) \leftarrow \mathcal{H}(\boldsymbol{p}_j, \boldsymbol{p}'_j)$
10                       $\Delta(W_j \mid M_j) \leftarrow \sum_{\boldsymbol{x}_{M_j}, x_j} P(\boldsymbol{x}_{M_j}, x_j) D(\boldsymbol{x}_{M_j}, x_j)$
11               $j^\star \leftarrow \arg\min_j \Delta(W_j \mid M_j)$
12               $G \leftarrow G \setminus \{W_{j^\star}\}$
13   **until** $|G| = n'$
14   **return** G

At each step, the algorithm computes for each feature W_j the set $M_j \subset (G \setminus \{F_j\})$ containing the k features that have the highest correlation with W_j, where k is a parameter of the algorithm and where correlation is measured by the Pearson correlation coefficient in line 5. Then, in line 8, the quantity $\Delta(W_j \mid M_j)$ is computed as the average cross entropy between the conditional distributions of the class that result from the inclusion and the exclusion of feature W_j. This quantity is clearly zero if M_j is a Markov blanket. Thus, picking $j^\star$ that minimizes it (line 9) selects a feature for which an approximate Markov blanket exists. Such a feature is removed and the process iterated until n' features remain in G.

Several other filter approaches have been proposed in the context of text categorization, including the use of minimum description length (Lang 1995), and symbolic rule learning (Raskinis and Ganascia 1996). A comparison of alternative techniques is reported in Yang (1999).

4.6.5   Measures of performance

The performance of a hypothesis function $h(\cdot)$ with respect to the true classification function $f(\cdot)$ can be measured by comparing $h(\cdot)$ and $f(\cdot)$ on a set of documents D_t whose class is known (test set). In the case of two categories, the hypothesis can be completely characterized by the confusion matrix

Predicted Category	Actual category	
	−	+
−	TN	FN
+	FP	TP

where TP, TN, FP, and FN mean true positives, true negatives, false positives, and false negatives, respectively. In the case of balanced domains (i.e. where the unconditional probabilities of the classes are roughly the same) *accuracy* A is often used to characterize performance. Under the 0-1 loss (see Section 1.5 for a discussion), accuracy is defined as

$$A = \frac{\text{TN} + \text{TP}}{|D_t|}. \tag{4.42}$$

Classification error is simply $E = 1 - A$. If the domain is unbalanced, measures such as precision and recall are more appropriate. Assuming (without losing generality) that the number of positive documents is much smaller than the number of negative ones, precision is defined as

$$\pi = \frac{\text{TP}}{\text{TP} + \text{FP}} \tag{4.43}$$

and recall is defined as

$$\rho = \frac{\text{TP}}{\text{TP} + \text{FN}}. \tag{4.44}$$

A complementary measure that is sometimes used is *specificity*

$$\sigma = \frac{\text{TN}}{\text{TN} + \text{FN}}. \tag{4.45}$$

As in retrieval, there is clearly a trade-off between false positives and false negatives. For example, when using a probabilistic classifier like Naive Bayes we might introduce a decision function that assigns a document the class '+' if and only if $P(c \mid d) > t$. In so doing, small values of the threshold t yield higher recall and larger values yield higher precision. Something similar can be constructed for an SVM classifier using a threshold on the distance between the points and the separating hyperplane. Often this trade-off is visualized on a parametric plot where precision and recall values, $\pi(t)$ and $\rho(t)$ are evaluated for different values of the threshold (see Figure 4.9 for some examples). Sometimes these plots are also called ROC curves (from Receiver Operating Characteristic) neglecting the caveat that the original name was coined in clinical research for diagrams plotting specificity versus sensitivity (an alias for precision). A very common measure of performance that synthesizes a precision-recall curve is the *breakeven point*, defined as the best[4] point where $\pi(t^\star) = \rho(t^\star)$ and can be seen as an alternative to the F_β measure (discussed earlier in Section 4.3.3) for reducing performance to a single number.

In the case of multiple categories we may define precision and recall separately for each category c, treating the remaining classes as a single negative class (see Table 4.2 for an example). Interestingly, this approach also makes sense in domains where the same document may belong to more than one category. In the case of multiple categories, a single estimate for precision and a single estimate for recall can

---

[4] Since in general there may be more than one value of the threshold τ at which $\pi(t) = \rho(t)$, we take the value $t^\star$ that maximizes $\pi(t^\star) = \rho(t^\star)$.

be obtained by averaging results over classes. Averages, however, can be obtained in two ways. When *microaveraging*, correct classifications are first summed individually:

$$\pi^\mu = \frac{\sum_{c=1}^K \text{TP}_c}{\sum_{c=1}^K \text{TP}_c}, \tag{4.46}$$

$$\rho^\mu = \frac{\sum_{c=1}^K \text{TP}_c + \text{FP}_c}{\sum_{c=1}^K \text{TP}_c + \text{FN}_c}. \tag{4.47}$$

When *macroaveraging*, precision and recall are averaged over categories:

$$\pi^M = \frac{1}{K} \sum_{c=1}^K \pi_c, \tag{4.48}$$

$$\rho^M = \frac{1}{K} \sum_{c=1}^K \rho_c. \tag{4.49}$$

Compared to microaverages, macroaverages tend to assign a higher weight to classes having a smaller number of documents.

4.6.6   Applications

Up to this point we have largely discussed the classification of generic text documents as represented by a 'bag-of-words' vector representation. From this viewpoint it did not really matter whether the bag of words represented a technical article, an email, or a Web page. In practice, however, when applying text classification to Web documents there are several domain-specific aspects of the Web that can be leveraged to improved classification performance. In this section we review some typical examples that demonstrate how ideas from text classification can be applied and extended in a Web context.

Classification of Web pages

Craven *et al.* (2000) have tackled the difficult task of extracting information from Web documents in order to automatically generate knowledge bases (KB). In their WEB → KB system, information extraction is carried out by first training machine-learning subsystems to make predictions about classes and relations, and then using the trained subsystems to populate an actual KB starting from data collected from the Web (see Chapter 6 for details on gathering collections of Web pages). Users are expected to provide as input an *ontology* (i.e. a formal description of classes and relations of interest) and training examples of actual instances of classes and relations. A sample ontology for modeling academic websites is shown in Figure 4.8.

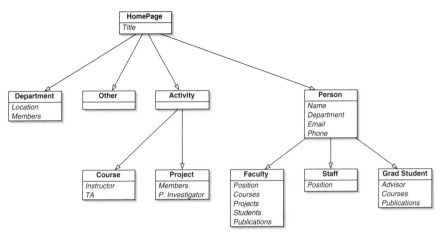

Figure 4.8 A sample ontology for representing knowledge about academic websites. The diagram indicates the class hierarchy and the main attributes of each class. Figure adapted from Craven *et al.* (2000).

Knowledge extraction consists essentially of

(1) assigning a new Web page to one node of the class hierarchy, and

(2) filling in the class attributes by extracting the relevant information from the document.

For example, once the system has gathered a home page of a professor at your university, it should first assign the page to the correct class (faculty) and then it should determine the value of attributes such as the email address, the courses taught, and the published papers. The first problem can be conveniently formulated as a text categorization task. In particular, Craven *et al.* (2000) study a Naive Bayes classifier to discriminate between the seven categories that appear as the leaves of the hierarchy shown in Figure 4.8. Their results are shown in Table 4.2 in the form of a confusion matrix. Each entry γ_{kl} for $k, l = 1, \ldots, K$ is the number of documents in class l that are assigned to class k, i.e. nondiagonal entries represent misclassifications. Precision for each class k is obtained by dividing γ_{kk} (the number of documents in class k that are correctly classified) by the total number of documents assigned to class k, $\sum_{l=1}^{K} \gamma_{kl}$. Similarly the recall for class l is obtained dividing γ_{kl} by the number of documents that actually belong to class l, i.e. $\sum_{k=1}^{K} \gamma_{kl}$.

Classification of news stories

The Reuters-21578 collection is perhaps the most widely use benchmark for text categorization systems (Lewis 1997). It consists of 21 578 stories that appeared in 1987 in the Reuters newswire, assembled and manually labeled by personnel from Reuters Ltd and Carnegie Group, Inc. There are 672 categories and each story may belong to more than one category. 148 of the categories contain at least 20 docu-

Table 4.2 Experimental results obtained in Craven *et al.* (2000) using the
Naive Bayes classifier on the WEB $\rightarrow$ KB domain.

Predicted category	Actual category							Precision
	Cou	Stu	Fac	Sta	Pro	Dep	Oth	
Cou	202	17	0	0	1	0	552	26.2
Stu	0	421	14	17	2	0	519	43.3
Fac	5	56	118	16	3	0	264	17.9
Sta	0	15	1	4	0	0	45	6.2
Pro	8	9	10	5	62	0	384	13.0
Dep	10	8	3	1	5	4	209	1.7
Oth	19	32	7	3	12	0	1064	93.6
Recall	82.8	75.4	77.1	8.7	72.9	100.0	35.0	

Table 4.3 Results reported by Joachims (1998) and Weiss *et al.* (1999) (last row) on 90 classes
of the Reuters-21578 collection. Performance is measured by microaveraged breakeven point.

Prediction method	Performance breakeven (%)
Naive Bayes (linear)	73.4
Rocchio (linear)	78.7
Decision tree C4.5	78.9
k-NN	82.0
Rule induction	82.0
Support vector (RBF)	86.3
Voting multiple decision trees	87.8

ments. The data set has been split into training and test data according to several
alternative conventions (see Lewis (1997) and Sebastiani (2002) for discussion). In
the so-called ModApte split, 9603 documents are reserved for training and 3299 for
testing (the rest being discarded). Ninety categories possess at least one training and
one test example. In this setting, Joachims (1998) experimented with several alternative classifiers, including those described in this chapter (see Table 4.3). Note that
the support vector classifier used in Joachims (1998) applies a radial basis function (RBF) kernel to learn nonlinear separation surfaces (see Schoelkopf and Smola
(2002) for a thorough discussion of kernel methods). It is worth noting that the simple k-NN classifier achieves relatively good performance and outperforms Naive
Bayes in this problem. On the same data set, Weiss *et al.* (1999) have reported better
results using multiple decision trees trained with AdaBoost (Freund and Schapire
1996; Schapire and Freund 2000) – specific results are provided in the last row of
Table 4.3.

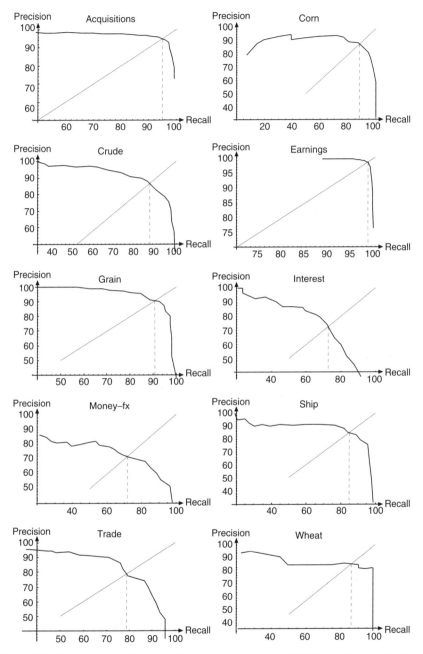

Figure 4.9 Precision-recall curves obtained by training an SVM classifier on the Reuters-21578 data set (ModApte split, 10 most frequent classes), where each plot is for one of the 10 classes.

Table 4.4   Experimental results obtained by Sahami *et al.* (1998)
in a junk email filtering problem.

Features	Junk		Not junk	
	Precision (%)	Recall (%)	Precision (%)	Recall (%)
Words	97.1	94.3	87.7	93.4
Terms	97.6	94.3	87.8	93.4
Terms and features	100.0	98.3	96.2	100.0

A simplified problem in the Reuters data set is obtained by removing all but the 10 most populated categories. In this setting, Dumais *et al.* (1998) report a comparison of several alternative learners obtaining their best performance (92% microaveraged breakeven point) with support vector machines. Figure 4.9 shows precision-recall curves we have obtained on this data set using an SVM classifier.

Email and news filtering

Sahami *et al.* (1998) have applied the Naive Bayes classifier to the problem of junk email filtering. This application is an example of a problem where hand-programming for data preprocessing (before learning) is particularly important. In this domain, the bag-of-words representation removes important information related to word proximity. Better features can be obtained by assessing the occurrence of terms such as 'confidential message', 'urgent and personal', or 'hot to make $', or perhaps by having features associated with regular expressions that detect unusual usage of punctuation. Sahami *et al.* (1998) manually identified a set of such terms and, in addition, enriched the textual representation with other features such as the TLD of the sender's address, attachments, timestamps, etc.

Feature selection in Sahami *et al.* (1998) was performed by eliminating rare words and then retaining the 500 most informative features as determined by mutual information. Results obtained on a data set of 1538 training messages and 251 test messages are reported in Table 4.4. The data set contained 1578 junk messages and 211 legitimate messages. Since the loss of false positives is clearly higher than the loss associated with false negatives, Sahami *et al.* (1998) proposed to classify a message as junk only if the probability predicted by Naive Bayes was greater than 99.9% (an empirically determined threshold). Good results with Naive Bayes have also been obtained by Androutsopoulos *et al.* (2000) on a publicly available data set (http://www.aueb.gr/users/ion/publications.html). A comparison of alternative learning methods, including SVM, is reported in Drucker *et al.* (1999).

NewsWeeder is a personalized filtering system for Usenet articles which uses relevance feedback to score news according to user interests (Lang 1995). The data set collected for training the NewsWeeder system contains 19 997 articles evenly sampled from 20 newsgroups and has been widely used as a benchmark for text categorization

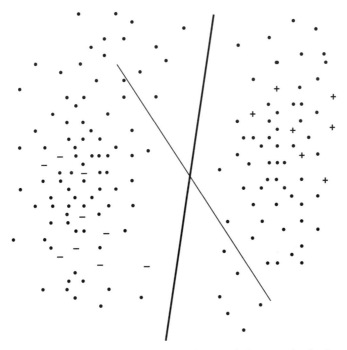

Figure 4.10   How unlabeled data (circles) may help categorization in
the case of large margin classifiers.

systems (the task being to predict which newsgroup a message was posted to). The
data set of 20 newsgroups is available at http://www.ai.mit.edu/people/
jrennie/20Newsgroups/.

4.6.7   Supervised learning with unlabeled data

The classifiers studied so far learn from labeled examples of text documents, i.e. the
category of the examples must be known. Labels must be assigned by human experts
and the process is clearly costly and time consuming. On the other hand, unlabeled
documents are abundantly available in many domains, particularly in the case of Web
documents. The available data can be written as $D \cup D'$, where $D = (d_i, f(d_i))$,
$i = 1, \dots n$, is the labeled data set and $D' = \{d'_j, \ j = 1, \dots, n'\}$ are unlabeled documents. Is there any useful information about classes in D'? Although it may seem
counterintuitive, the answer is yes, at least in principle, since unlabeled points allow
us to better characterize the shape of the data distribution.

   There are several intuitive explanations of why this is the case. First, suppose positive and negative examples are generated by two separate distributions, for example,
two Gaussians, and that a very large number of samples are available. In this case
we may expect that the parameters of the two distributions can be estimated well, for

example, by a training a mixture of two Gaussians. Then, just a few labeled points may be sufficient to decide which Gaussian is associated with the positive and negative class (see Castelli and Cover (1995) for a theoretical discussion). A different intuition, in terms of discriminant classifiers, can be gained by considering Figure 4.10, where unlabeled data points are shown as dots. The thicker hyperplane is clearly a better solution than the thin maximum margin hyperplane computed from labeled points only. Finally, as noted by Joachims (1999b), learning from unlabeled data is reasonable in text domains because categories can be guessed using term co-occurrences. For example, consider the 10 documents of Table 4.1 and suppose category labels are only known for documents d_1 and d_{10} ('Linux' and 'DNA', respectively). Then it should be clear that term co-occurrences allow us to infer categories for the remaining eight unlabeled documents. This observation links text categorization to LSI, a connection that has been exploited for example in Zelikovitz and Hirsh (2001).

None of the classification algorithms we have studied so far can deal directly with unlabeled data. We now present two approaches that develop the intuition above. The use of unlabeled data is further discussed in the context of co-training in Section 4.7.

EM and Naive Bayes

Nigam *et al.* (2000) propose a solution based on the Naive Bayes classifier and the EM algorithm for handling unlabeled data. In their setting, the class variable C for unlabeled documents is interpreted as a missing variable that can be filled in using EM. More precisely, the E step consists of computing, for each document $d_i \in D_u$, the conditional probability $P(c \mid d_i)$ (in the case of labeled documents this probability is clearly unity if $c = c_i$ and zero otherwise). This probability is then used to compute the expected sufficient statistics for the parameters, which are in turn used as 'soft' counts for parameters re-estimation. Under the word-based multinomial event model of Equation (4.25), the re-estimation equations in the M step are

$$\hat{\theta}_{c,j} = \frac{\alpha q_j + \sum_{i=1}^{n} n_{ij} P(c \mid d_i)}{\alpha + \sum_{k=1}^{|V|} \sum_{i=1}^{n} n_{ik} P(c \mid d_i)} \qquad (4.50)$$

and

$$\hat{\theta}_c = \frac{q_c' \alpha' + \sum_{i=1}^{n} P(c \mid d_i)}{\alpha' + n}. \qquad (4.51)$$

Nigam *et al.* (2000) validated the above method on three data sets: the 20 newsgroups, WEB → KB, and Reuters-21578 (ModApte split). For brevity we only report their results on the first data set in Figure 4.11. Note the significant difference of accuracy between Naive Bayes with no unlabeled documents and EM with 10 000 unlabeled documents when the number of labeled documents in the training set is small. As the number of labeled documents increases, the parameters of the Naive Bayes classifier are estimated more reliably and the two approaches tend to the same accuracy.

Transductive SVM

Support vector machines can be also be extended to handle nonlabeled data in the *transductive* learning framework of Vapnik (1998). In this setting, the optimization problem of Equation (4.33) (that leads to computing the optimal separating hyperplane for linearly separable data) becomes:

$$\min_{y'_1,\ldots,y'_{n'},\boldsymbol{w},w_0} \|\boldsymbol{w}\| \quad \text{subject to} \quad \begin{cases} y_i(\boldsymbol{w}^{\mathrm{T}}\boldsymbol{x}_i + w_0) \geqslant 1, & i = 1,\ldots,n, \\ y'_j(\boldsymbol{w}^{\mathrm{T}}\boldsymbol{x}'_j + w_0) \geqslant 1, & j = 1,\ldots,n'. \end{cases} \quad (4.52)$$

The solution is found by determining a label assignment $(y'_1,\ldots,y'_{n'})$ of the training examples so that the separating hyperplane maximizes the margin of both the data in D and D'. In other words missing values $(y'_1,\ldots,y'_{n'})$ are 'filled in' using maximum margin separation as a guiding criterion. A similar extension of Equation (4.38) for nonlinearly separable data is

$$\min_{y'_1,\ldots,y'_{n'},\boldsymbol{w},w_0} \|\boldsymbol{w}\| + C \sum_{i=1}^{n} \xi_i + C' \sum_{j=1}^{n'} \xi'_j$$

$$\text{subject to} \quad \begin{cases} y_i(\boldsymbol{w}^{\mathrm{T}}\boldsymbol{x}_i + w_0) \geqslant 1 - \xi_i, & i = 1,\ldots,n, \\ y'_j(\boldsymbol{w}^{\mathrm{T}}\boldsymbol{x}'_j + w_0) \geqslant 1 - \xi'_j, & j = 1,\ldots,n', \\ \xi_i \geqslant 0, & i = 1,\ldots,n, \\ \xi'_j \geqslant 0, & j = 1,\ldots,n'. \end{cases} \quad (4.53)$$

Clearly an exact solution of the above transductive inference problem is intractable, as it involves searching in the space of all possible label assignments on n' points.

Joachims (1999b) proposes a heuristic algorithm that starts assigning points in D', using the labeling induced by the classifier obtained by training a standard SVM on D. The algorithm then iteratively improves the solution by switching the labels of a positive and negative example in D' if the switch leads to a reduction of the objective function. Joachims (1999b) reports empirical results on Reuters-21578 (ModApte split with 10 categories) and on a subset of WEB → KB data. The results show that using a very small number of labeled documents still yields acceptable performance. For example, Joachims found that the average breakeven point of the transductive SVM was 60.8% on the Reuters-21578 data set (ModApte split, 10 most frequent classes) when using a data set D of 17 labeled documents and a test set D' of 3299 documents. By comparison, training a standard SVM on the same D and testing on the same D' yields an average breakeven score of 48.4%. Similarly, in the WEB → KB domain the average breakeven score was found to improve from 48.6% to 53.5% using a training set $|D| = 9$ and a test set $|D'| = 3957$. Although positive, these results are obtained in a quite specific experimental setup: the labeled data set D is a subset of the training data specified in the ModApte split. However, the unlabeled data D' are the ModApte test documents, i.e. the same documents where precision and recall are later estimated. Zhang and Oles (2000) repeated the same experiment using a different data split and

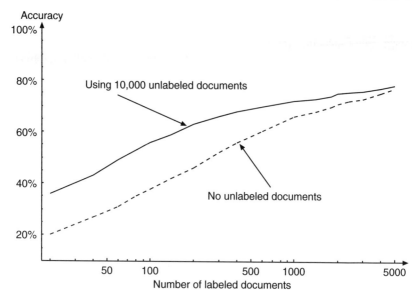

Figure 4.11 Classification accuracy as a function of the number of labeled documents using EM and 10 000 unlabeled documents (solid line) and using no unlabeled documents and the Naive Bayes classifier (dotted line). Experimental results reported in (Nigam *et al.* 2000) for the 20 newsgroups data set.

found that the transductive SVM actually maximizes the 'wrong margin' i.e. pushes toward a large separation of unlabeled data but achieve this by actually mislabeling the unlabeled data. In the setting of Zhang and Oles (2000) the performance of the transductive SVM was found to be worse than that of the inductive SVM.

4.7 Exploiting Hyperlinks

Text analysis in the case of Web documents can take advantage of the hyperlinked structure to gain additional information about a document that is not contained in the document itself. For example, the text in the source anchor text is very often a synthetic description of the contents of the page being linked to. Even if we did not observe a target document $d(v)$, we can gain information about this document from source anchors text in all the documents $d(w)$ such that $w \in \text{pa}[v]$. In the case of retrieval, for example, anchor text has been very usefully exploited to score target pages (Brin and Page 1998).

4.7.1 Co-training

Links in hypertexts offer additional information that can also be exploited for improving categorization and information extraction. The co-training framework introduced by Blum and Mitchell (1998) especially addresses this specific property of the Web, although it has more general applicability. In co-training, each instance is observed through two alternative sets of attributes or 'views' and it is assumed that each view is sufficient to determine the class of the instance. More precisely, the instance space $\mathcal{X}$ is factorized into two subspaces $\mathcal{X}_1 \times \mathcal{X}_2$ and each instance x is given as a pair (x_1, x_2). For example, x_1 could be the bag of words of a document and x_2 the bag of words obtained by collecting all the text in the anchors pointing to that document. Blum and Mitchell (1998) assume that

(1) the labeling function that classifies examples is the same as applied to x_1 or to x_2, and

(2) x_1 and x_2 are conditionally independent given the class.

Under these assumptions, unlabeled documents can be exploited in a special way for learning. Specifically, suppose two sets of labeled and unlabeled documents D and D' are given. The iterative algorithm described in Blum and Mitchell (1998) then proceeds as follows. Labeled data are used to infer two Naive Bayes classifiers, one that only considers the x_1 portion of x and one that only considers the portion x_2. Then these two classifiers are used to guess class labels of documents in a subset of D'. A fixed amount of such instances that are classified as positive with highest confidence is then added as positive examples to D. A similar procedure is used to add 'self-labeled' negative examples to D and the process iterated, retraining the two Naive Bayes classifiers on both labeled and self-labeled data. Blum and Mitchell (1998) report a significant error reduction in an empirical test on WEB $\rightarrow$ KB documents. Nigam and Ghani (2000) have shown that co-training offers significant advantages over EM if there is a true independence between the two feature sets.

4.7.2   Relational learning

Relational learning is another interesting approach for exploiting hypertextual information. In this setting, data are assumed to be available in a relational format and a learning algorithm can exploit relations among data items. In the case of Web data, there are obvious relations that are encoded in the hyperlinked structure of Web pages, but also local relations associated with the semi-structured organization of text as determined by the descriptive markup in HTML. As in the co-training framework described above, the general intuition is that, as well as the text on the pages being relevant to the classification of the page, the hyperlinks also contain useful information about the category a page belongs to.

FOIL (Quinlan 1990) is a classic algorithm for dealing with relational data and learning first-order clauses such as those used in the Prolog programming language. Although a description of FOIL is out of the scope of this book, it is interesting to remark that these approaches have been exploited with success to take hypertextual information into account. Craven et al. (2000) applied FOIL to learn classification rules in the WEB $\rightarrow$ KB domain (Figure 4.8 shows an ontology for that domain).

Table 4.5   Examples of first-order logic classification rules learned from WEB → KB data (Craven *et al.* 2000). Note that terms such as *jame* and *faculti* result from text conflation.

student(A)	:- not(has_ *data*(A)), not(has_ *comment*(A)), link_to(B,A), has_*jame*(B), has_*paul*(B), not(has_ *mail*(B)).
faculty(A)	:- has_ *professor*(A), has_ *ph*(A), link_to(B,A), has_*faculti*(B).

Two sample rules are reported in Table 4.5. For example, a Web page is classified as a 'student' page if it does not contain the terms *data* and *comment* and it is linked by another page that contains the terms *jame* and *paul* but not the term *mail*. The method was later refined by characterizing documents and links in a statistical way using the Naive Bayes model in combination with FOIL Craven and Slattery (2001).

Graphical models such as Bayesian networks that are traditionally conceived for propositional data (i.e. each instance of case is a tuple of attributes) have been more recently generalized to deal with relational data as well. Relational Bayesian networks have been introduced in Jaeger (1997) and learning algorithms for probabilistic relational models are presented in Friedman *et al.* (1999). Taskar *et al.* (2002) propose a probabilistic relational model for classifying entities, such as Web documents, that have known relationships. This leads to the notion of *collective classification*, where classification of a set of documents is viewed as a global classification problem rather than classifying each page individually. Taskar *et al.* (2002) use a general probabilistic Markov network model for this class of problems and demonstrate that taking relational information (such as hyperlinks) into account clearly improves classification accuracy. Cohn and Hofmann (2001) describe a probabilistic model that jointly accounts for contents and connectivity integrating ideas from PLSA (see Section 4.5.2) and PHITS (see Section 5.6.2).

Classification and probabilistic modeling of relational data in general is still a relatively less explored research area at this time but is clearly a promising methodology for classification and modeling of Web documents.

4.8   Document Clustering

4.8.1   Background and examples

As discussed in Chapter 1, clustering is the process of finding natural groups in data where there are no class labels present, i.e. all of the training data are unsupervised. In document clustering the objects to be clustered are typically represented using a bag-of-words representation. Document clustering is useful across a variety of applications related to the Web. We may for example want to automatically group Web pages on a website into clusters based on their content for the purposes of automatically generating links in real time to different clusters. If we have a relatively small number of pages, it may be possible to manually assign pages to clusters. However, many websites have tens of thousands of pages. In these cases it is clearly impractical to manually label all of the Web pages and automated clustering is an attractive approach.

Another application of clustering of Web documents is the automated grouping of the results of search engine queries. Search engines typically return a ranked list of URLs relative to a specified query. For example, issuing the query 'World Cup' to the Google search engine on 2 December 2002 yielded the following results.

- The top ranked URL was `http://www.fifaworldcup.com/` and indeed the next 10 ranked URLs, as well as many of the top 100, were also about soccer. Clearly 'soccer' is the largest cluster of URLs for this query.

- However, the 12th ranked URL was `http://www.dubaiworldcup.com`, a Web page about 'the richest horse race in the world' and obviously a different topic from soccer. As such it could be assigned to a different cluster.

- Further down the list we find items such as `http://www.wcsk8.com/` for skateboarding, `http://www.cricketworldcup.com/` for cricket, `http://www.pwca.org/` for paragliding, `http://www.rugby2003.com.au/` for rugby, and so on. Each of these could reasonably be expected to be in different small clusters.

- Robot soccer also appears on the list, `http://www.robocup.org/`, and could perhaps be a subcluster of the main soccer cluster.

- Skiing appears multiple times, both downhill skiing with `http://www.fis-ski.com/` and `http://www.skiworldcup.org/` and cross-country skiing with `http://www.uwasa.fi/~hkn/WC-Skiing/`. Again perhaps these could be viewed as two subclusters of a cluster about skiing.

Naturally it would be quite useful to automatically group these URLs in real time into meaningful clusters. In fact some search engines already support this feature, one example being `http://www.vivisimo.com/`. For the same query of 'World Cup' on 2 December 2002, the Vivisimo engine returned 175 URLs with the following cluster titles and number of URLs in each cluster in parentheses: FIFA World Cup (44); Soccer (42); Sports (24); History (19); Rugby World Cup (13); Women's World Cup (10); Betting (8); Cricket (6); Skiing World Cup (3); and so on. The details of this particular algorithm are not published but it can cluster about 200 URLs in response to a query in 'human real time' and automatically assign labels to the clusters (such as those assigned to the World Cup clusters above).

4.8.2   Clustering algorithms for documents

There are a variety of classic clustering algorithms that can be applied to document clustering including hierarchical clustering, K-means type clustering, or probabilistic clustering (see the earlier discussion in Chapter 1 on clustering). In what follows we provide a brief discussion of applying these algorithms to document clustering.

Hierarchical clustering

Hierarchical clustering algorithms do not presume a fixed number of clusters K in advance but instead produce a binary tree where each node is a subcluster of the parent node. Assume that we are clustering n objects. The root node consists of a 'cluster' containing all objects and the n leaf nodes correspond to 'clusters' where each contains one of the n original objects. This tree structure is known as a dendrogram.

Agglomerative hierarchical clustering algorithms start with a pairwise matrix of distances or similarities between all pairs of objects. At the start of the algorithm all objects are considered to be in their own cluster. The algorithm operates by iteratively and greedily merging the two closest clusters into a new cluster. The resulting merging process results in the gradual bottom-up construction of a dendrogram. The definition of 'closest' depends on how we define distance between two sets (clusters) of objects in terms of their individual distances. For example, if we define closest as being the smallest pairwise distance between any pair of objects (where the first is in the first cluster, and the second is in the second cluster) we will tend to get clusters where some objects can be relatively far away from each other. This leads to the well-known 'chaining effect' if, for example, the distances correspond to Euclidean distance in some d-dimensional space, where the cluster shapes in d-space can become quite elongated and 'chain like.' This minimum-distance algorithm is also known as single-link clustering.

In contrast we could define closest as being the maximum distance between all pairs of objects. This leads naturally to very compact clusters if viewed in a Euclidean space, since we are ensuring at each step that the maximum distance between any pair of objects in the cluster is minimized. This maximum-distance method is known as complete-link clustering. Other definitions for 'closest' are possible, such as computing averages of pair-wise distances, providing algorithms that can be thought of as between the single-line and complete-link methods.

A useful feature of hierarchical clustering for documents is that we can build domain-specific knowledge into the definition of the pairwise distance measure between objects. For example, the cosine distance in Equation (4.1) or weighting schemes such as TF–IDF (Equation (4.5)) could be used to define distance measures. Alternatively, for HTML documents where we believe that some documents are structurally similar to others, we could define an edit distance to reflect structural differences.

However, a significant disadvantage of hierarchical clustering is that the agglomerative algorithms have a time complexity between $O(n^2)$ and $O(n^3)$ depending on the particular algorithm being implemented. All agglomerative algorithms are at least $O(n^2)$ due the requirement of starting with a pairwise distance matrix between all n objects. For small values of n, such as the clustering of a few hundred Web pages that are returned by a search engine, an $O(n^2)$ algorithm may be feasible. However, for large values of n, such as 1000 or more, hierarchical clustering is somewhat impractical from a computational viewpoint.

Probabilistic model-based clustering

An alternative approach to hierarchical clustering is that of probabilistic model-based clustering. Earlier in Chapter 1 we briefly reviewed clustering using the Naive Bayes model, and earlier in this chapter we discussed using the same model for classification. Recall from Chapter 1 that model-based clustering assumes the existence of a generative probabilistic model for the data, typically in the form of a mixture model with K components, where each component corresponds to a probability distribution model for one of the clusters. The problem of clustering in this context amounts to one of learning the parameters of this mixture model, specifically, the parameters of each component model and the mixture weights for each component.

As we saw with classification earlier in this chapter, the Naive Bayes model is attractive as a component model for document clustering, since it contains only one parameter per dimension and the dimensionality of our document vectors is likely be high, e.g. as high as 5000–50 000 if we are using the bag-of-words representation. Even with a model such as the Naive Bayes that is linear in the number of parameters, these dimensionalities are still rather high. In practice a number of heuristics can be used to reduce dimensionality, such as removing very common words as well as very rare words. We cannot use standard class-based feature selection criteria directly (as we did for classification earlier in this chapter), since in clustering we do not know *a priori* what the classes (clusters) are.

4.8.3   Related approaches

Tantrum *et al.* (2002) discuss various strategies for combining ideas from hierarchical clustering and model-based clustering with applications to sets of documents. In this work the original bag-of-words representation is reduced to a 50-dimensional space by applying latent semantic indexing before clustering, and then performing clustering in the reduced-dimensional space. Dhillon and Modha (2001) investigate a spherical K-means algorithm for clustering of sparse high-dimensional document vectors and again use ideas from linear algebra to reduce dimensionality improve the quality of clustering. The motivation in both of these approaches is to combine dimension reduction with clustering, with the promise of being able to achieve more reliable clustering in the lower-dimensional space. A caveat here is that dimension-reduction techniques could in fact destroy the very cluster structure that we are seeking, unless the objective function for dimension reduction somehow incorporates the notion of what constitutes a good clustering of the data.

From a computational viewpoint, Dhillon *et al.* (2001) describe a number of practical methods for efficient implementation of the spherical K-means algorithm for document clustering. McCallum *et al.* (2000b) propose the use of approximate distance measures that are cheap to compute and that can divide a data set into subsets called 'canopies.' The general idea is that only objects within the same canopy need to be considered when updating clusters associated with that canopy and this is particularly efficient when both the dimensionality of the problem and the number of objects are both very large. The authors describe speed-ups of up to an order of magnitude for greedy agglomerative clustering, K-means clustering, and probabilistic model-based clustering, using their approach.

A number of other representations (besides hierarchical clustering, K-means, or mixtures models) have also been proposed for clustering documents. Taskar *et al.* (2002) describe how their Markov network model for relational data can be used to incorporate both text on the page as well as hyperlink structure for clustering of Web pages. Slonim and Tishby (2000) propose a very general information-theoretic technique for clustering called the *information bottleneck*. The technique appears to work particularly well for document clustering (Slonim *et al.* 2002).

Zamir and Etzioni (1998) describe a specific algorithm for the problem discussed earlier of clustering the results of search engines that was . They use 'snippets' returned by Web search engines as the basis for clustering and propose a clustering algorithm that uses suffix tree data structures based on phrases between algorithms. In the data sets used in their experiments they show that the resulting algorithm is both computationally efficient (linear in the number of documents) and finds clusters that are approximately as good as those obtained from clustering the full text of the Web pages.

4.9   Information Extraction

An interesting application of machine-learning and artificial intelligence techniques to text documents in general (and to text data on the Web in particular) is that of information extraction (Cohen *et al.* 2000). The general idea is to automatically extract unstructured text data from Web pages and to represent this extracted information in some well-defined schema (e.g. in a form suitable to enter into a relational database table). For example, we might want to extract information on authors and books from various online bookstore and publisher pages. Furthermore we might want to search for online reviews of these books, in newspaper and magazine articles, at online stores, etc. In another type of application a company might want to continuously crawl the Web searching for information about certain technologies or products of interest.

In general the problem of information extraction can often be characterized as that of detecting information about 'entities' (individuals, organizations, objects, etc.) in Web pages and then performing some form of 'parsing' to extract the relevant information such as the name of the entity, attributes of the entity, and relationships among multiple entities. This is a rich research topic – here we only provide the briefest of overviews but hopefully this will whet the reader's interest in exploring some of the references below and learning more about this topic. Our overview is loosely based on Cohen and McCallum (2002).

Early efforts in information extraction relied on hand-built models, for example the FASTUS system (Appelt *et al.* 1995) that uses 'cascades' of finite-state machines to parse text sequences into lexical units, entity names, groups of words associated with verbs, and so forth. More recent work, including, for example, the Web-KB system

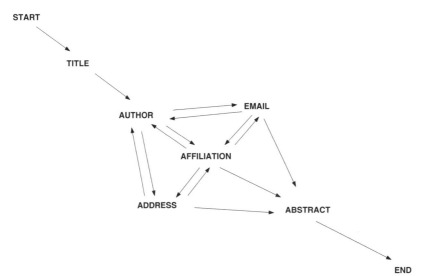

Figure 4.12 A toy example of states and transitions in an HMM for extracting various fields from the beginning of research papers. Not shown are various possible self-transition probabilities (e.g. multiple occurrences of the state 'author'), transition probabilities on the edges, and the probability distributions of words associated with each state. See Figure 7 in McCallum *et al*. (2000b) for a more detailed and realistic example of an HMM for this problem.

described earlier in this chapter, has focused largely on the use of machine-learning and statistical techniques that leverage human-labelled data to automate the process of constructing information extractors. In effect these systems use the human-labelled text to learn how to extract the relevant information (Cardie 1997; Kushmerick *et al*. 1997).

One general approach is to define information extraction as a classification problem. For example, we might want to classify short segments of text in terms of whether they correspond to the title, author names, author addresses and affiliations, and so forth, from research papers that are found by crawling the Web (for example this is one component of the functionality in the CiteSeer digital library system (Lawrence *et al*. 1999)). A classification approach to this problem would be to represent each document as a sequence of words and then use a 'sliding window' of fixed width k as input to a classifier – each of the k inputs to the classifier is a word in a specific position. The system can be trained on positive and negative examples that are typically manually labeled. A variety of different classifiers such as Naive Bayes, decision trees, and relational rule representations have been used for sliding-window based classification (Baluja *et al*. 2000; Califf and Mooney 1998; Freitag 1998; Soderland 1999).

A limitation of the sliding window is that it does not take into account sequential constraints that are naturally present in the data, e.g. the fact that the 'author' field almost always precedes the 'address' field in the header for a research paper. To take this type of structure into account, one approach is to train stochastic finite-state machines that can incorporate sequential dependence. One popular choice has been hidden Markov models (HMMs), which were mentioned briefly in Chapter 1 during our discussion of graphical models. An HMM contains a finite-state Markov model, where each state in the model corresponds to one of the fields that we wish to extract from the text. For example, in parsing headers of research papers we could have states corresponding to the title of the paper, author name, etc. An example of a simple Markov state diagram is shown in Figure 4.12. The key idea in HMMs is that the true Markov state sequence is unknown at parse-time – instead we see noisy observations from each state, in this case the sequence of words from the document. Each state has a characteristic probability distribution over the set of all possible words, e.g. the distribution of words from the state 'title' will be quite different from the distribution of words for the state 'author names.' Thus, given a sequence of words, and an HMM, we can 'parse' the observed sequence into a corresponding set of inferred states – the Viterbi algorithm provides an efficient method (linear in the number of observed words) for producing the most likely state-sequence, given the observations.

For information extraction, the HMM can be trained in a supervised manner with manually labeled data, or bootstrapped using a combination of labeled and unlabeled data (where the EM algorithm is used for training HMMs on unlabeled data). An early application of this approach to the problem of named entity extraction (automatically finding names of people, organizations, and places in free text) is described in Bikel *et al.* (1997). A variety of related ideas and extensions are described in Leek (1997), Freitag and McCallum (2000), McCallum *et al.* (2000a) and Lafferty *et al.* (2001). For a detailed description of the use of machine-learning methods (and HMMs in particular) to automatically create a large-scale digital library see McCallum *et al.* (2000c).

Information extraction is a broad and growing area of research at the intersection of Web research, language modeling, machine learning, information retrieval, and database theory. In this section we have barely scratched the surface of the many research problems, techniques, and applications that have been proposed, but nonetheless, we hope that the reader has gained a general high-level understanding of some of the basic concepts involved.

4.10   Exercises

Exercise 4.1. Given a collection of m documents to be indexed, compare the memory required to store pointers as integer document identifiers and as the difference between consecutive document identifiers using Elias's γ coding. (Hint: use Zipf's Law for term frequency and note that the most frequent term occurs in n documents, the second most frequent in $n/2$ documents, the third in $n/4$ documents and so on.)

Exercise 4.2. A vector $x \in \mathbb{R}^m$ is said to be *sparse* if

$$\ell = |\{j = 1, \ldots, m : x_j \neq 0\}| \ll m.$$

The dot product of two vectors x and x' normally requires $\Omega(m)$ time. In the case of sparse vectors, this can be reduced to $\Omega(\ell + \ell')$. Describe an efficient algorithm for computing the dot product of sparse vectors. (Hint: use linear lists to store the vectors in memory.)

Exercise 4.3. Consider the vector-space representation of documents and compare the cosine distance (Equation (4.1)) to the ordinary Euclidean distance. Show that for vectors of unit length the ranking induced by the two distances is the same.

Exercise 4.4. Show that a data set of n vectors in $\mathbb{R}^m$ is linearly separable if $n \leqslant m+1$.

Exercise 4.5. Consider a data set of n text documents. Suppose L is the average number of words in each document. How many documents must be collected, on average, before the sufficient condition of Exercise 4.4 fails? (Hint: use Heap's law to estimate vocabulary growth.)

Exercise 4.6. Collect subjects of your email mailbox and build a term–document matrix X using a relatively small set of frequent terms. Compute the SVD decomposition of the matrix using a software package capable of linear algebra computations (e.g. octave) and examine the resulting reconstructed matrix $\hat{X}$. Try querying your messages and compare the results obtained with a simple Boolean model where you simply retrieve matching subjects and the LSI model. Experiment with different values of k (the dimension of the latent semantic space).

Exercise 4.7. Build your own version of the Naive Bayes classifier and evaluate its performance on the Reuters-21578 data set. In a first setting, keep all those classes having at least one document in the training set and one in the test set. Repeat the experiment using only the top 10 most frequent categories. Beware that categories in the collection are not mutually exclusive (some documents belong to several categories).

Exercise 4.8. Write a program that selects the most informative features using mutual information. Test your program on the 20 Newsgroups corpus. Train your Naive Bayes classifier using the most k informative terms for different values of k and plot your generalization accuracy. What is your best value of k?

Repeat the exercise by using Zipf's Law on text, removing words that are either too frequent or too rare. Try different cutoff values for the k_1 most frequent words and the k_2 most rare words. Compare your results to those obtained with mutual information.

Finally combine the two methods above and compare your results.

Exercise 4.9. Obtain a public domain implementation of Support Vector Machines (e.g. Joachims's SVM^{light}, `http://svmlight.joachims.org/`) and set up an experiment to classify documents in the WEB $\rightarrow$ KB corpus. Since SVMs are binary classifiers you need to devise some coding strategy to tackle the multiclass problem. Compare results obtained using the two easiest strategies: one-against-all and all-pairs.

5

Link Analysis

Computing the relevance of a document is a major issue in Web-based information retrieval. As we have seen in Chapter 4, when considering unstructured collections of documents, it is possible to compute relevance with respect to user queries such as those involving keyword-based Boolean expressions, or those involving measures of similarity between documents. In each of these cases the score is a function of the document and the query. Large collections of hypertext such as the Web are more interesting in this respect since they allow us to compute scoring functions that include topological information about the hypertext graph. The basic assumption is that hyperlinks contain information about the human judgment of a document. To a first approximation, the more incoming links that exist for a document, the more likely it is that the document was judged to be 'important' by the authors of other documents linking to it.

Incoming links embody a common notion of popularity that also exists in other domains or in other webs. Networks of interaction have been studied for a long time in social sciences (Wasserman and Faust 1994), where nodes correspond to persons or organizations, and edges represent some type of social interaction. Intuitively, increasing the number of incoming links to a node should increase a common-sense measure of standing or popularity or prestige for that node. However, it should be also common sense that just counting the number of links does not necessarily provide an accurate measure of standing. For example, measuring the prestige of an enterprise by the mere number of its clients could be misleading, since different clients may have very different weights.

An important example is the network obtained by considering the scientific literature. Nodes in this case are papers, books, or entire journals, and edges correspond to citations. It makes sense to assume that the more citations a paper or a book receives, the more it can be assumed to be important, since it has been judged as useful by other scientists. The systematic construction of such networks through citation indexes was introduced by Garfield (1955), who later proposed a measure of standing for journals that is still in use. This measure, called *impact factor* (Garfield 1972), is defined as the average number of citations per recently published item. More precisely, if C is the total number of citations in a given time interval $[t, t + t_1]$ to articles published

Modeling the Internet and the Web   P. Baldi, P. Frasconi and P. Smyth
© 2003 P. Baldi, P. Frasconi and P. Smyth   ISBN: 0-470-84906-1

by a given journal during $[t - t_2, t]$, and N is the total number of articles published by that journal in $[t - t_2, t]$, the impact factor is defined as C/N (typically $t_1 = 1$ year and $t_2 = 2$ years). Thus, the impact factor is a very simple measure, since it basically corresponds to the normalized indegree of a journal in a subgraph of the citation network. Graph-based link analysis for the scientific literature goes back to the 1960s (Garner 1967). However, these ideas were not exploited in the development of first-generation Web search tools.

The paper by Bray (1996) reports an early attempt to apply social networks concepts to the Web. He suggested a Web visualization approach where the '... appearance of a site should reflect its *visibility*, as measured by the number of other sites that have pointers to it . . . ' and '. . . its *luminosity*, as measured by the number of pointers with which it casts navigational light off-site. . . '. Visibility and luminosity defined in this way are directly related to the indegree and the outdegree of websites, respectively. More recently, toward the end of the 1990s, link analysis methods became more widely known and used in a search engine context, leading to what is sometimes called the *second generation* of Web searching tools.

This chapter reviews the most common approaches to link analysis and how these techniques are be applied to compute the popularity of a document or a site. The algorithms presented in this chapter extract emergent properties from a complex network of interconnections, attempting to model (indirectly) subjective human judgments. It remains debatable as to whether popularity (as implied by the mechanism of citations) captures well the notions of relevance and quality as they are subjectively perceived by humans, and whether link analysis algorithms can successfully model human judgments.

5.1   Early Approaches to Link Analysis

Our notation for hypertext will be straightforward. For each vertex v in the hypertext graph $G = (V, E)$, $d(v)$ denotes the contents of the document at vertex v. If $d(v)$ is considered to be an isolated document, then its score with respect to a query q is $s(v \mid d(v), q)$. When considering $d(v)$ in its hypertext context, the score should also depend on G and will be denoted as $S(v \mid d(v), q, G)$. In the following, we will simplify the notation of these scores by just writing $s(v)$ and $S(v)$ if the dependencies on $d(v)$, q, and G are obvious from the context.

To quantify visibility (luminosity) $S(v)$ could simply be designed to grow with the indegree (outdegree) of v as hinted by Bray (1996). Clearly, however, such an approach suffers from a fundamental limitation: it would fail to capture the relative importance of different parents (children) in the graph. For example, a Web page with a small number of links coming from important sites should be considered more popular than a Web page with a larger number of links and whose sources are all from unimportant or irrelevant sites. Hence, rather than a mere count, popularity should be computed as a weighted sum of the citations a document receives through hypertext links. Ideas having this flavor are less recent than we might expect.

The use of hypertext information in information retrieval is older than the Web. Mark (1988), for example, was concerned with retrieving hypertext cards in a medical domain and noted that '... often cards do not even mention what they are about, but assume that the reader understands the context because he or she has read *earlier* cards.' He then proposed a simple algorithm for scoring documents where relevance information was transmitted from documents to their parents in the hypertext graph G. More precisely, the 'global' score of v given the query and the topology of G was computed as:

$$S(v) = s(v) + \frac{1}{|\text{ch}[v]|} \sum_{w \in |\text{ch}[v]|} S(w). \tag{5.1}$$

This simple algorithm somewhat resembles message passing schemes that are very common in connectionism (McClelland and Rumelhart 1986) or in graphical modeling (Pearl 1988). As such, it requires G to be a DAG so that a topological sort[1] can be chosen for updating the global scores S. The DAG assumption is reasonable in small hypertexts with a root document and a relatively strong hierarchical structure (in this case, even if G is not acyclic, not much information would be lost by replacing it with its spanning tree). The Web, however, is a large and complex graph. This may explain why search engines largely ignored its topology for several years.

The paper by Marchiori (1997) was probably the first one to discuss the quantitative concept of *hyper information* to complement *textual information* in order to obtain the *overall information* contained in a Web document. The idea somewhat resembles Frisse's approach. Indeed, if we rewrite Equation (5.1) as

$$S(v) = s(v) + h(v), \tag{5.2}$$

then $s(v)$ can be thought of as the textual information (that only depends on the document and the query), $h(v)$ corresponds to the hyper information that depends on the link structure where v is embedded, and $S(v)$ is the overall information. Marchiori (1997) did not cite Mark (1988), but nonetheless he identified a fundamental problem with Equation (5.1). If an irrelevant page v has a single link to a relevant page w, Equation (5.1) implies that $S(v) \geq S(w)$. The scenario would be even worse in a chain of documents $v_0, v_1, \ldots, v_k$. Here if $S(v_k)$ is very high but $S(v_0), \ldots, S(v_{k-1})$ are almost zero, then v_0 would receive a global score higher than v_k, even though a user would need k clicks to reach the important document.

As a remedy, Marchiori suggested that in this case the hyper information of v_0 should be computed as

$$h(v) = \sum_{w \in \text{ch}[v]} F^{r(v,w)} S(w), \tag{5.3}$$

where $F \in (0, 1)$ is a fading constant and $r(v, w) \in \{1, \ldots, |\text{ch}[v]|\}$ is the rank of w after sorting (in ascending order) the children of v according to the value of

---

[1] A topological sort is an ordering '$<$' of the vertices such that $v < v'$ if and only if there is a directed path from v' to v (Cormen *et al.* 2001).

$S(w)$. When applying Equation (5.3) to a linear chain $v_0, v_1, \ldots, v_k$ of documents, one would get $S(v_0) = \sum_{i=1}^{k} F^i S(v_i)$, so the contribution of the score of v_k fades exponentially as one moves back in the graph. As it turns out, Equation (5.3) in general implies a recursive form of computation that cannot be carried out in a cyclic graph. Marchiori (1997) suggested a solution to this problem assuming a finite horizon of propagation, i.e. $S(v)$ was computed on the tree rooted at v and having a fixed small depth k.

Before we discuss link analysis of the Web graph, it will first be useful to establish certain basic mathematical results relating to nonnegative matrices, graphs, and Markov chains. The reader familiar with these topics can safely skip the next section.

5.2 Nonnegative Matrices and Dominant Eigenvectors

A square matrix A is said to be *nonnegative*, written $A \geqslant 0$, if all its elements are nonnegative. Important examples of nonnegative matrices include graph incidence (of adjacency) matrices and stochastic matrices. Given a directed graph $G = (V, E)$, the incidence matrix A of G is defined as the 0-1 matrix with $a_{ij} = 1$ if and only if $(i, j) \in E$. Note that for simplicity we have assumed that the vertices in V are identified by the integers $1, 2, \ldots, n$, where $n = |V|$. Stochastic matrices are commonly used to describe first-order Markov chains as discussed in Appendix A. In a system with n discrete states, each entry of a stochastic matrix A contains the transition probability $a_{ij} = P(S_t = j \mid S_{t-1} = i)$. In this case, the probability axioms imply that each $a_{ij} \geqslant 0$ and that the elements of each row i should sum to unity.

A nonnegative $n \times n$ matrix A is said to be *irreducible* if, for each pair of indices (vertices) i and j, there exists a corresponding integer t such that $(A^t)_{ij} > 0$. If A is the adjacency matrix of a (directed) graph, this property tells us that the graph is (strongly) connected. By contrast, a reducible matrix is associated with a graph with more than one (strongly) connected component. In this case, if there exists a path of length t from a node i to itself, $(A^t)_{ii} > 0$. The greatest common divisor (gcd) of the set $\{t : (A^t)_{ii} > 0\}$ is called the *period* of i. If A is irreducible, then the period is the same for all indices (nodes) i. The common period is the gcd of the lengths of all the cycles in the graph. Interestingly, these topological properties of a graph have a correspondence in the spectral structure of its adjacency matrix, as shown by the Perron–Frobenius theorem (Seneta 1981).

For a nonnegative irreducible primitive matrix A, the Perron–Frobenius theorem allows us to conclude that there exists an eigenvalue λ of A such that

(1) λ is real and positive, and $\lambda \geqslant |\lambda'|$ for every other eigenvalue $\lambda' \neq \alpha$;

(2) λ corresponds to a strictly positive eigenvector;

(3) λ is a simple root of the characteristic equation $(A - \alpha I_n) = 0$.

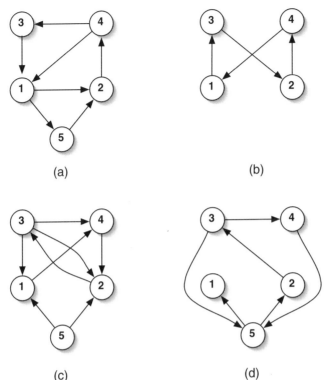

Figure 5.1   Graphs with different types of incidence matrices. (a) is primitive,
(b) is irreducible (with period 4) but not primitive, (c) and (d) are reducible.

In this case, λ is called the *dominant eigenvalue* of A and the associated eigenvector
is called the *dominant eigenvector*. Denoting by $(\lambda_1, \ldots, \lambda_n)$ the eigenvalues of A,
in the following we will assume that the dominant eigenvalue is always λ_1.

Note, however, that although there cannot be multiple roots there may be some
other eigenvalue $\lambda_j \neq \lambda_1$ such that $|\lambda_j| = |\lambda_1|$. It can be shown that if there are k
eigenvalues having the same magnitude as the dominant eigenvalue, then they are
equally spaced in the complex circle of radius λ_1. Moreover, if A is the adjacency
matrix of a graph, k is the gcd of the lengths of all the cycles in the graph. In order
to get a dominant eigenvalue that is strictly greater than all other eigenvalues further
conditions are necessary.

A matrix A is said to be *primitive* if there exists a positive integer t such that
$A^t > 0$ (note the strict inequality). A primitive matrix is also irreducible, but the
converse is not true in general. For a primitive matrix, condition 1 of the Perron–
Frobenius theorem holds with strict inequality. This means that all the remaining
eigenvalues are smaller in modulus than the dominating eigenvalue. Moreover, if the
adjacency matrix of a graph is primitive, then the gcd of the lengths of all cycles is
unity. Figure 5.1 illustrates some examples.

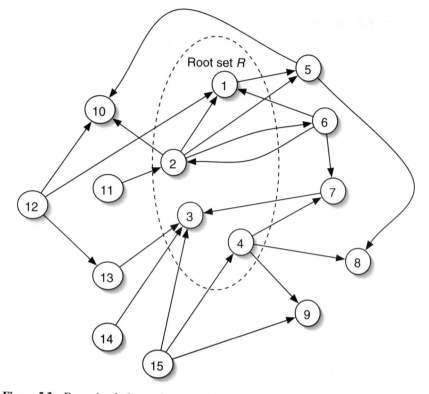

Figure 5.2   Example of a base subgraph obtained by starting from the vertex set $\{1, 2, 3, 4\}$.

The fundamental property of primitive matrices also suggests a simple iterative algorithm for computing the dominant eigenvalue and the associated eigenvector. Let $x \in \mathbb{R}^n$ and let $(a_1, \ldots, a_n)$ denote the coordinates of x in the basis formed by the eigenvectors $(v_1, \ldots, v_n)$. If we expand the product $A^t x$, remembering that $A v_i = \lambda_i v_i$, we obtain

$$A^t x = \sum_i a_i \lambda_i^t v_i, \qquad (5.4)$$

but since $\lambda_1 > |\lambda_i|, i > 1$, the first term dominates the above sum as t gets large, i.e. $a_i \lambda_i v_i \approx A^t$ for large t. This gives us a vector proportional to the dominant eigenvector, provided that x is not orthogonal to v_1. Since the Perron–Frobenius theorem tells us that v_1 is strictly positive, any random positive vector will yield the correct solution, for example, $x = \mathbf{1} = (1, 1, \ldots, 1)^{\mathrm{T}}$.

In the special case of primitive stochastic matrices, it is easy to see that $\lambda_1 = 1$ since, by the definition of a stochastic matrix, $A\mathbf{1} = \mathbf{1}$. Moreover, since all the remaining eigenvalues are strictly smaller than one in modulus, the sequence A^t converges at an exponential rate and it can be shown that

$$\lim_{t \to \infty} A^t = \mathbf{1}^{\mathrm{T}} r. \qquad (5.5)$$

In this case, r is known as the *stationary distribution* of the Markov chain.

5.3   Hubs and Authorities: HITS

Kleinberg's algorithm, called 'Hypertext Induced Topic Selection' (HITS), simultaneously computes a pair of scoring values associated with hypertext documents (Kleinberg 1998, 1999). The semantics attached to these quantities essentially match Bray's concepts of visibility and luminosity.

HITS works on a small graph associated with a selected collection of hypertext documents, for example a focused portion of the Web that is expected to be related to a given topic of interest. In the original formulation of HITS, the subgraph of interest (also known as the *base subgraph*) is computed by selecting the neighbors of a root set R of Web pages that are known to be relevant with respect to the topic. The root set is expanded to include all the children and a fixed number of parents of nodes in R. Details of the procedure are given in the following algorithm (see Figure 5.2 for an illustration).

BASESUBGRAPH(R, d)
1   $S \leftarrow R$
2   **for each** v **in** R
3   **do** $S \leftarrow S \cup \text{ch}[v]$
4       $P \leftarrow \text{pa}[v]$
5   **if** $|P| > d$
6   **then** $P \leftarrow$ arbitrary subset of P having size d
7                 $S \leftarrow S \cup P$
8   **return** S

Note that the children of a given node (line 3) are forward links and can be obtained directly from each page v. Parents (line 4) correspond to backlinks and can be obtained from a representation of the Web graph obtained, for example, through a crawl. Several commercial search engines currently support the special query link:url that returns the set of documents containing url as a link. In the case of small scale applications, this approach can be used to obtain the set of parents in line 4. Parameter d is the maximum number of parents of a node in the root set that can be added. As we know (see Chapter 3), some pages may have a very large indegree. Thus, bounding the number of parents is crucial in practical applications. Algorithm BASESUBGRAPH returns a set of nodes S. In what follows, HITS is assumed to work on the subgraph of the Web induced by S.

Let $G = (V, E)$ denote the subgraph of interest, where $V = S$. For each vertex $v \in V$, let us introduce two positive real numbers $a(v)$ and $h(v)$. These quantities are called the *authority* and the *hubness* weights of v, respectively. Intuitively, a document should be very authoritative if it has received many citations. As discussed above, citations from important documents should be weighted more than citations from less-important documents. In the case of HITS, the importance of a document as a source of citations is measured by its hubness. Intuitively, a good hub is a document that allows us to reach many authoritative documents through its links. The result is that the hubness of a document depends on the authority of the cited documents, and the authority of a document depends on the hubness of the citing documents. We are apparently stuck in a loop, but let us observe that this recursive form of dependency between hubs and authority weights naturally leads to the definition of the following operations:

$$a(v) \leftarrow \sum_{w \in \text{pa}[v]} h(w), \tag{5.6}$$

$$h(v) \leftarrow \sum_{w \in \text{ch}[v]} a(w). \tag{5.7}$$

The two operations above can be carried out to update authority and hubness weights starting from initial values. This approach is meaningful because Kleinberg (1999) showed that iterating Equations (5.6) and (5.7), intermixed with a proper normalization step, yields a convergent algorithm. The output is a set of weights that can be therefore considered to be globally consistent. Kleinberg's algorithm is listed below. For convenience, weights are collected in two n-dimensional vectors a and h.

HubsAuthorities(G)
1    $\mathbf{1} \leftarrow [1, \ldots, 1] \in \mathbb{R}^{|V|}$
2    $a_0 \leftarrow h_0 \leftarrow \mathbf{1}$
3    $t \leftarrow 1$
4   **repeat**
5   **for each** v **in** V
6   **do** $a_t(v) \leftarrow \sum_{w \in \text{pa}[v]} h_{t-1}(w)$
7                  $h_t(v) \leftarrow \sum_{w \in \text{ch}[v]} a_{t-1}(w)$
8              $a_t \leftarrow a_t / \|a_t\|$
9              $h_t \leftarrow h_t / \|h_t\|$
10             $t \leftarrow t + 1$
11   **until** $\|a_t - a_{t-1}\| + \|h_t - h_{t-1}\| < \varepsilon$
12   **return** (a_t, h_t)

To show that HubsAuthorities terminates, we need to prove that for each $\varepsilon > 0$ the condition controlling the outer loop will be met for t large enough. Formally, this means that the sequences $\{a_t\}_{i \in \mathbb{N}}$ and $\{h_t\}_{t \in \mathbb{N}}$ converge to limits $a^\star$ and $h^\star$, respectively. The proof of this result is based on rewriting HITS using linear algebra. In particular, if we denote by A the incidence matrix of G, it can be easily verified that the updating operations can be written compactly in vector notation as $a_t = A^\mathrm{T} h_{t-1}$

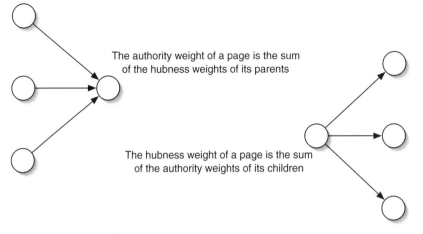

The authority weight of a page is the sum
of the hubness weights of its parents

The hubness weight of a page is the sum
of the authority weights of its children

Figure 5.3   Graphical explanation of the basic operations in HITS
(see Equations (5.6) and (5.7)).

and $h_t = A h_{t-1}$. As a result, after t iterations,

$$a_t = \alpha_t (A^T A)^{t-1} A^T 1, \tag{5.8}$$

$$h_t = \beta_t (A A^T)^t 1, \tag{5.9}$$

where α_t and β_t are scalar normalization factors. Thus, a sufficient condition for
HUBSAUTHORITIES to terminate is that the sequences of vectors $\{\alpha_t (A^T A)^{t-1} A^T 1\}$
and $\beta_t (A A^T)^t 1$ converge for $t \to \infty$. It is possible to prove that these sequences
converge under fairly unrestrictive hypotheses. For example, it can be shown that a
sufficient condition is that M be a nonnegative nonsingular symmetric matrix. In this
case, the dominant eigenvector $\omega_1(M)$ (i.e. the eigenvector associated with the largest
eigenvalue $\lambda_1(M)$) is nonnegative, and for every vector x such that $\omega_1(M)^T x \neq 0$
we have

$$\lim_{t \to \infty} \frac{M^t x}{\| M^t x \|} \propto \omega_1(M). \tag{5.10}$$

Since 1 cannot be orthogonal to a nonnegative vector, the sequences $\{a_t\}$ and $\{h_t\}$
converge to $\omega_1(A^T A)$ and $\omega_1(A A^T)$, respectively.

As an example, in Figure 5.4 we show the authority and hubness weights computed
by HUBSAUTHORITIES on the graph of Figure 5.2. We can note some unobvious
weight assignments. For example, vertex 3 has the largest indegree in the graph but
nonetheless its authority is rather small because of the low hubness weight of its
parents.

Bharat and Henzinger (1998) have suggested an improved version of HITS that
addresses some specific problems that are encountered in practice. For example, a
mutual reinforcement effect occurs when the same host (or document) contains many
identical links to the same document in another host. To solve this problem, Bharat
and Henzinger (1998) modified HITS by assigning weight to these multiple edges that

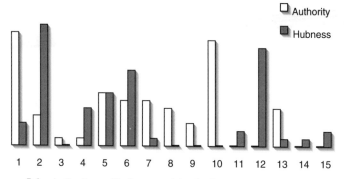

Figure 5.4   Authority and hubness weights in the example graph of Figure 5.2.

are inversely proportional to their multiplicity. The method presented by Bharat and Henzinger (1998) also addresses the problem of links that are generated automatically, for example, by converting messages posted to Usenet news groups into Web pages. Finally, they address the so-called *topic drift* problem, i.e. some nodes in the base subgraph may be irrelevant with respect to the user query and documents with highest authority or hubness weights could be about different topics. This problem can be addressed either by pruning irrelevant nodes or by regulating the influence of a node with a relevance weight.

5.4   PageRank

The theory developed in this section was introduced by Page *et al.* (1998) and resembles in many ways the recursive propagation idea we have seen in HITS. However, unlike HITS, only one kind of weight is assigned to Web documents. Intuitively, the rank of a document should be high if the sum of its parents' ranks is high. To a first approximation, this intuition might be embodied in the equation

$$r(v) = \alpha \sum_{w \in \text{pa}[v]} \frac{r(w)}{|\text{ch}[w]|}, \tag{5.11}$$

where $r(v)$ is the rank assigned to page v and α is a normalization constant. Note that each parent w contributes by a quantity that is proportional to its rank $r(w)$ but inversely proportional to its outdegree. This is a fundamental difference with respect to authority in HITS. The endorsement signal that flows from a given page w to each of its children decreases as the number of outgoing links (and, therefore, the potential of being a good hub) increases. Equation (5.11) can also be written in matrix notation, as

$$r = \alpha B r = M r. \tag{5.12}$$

The matrix B is obtained from the adjacency matrix A of the graph by dividing each element by the corresponding row sum, i.e.

$$b_{uv} = \begin{cases} \dfrac{a_{uv}}{\sum_w a_{uw}}, & \text{if } ch[u] \neq \emptyset, \\ a_{u,v} = 0, & \text{otherwise.} \end{cases} \tag{5.13}$$

As implied by Equation (5.12), the vector r is a right eigenvector of B with an associated eigenvalue α.

An interesting property of this solution can be seen by interpreting the computation expressed by Equation (5.12) as the description of a random walk through the Web graph. More precisely, suppose each vertex in the graph is associated with a realization of a discrete random variable S_t that models the position of a hypothetical surfer at a given time t. The rank of a page v could be then thought of as the asymptotic probability that the surfer is currently browsing that page, i.e. $P(S_t = v)$. Under this perspective, M is interpreted as the transition matrix for a first-order Markov chain:

$$r_t(v) = P(S_t = v) = \sum_w P(S_t = v \mid S_{t-1} = w) P(S_{t-1} = w)$$

$$= \sum_w m_{wv} r_{t-1}(w). \tag{5.14}$$

This equation updates the probability that our random surfer will browse page v at time t, given the vector of probabilities at time $t-1$ and the transition probabilities m_{wv}. In matrix notation this can be written as

$$r_t = M r_{t-1}. \tag{5.15}$$

To satisfy probability axioms, M must be a stochastic matrix, i.e. its rows should sum to one. Since the rows of B are normalized, the probability axioms are satisfied if $\alpha = 1$. This simply means that the random surfer picks one of the outlinks in the page being visited according to the uniform distribution.

A fundamental question is whether iterating Equation (5.14) converges to some sensible solution regardless of the initial ranks r_0. To answer this we need to inspect different cases in the light of the theory of nonnegative matrices developed in Section 5.2.

Four interesting cases are illustrated in Figure 5.5. The first graph has a primitive adjacency matrix. Therefore Equation (5.5) holds, and values of $r(v)$ corresponding to the steady state are indicated inside each node (below the node index). In the same figure, arcs are labeled by the amount of rank that is passed from a node to its children. As expected, Equation (5.11) holds everywhere. The second graph is more problematic, since its adjacency matrix is irreducible but not primitive. In this case, passing ranks from nodes to their children results in a cyclic updating. The random walk recursion of Equation (5.14) converges in this case to a limit cycle rather than to a steady state, and the periodicity of the limit cycle is the period of the matrix, or, as

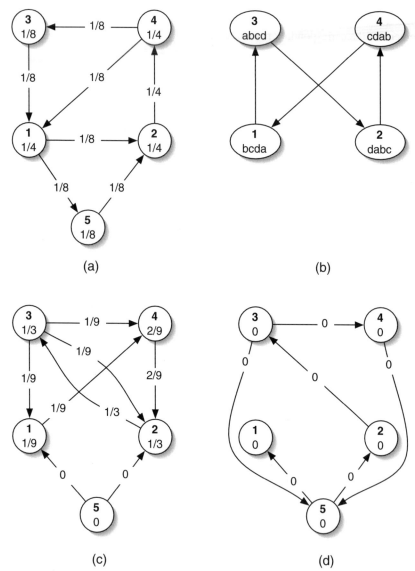

Figure 5.5 Rank propagation on graphs with different types of incidence matrices. Equation (5.14) converges to a nontrivial steady state in case (a) and (c), to a limit cycle in case (b), and to zero in case (d).

we know, the gcd of the lengths of the cycles (4 in the example). This is indicated in Figure 5.5b by four values a, b, c, d of rank that cyclically bounce along the nodes. The third graph of Figure 5.5 has a reducible adjacency matrix. In this case M^t converges to a matrix whose last column is all zero, reflecting the fact that the node should have zero rank as it has no parents. Finally, the fourth graph has also a reducible adjacency matrix but this time the maximum eigenvalue is less than one, so M^t converges to the zero matrix. This is due to the existence of a node (1) with no children that effectively acts as a rank sink.

The situation in Figure 5.5d is of course undesirable but is very common in the actual Web. Many pages have no outlinks at all. Furthermore, pages that remain on the crawling frontier and are never fetched will likely produce dangling edges in the graph that is obtained from crawling. To solve this difficulty, observe that the connectivity of node 1 should be defined as illegal, since it violates the basic hypothesis underlying the random walk model: the sum of the probabilities of the available actions should be one in each node, but once in the sink nodes our random surfer would be left with no choices. A sensible correction consists of giving the random surfer a 'method of escape' by adding allowable actions. One possibility is to assume that the surfer, who cannot possibly follow any link, will restart browsing by picking a new Web page at random. This is the same as adding a link from each sink to each other vertex, i.e. introducing an escape matrix E defined as $e_{vw} = 0$ if $|\text{ch}[v]| > 0$ and $e_{vw} = 1/n$ otherwise, for each w. Then the transition matrix becomes

$$M = (B + E).$$

M is now a stochastic matrix and the Markov chain model for a Web surfer is sound. In general, however, there is no guarantee that M is also primitive (if there are cycles with zero outdegree as in Figure 5.1b, these bring irreducible but periodic components). This difficulty will be addressed shortly and for now let us assume that M is primitive.

The following iterative algorithm was suggested in the original paper on PageRank (Page *et al.* 1998). It takes as input a nonnegative square matrix M, its size n, and a tolerance parameter ϵ.

PAGERANK(M, n, ϵ)
1   $\mathbf{1} \leftarrow [1, \dots, 1] \in \mathbb{R}^n$
2   $z \leftarrow \frac{1}{n}\mathbf{1}$
3   $x_0 \leftarrow z$
4   $t \leftarrow 0$
5   **repeat**
6         $t \leftarrow t + 1$
7         $x_t \leftarrow M^T x_{t-1}$
8         $d_t \leftarrow \|x_{t-1}\|_1 - \|x_t\|_1$
9         $x_t \leftarrow x_1 + d_t z$
10        $\delta \leftarrow \|x_{t-1} - x_t\|_1$
11   **until** $\delta < \epsilon$
12   **return** x_t

The quantity d_t is the total rank being lost in sinks. Adding $d_t z$ to $M^T x_{t-1}$ is basically a normalization step. As it turns out, if M is a stochastic primitive matrix, then $d_t = 0$ in each iteration (no normalization is necessary) and PageRank converges to the stationary distribution of M. Otherwise, the above algorithm implicitly 'repairs' the matrix M into a stochastic matrix and converges to the corresponding stationary distribution (see Exercise 5.4).

Now we address the problem of irreducible but periodic components. These also act as rank sinks because they never pass rank to other parts of the graph. Moreover, periodicity may hurt convergence and the algorithm PAGERANK above is not anymore guaranteed to terminate. The solution suggested in Page *et al.* (1998) consists of forcing some source or rank by introducing a 'static' stochastic process that models the 'distribution of Web pages that a random surfer periodically jumps to.' This distribution can be any nonnegative vector e such that $\|e\|_1 = 1$. The probability distribution that results from combining the Markovian random walk distribution x and the static rank source distribution is a mixture model with parameter ε:

$$r = \varepsilon e + (1 - \varepsilon)x.$$

The simplest choice for e is a uniform distribution, i.e. $e = (1/n)\mathbf{1}$. Intuitively, this approach can be motivated by the metaphor that browsing consists of following existing links with some probability $1 - \varepsilon$ or selecting a nonlinked page with probability ε. When the latter choice is made, each page in the entire Web is sampled according to the probability distribution e. In the case of the uniform distribution, Equation (5.15) will be rewritten as

$$r_t = [\varepsilon H + (1 - \varepsilon)M]^{\mathrm{T}} r_{t-1}, \tag{5.16}$$

where H is a square matrix with $h_{uv} = 1/n$ for each u, v. In this way we have obtained an ergodic Markov chain whose underlying transition graph is fully connected. The associated transition matrix $\varepsilon H + (1 - \varepsilon)M$ is primitive and therefore the sequence r_t converges to the dominant eigenvector. The stationary distribution r associated with the Markov chain described by Equation (5.16) is known as *PageRank*. In practice, ε is typically chosen to be between 0.1 and 0.2 (Brin and Page 1998).

5.5 Stability

An important question is whether the link analysis algorithms based on eigenvectors (such as HITS and PageRank) are stable in the sense that results do not change significantly as a function of modest variations in the structure of the Web graph. More precisely, suppose the connectivity of a portion of the graph is changed arbitrarily, i.e. let $G = (V, E)$ be the graph of interest and let us replace it by a new graph $\tilde{G} = (V, \tilde{E})$, where some edges have been added or deleted. How will this affect the results of algorithms such as HITS and PageRank?

Ng *et al.* (2001) proved two interesting results about the stability of algorithms based on the computation of dominant eigenvectors.

5.5.1 Stability of HITS

First, Ng *et al.* (2001) derived a bound on the number of hyperlinks that can added or deleted from one page without significantly affecting the authority (or hubness) weights computed by HITS. The bound essentially depends on the *eigengap*, namely the difference $\delta \doteq \lambda_1 - \lambda_2$ between the two largest eigenvalues of $M = A^T A$ (since this matrix is symmetric the eigengap is a real number).

The result can be formally stated as follows. For every $\alpha > 0$, suppose G is perturbed by adding or deleting at most k hyperlinks from one page,

$$k \leqslant \left(\sqrt{d + \frac{\alpha \delta}{4 + \sqrt{2}\alpha}} - \sqrt{d} \right)^2, \tag{5.17}$$

where d is the maximum outdegree of G. The principal eigenvector associated with the perturbed graph $\tilde{G}$ then satisfies

$$\|a - \tilde{a}\|_2 \leqslant \alpha. \tag{5.18}$$

Moreover, Ng *et al.* (2001) show that it is possible to perturb a symmetric matrix, by a quantity that grows as δ, that produces a constant perturbation of the dominant eigenvector. Thus, matrices with small eigengap can have low robustness with respect to perturbation.

In practice, it is not difficult to construct graphs where even adding or deleting even a single edge results in large variations in the authority and hubness weights. For example, consider two isolated communities, one possessing a hub h but no emerging authority (so the authority weight is dispersed among all the nodes), and the other possessing a well recognized authority a but no important hubs. Adding an edge from h to a would result in a large change in weight assignments, since the authority weight of a would be increased at the expense of some amount of authority weights contributed by all the nodes in the first community.

5.5.2 Stability of PageRank

In this case suppose r is the stationary distribution associated with matrix

$$\varepsilon H + (1 - \varepsilon)M$$

(see Equation (5.16)). If the adjacency matrix A is perturbed to a new matrix $\tilde{A}$, then Ng *et al.* (2001) show that

$$\|\tilde{r} - r\| \geqslant \frac{2 \sum_{j \in \tilde{V}} r(j)}{\varepsilon}, \tag{5.19}$$

where $\tilde{V}$ denotes the set of vertices touched by the perturbation. This demonstrates two interesting facts: first, the parameter ε of the mixture model in Equation (5.16) has a stabilization role; second, if the set of pages affected by the perturbation have a

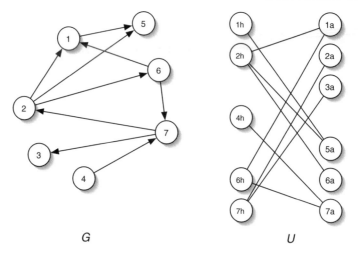

Figure 5.6   Forming a bipartite graph in SALSA.

small rank, the overall change will also be small. Bianchini *et al.* (2001) later proved
the tighter bound

$$\|\tilde{r} - r\| \geqslant \frac{1 - \varepsilon}{\varepsilon} 2 \sum_{j \in \tilde{V}} \delta(j) r(j), \qquad (5.20)$$

where $\delta(j) \geqslant 2$ depends on the edges incident on j affected by the perturbation.

5.6   Probabilistic Link Analysis

The probabilistic interpretation of PageRank based on random walks can be extended
to link analysis algorithms that, like HITS, distinguish the importance of a node as
an authority or as a hub.

5.6.1   SALSA

Lempel and Moran (2001) have proposed a probabilistic extension of HITS called the
'Stochastic Approach for Link Structure Analysis' (SALSA). Similar extensions have
been proposed independently by Rafiei and Mendelzon (2000) and Ng *et al.* (2001).
In all of these proposals, the random walk is carried out by following hyperlinks both
in the forward and in the backward direction.

SALSA starts from a graph $G = (V, E)$ of topically related pages (like the base
subgraph of HITS) and constructs a bipartite undirected graph $U = (\hat{V}, \hat{E})$ as (see
Figure 5.6)

$$\hat{V} = V_h \cup V_a,$$

where

$$V_h \doteq \{v_h : v \in V, \ ch[v] \neq \emptyset\},$$
$$V_a \doteq \{v_a : v \in V, \ pa[v] \neq \emptyset\},$$
$$\hat{E} \doteq \{(u_h, v_a) : (u, v) \in E\}.$$

The sets V_h and V_a are called the *hub side* and the *authority side* of U, respectively.

Two separate random walks are then introduced. In the 'hub' walk, each step consists of

(1) following a Web link from a page u_h to a page w_a, and

(2) immediately afterward following a backlink going back from w_a to v_h, where we have assumed that $(u, w) \in E$ and $(v, w) \in E$.

For example, jumping from 1_h to 5_a and then back from 5_a to 2_h in Figure 5.6. In the 'authority' walk, a step consists of following a backlink first and a forward link next. In both cases, a step translates into following a path of length exactly two in U. Note that, by construction, each walk starts on one side of U, either the hub side or the authority side, and will remain confined to the same side. The Markov chains associated with the two random walks have transition matrices $\tilde{H}$ and $\tilde{T}$, respectively, defined as follows:

$$\tilde{h}_{uv} = \sum_{\substack{w:(u,w)\in E, \\ (v,w)\in E}} \frac{1}{\deg(u_h)} \frac{1}{\deg(w_a)},$$

$$\tilde{t}_{uv} = \sum_{\substack{w:(w,u)\in E, \\ (w,v)\in E}} \frac{1}{\deg(v_a)} \frac{1}{\deg(w_h)}.$$

The hub and authority weights are then obtained as principal eigenvectors of the matrices $\tilde{H}$ and $\tilde{T}$. Note that these two matrices could also be defined in an alternative way. Let A be the adjacency matrix of G, A_r the row-normalized adjacency matrix (as in Equation (5.13)) and let A_c the column-normalized adjacency matrix of G (i.e. dividing each nonzero entry by its column sum). Then $\tilde{H}$ consists of the nonzero rows and columns of $A_r \cdot A_c^T$, while $\tilde{T}$ consists of the nonzero rows and columns of $A_c^T \cdot A_r$.

Note that $\tilde{h}_{uv} > 0$ implies that there exists at least one page w that has links to both u and v. This is known as co-citation in bibliometrics (Kessler 1963) (see Figure 5.9 and Exercise 5.2). Similarly, $\tilde{t}_{uv} > 0$ implies there exists at least one page that is linked to by both u and v, a bibliographic coupling (Small 1973).

Lempel and Moran (2001) showed theoretically that SALSA weights are more robust that HITS weights in the presence of the Tightly Knit Community (TKC) Effect. This effect occurs when a small collection of pages (related to a given topic) is connected so that *every* hub links to *every* authority and includes as a special case the mutual reinforcement effect identified by Bharat and Henzinger (1998) (see

Section 5.3). It can be shown that the pages in a community connected in this way can be ranked highly by HITS, higher than pages in a much larger collection where only *some* hubs link to *some* authorities. Clearly the TKC effect could be deliberately created by spammers interested in pushing the rank of certain websites. Lempel and Moran (2001) constructed examples of community pairs C_s connected in a TKC fashion, and C_l sparsely connected, and proved that authorities of C_s are ranked above the authorities of C_l by HITS but not by SALSA.

In a similar vein, Rafiei and Mendelzon (2000) and Ng *et al.* (2001) have proposed variants of the HITS algorithm based on a random walk model with reset, similar to the one used by PageRank. More precisely, a random surfer starts at time $t = 0$ at a random page and subsequently follows links from the current page with probability $1 - \varepsilon$, or (s)he jumps to a new random page with probability ε. Unlike PageRank, in this model the surfer will follow a forward link on odd steps but a backward link on even steps. For large t, two stationary distributions result from this random walk, one for odd values of t, that corresponds to an authority distribution, and one for even values of t that correspond to a hubness distribution. In vector notation the two distributions are proportional to

$$a_{2t+1} = \varepsilon \mathbf{1} + (1 - \varepsilon) A_r h_{2t}, \qquad (5.21)$$

$$h_{2t} = \varepsilon \mathbf{1} + (1 - \varepsilon) A_c^T a_{2t-1}. \qquad (5.22)$$

The stability properties of these ranking distributions are similar to those of PageRank (Ng *et al.* 2001).

Some further improvements of HITS and SALSA, as well as theoretical analyses on the properties of these algorithms can be found in Borodin *et al.* (2001).

5.6.2 PHITS

Cohn and Chang (2000) point out a different problem with HITS. Since only the principal eigenvector is extracted, the authority along the remaining eigenvectors is completely neglected, despite the fact that it could be significant. An obvious approach to address this limitation consists of taking into account several eigenvectors of the co-citation matrix, in the same spirit as PCA is used to extract several factors that are responsible for variations in multivariate data. As we have discussed in Section 4.5.2, however, the statistical assumptions underlying PCA are not sound for multinomial data such as term–document occurrences or bibliographical citations. PHITS can be viewed as probabilistic LSA (see Section 4.5.2) applied to co-citation and bibliographic coupling matrices. In this case citations replace terms. As in PLSA, a document d is generated according to a probability distribution $P(d)$ and a 'latent' variable z is then attached to d with probability $P(z \mid d)$. Here z could represent research areas (in the case of bibliographic data) or a (topical) community in the case of Web documents. Citations (links) are then chosen according to a probability distribution $P(d' \mid z)$.

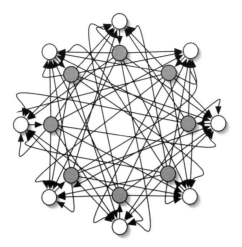

Figure 5.7 A link farm. Shaded nodes are all copies of the same page.

5.7   Limitations of Link Analysis

Search engines can be 'spammed' by websites faking high relevance with respect to some topics for the sole purpose of attracting visitors. Marchiori (1997) called this phenomenon Search Engine Persuasion (SEP). First-generation engines that based their ranking on classical information retrieval measures were clearly very sensitive to SEP. Common early techniques used to fool these engines included the use of inappropriate site titles or descriptions, the inclusion of META keywords in the HTML code, or even the use of extra text invisible to human surfers. Link analysis was immediately recognized as a solid defense against SEP. In their seminal paper on PageRank, Page *et al.* (1998) stated that

> . . . for a page to get a high PageRank, it must convince an important page, or a lot of non-important pages to link to it. At worst, you can have manip-ulation in the form of buying advertisements (links) on important sites. But this seems well under control since it costs money. This immunity to manipulation is an extremely important property.

In the intervening time period, ranking pages using link analysis has become the stan-dard approach followed by all the 'second generation' search engines. Consequently, website owners have realized the enormous importance of maximizing PageRank or similar link-based scoring functions in order to increase visibility. For some websites, having high search engine ranking (with respect to some keywords) is so important that it justifies considerable financial investments. Buying links as a form of adver-tisement has become reality and perhaps in some cases it is a sensible alternative to paying search engine companies directly for advertised links.

A *link farm* is a densely connected Web subgraph artificially built for the purpose of accumulating PageRank (or similar measures of popularity). Link farms can be built in many ways but link exchange is a common approach. A website owner who

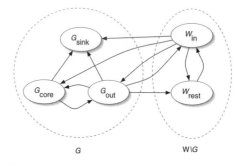

Figure 5.8 Canonical decomposition of the Web relative to a subgraph G.

decides to join the farm agrees to store a copy of a 'hub' page on her server and to link it from the root of her site. In return, the main URL of her site is added to the hub page, which is in turn redistributed to the sites participating in the link exchange. The result is a densely connected subgraph like the one shown in Figure 5.7.

It may appear that, since structures of this kind are highly regular, they should be relatively easy to detect (see Exercise 5.8) and thus link farming should not be a serious concern for search engines. However, it is possible to build farms that are more tightly entangled in the Web and are therefore more difficult to detect by simple topological analyses. This problem has been recently pointed out by Bianchini *et al.* (2001), who have shown that every community, defined as an arbitrary subgraph G of the Web, must satisfy a special form of 'energy balance'. The overall PageRank assigned to pages in G grows with the 'energy' that flows in from pages linking to the community and decreases with the energy dispersed in sinks and passed to pages outside the community. With reference to Figure 5.8, let G_{out} denote the subgraph of G induced by pages that contain hyperlinks pointing outside to G and let G_{sink} denote the sink subgraph of G. Also, let W_{in} be the subgraph induced by the pages outside G that link to pages in G. Then the equation

$$\|r_G\| = \alpha|G| + E_G^{\text{in}} - E_G^{\text{sink}} - E_G^{\text{out}}. \tag{5.23}$$

holds, where $\alpha|G|$ is the 'default' energy that is assigned to the community,

$$E_G^{\text{in}} = \frac{1-\varepsilon}{\varepsilon} \sum_{w \in W_{\text{in}}} f_G(w)\, r(w)$$

is the energy flowing in from outside, where $f_G(w)$ is the fraction of links in w that point to pages in G,

$$E_G^{\text{out}} = \frac{1-\varepsilon}{\varepsilon} \sum_{w \in G_{\text{out}}} (1 - f_G(w))\, r(w)$$

is the energy flowing out to pages outside the community, and

$$E_G^{\text{sink}} = \frac{1-\varepsilon}{\varepsilon} \sum_{w \in G_{\text{sink}}} r(w).$$

Suppose a set of 'sponsoring' pages S is generated artificially to boost the rank of a target website. The above energy balance analysis shows that the PageRank of the target can be increased linearly with $|S|$ regardless of the topology in S, making it difficult to detect the origin of spam using methods that are only based on graph topology. The fact that spamming activities of this kind are indeed possible is borne out, for example, by some recent anecdotes such as the popularity battle of the Church of Scientology against its main opponent Xenu.net.[2]

Websites that allow their users to post HTML code, for example, Weblogs (Walker 2002), potentially offer a cheap way for constructing artificial rank-boosting communities. Specialized algorithms that exploit more information than just topology are likely to be needed in the near future in order to prevent these spamming activities. This may also help prevent the future topology of the Web (and associated notions of popularity) from becoming significantly controlled by economic interests.

One example in this direction is the paper by Davison (2000a) that describes a machine-learning method for the automatic discrimination of links that are unrelated to the intrinsic merit of the pages. In this case, hyperlinks are described by several binary features that include, for example, tests about the structure of the URL, identity of source and target host or domain, or the total number of outlinks found on the source page.

Finally, we note that assessing the *quality* of pages returned by a search engine is difficult because quality is ultimately defined by human judgement. Amento *et al.* (2000) have reported an empirical study in which 16 human experts in five popular topics related to TV entertainment and music were asked to rank the quality of a set of Web pages. The experiment was aimed at testing whether expert judgements were correlated with scores based on link analysis. The study revealed that ranking documents according to the authority score (as computed by HITS) or according to PageRank yields high precision for the top 5 or 10 ranked documents. However, it was also found that alternative metrics such as page indegree or total number of pages in the website perform equally well.

Exercises

Exercise 5.1. Draw a graph of reasonable size, connecting vertices at random, and compute the principal eigenvectors of the matrices $A^{\mathrm{T}}A$ and AA^{T} to get authority and hubness weights. A very rapid way of doing this is by using linear algebra software such as Octave. Now select a vertex having nonzero indegree but small authority and try to modify the graph to increase its authority without increasing its indegree nor the indegree of its parents.

Exercise 5.2. The two matrices involved in Equations (5.8) and (5.9) were introduced several years before in the field of bibliometrics. In particular, $C = A^{\mathrm{T}}A$ is known as the *co-citation matrix* (Kessler 1963) and $B = AA^{\mathrm{T}}$ is known as the *bibliographic*

---

[2] See http://www.operatingthetan.com/google/ for details.

Co-citations (Small, 1973)

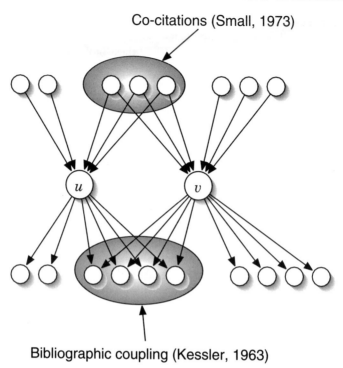

Bibliographic coupling (Kessler, 1963)

Figure 5.9 Co-citations and bibliographic coupling.

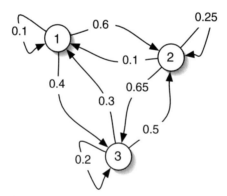

Figure 5.10 Markov chain for Exercise 5.3.

coupling matrix (Small 1973). Show that c_{uv} is the number of documents that cite both documents u and v, while b_{uv} is the number of pages that are cited by both u and v (see Figure 5.9).

Exercise 5.3. Consider the Markov chain in Figure 5.10 (where arcs are labeled by transition probabilities). Is it ergodic? What is the steady-state distribution?

Exercise 5.4. Let N be the $n \times n$ row normalized adjacency matrix of a graph as in Section 5.4. Suppose N is not stochastic and let R be any matrix such that $N + R$ is a stochastic matrix. Show that PAGERANK(N, n, ϵ) and PAGERANK$(N + R, n, \epsilon)$ converge to the same solution.

Exercise 5.5. In Equation (5.21), the stationary distributions of authority and hubness for randomized HITS are defined within a proportionality constant. Determine what this constant is.

Exercise 5.6. Use the stability results in Section 5.5.2 to estimate how often you should crawl the Web (and recompute PageRank) in order to guarantee that

$$\|\tilde{r} - r\| < \epsilon,$$

ϵ being an assigned tolerance. Assume for simplicity that a constant number of pages are changed in a given unit of time.

Exercise 5.7. Implement the PageRank computation and simulate the results on a relatively large artificial graph (build the graph using ideas from Chapter 3). Then introduce link farms in your graph and study the effect they have on the PageRank vector as a function of the number and the size of the farms.

Exercise 5.8. Propose an efficient algorithm to detect link farms structured as in Figure 5.7 in a large graph.

6

Advanced Crawling Techniques

In this chapter we refine the basic design of Web crawlers, building on the material presented in Section 2.5.

6.1 Selective Crawling

The search engine coverage experiments reported in Section 2.5.2 suggest that no crawler can realistically harvest the entire Web. An important fact is that the time required to complete a large crawl can be significant, independent of whatever technology is available at the site where the search engines operate. Moreover, fetching and indexing a larger set of documents has significant implications on the scalability of the overall system and, consequently, on the cost of the required hardware and maintenance services. Thus, to optimize available resources, a crawler should ideally be capable of recognizing the relevance or the importance of sites or pages, and limiting fetching to the most important subset of pages that can be downloaded in a given amount of time.

Relevance can be estimated with respect to several different criteria. To formalize the intuitive idea of selective crawling, we need to introduce, for each URL u, a scoring function $s_\theta^{(\xi)}(u)$, with respect to some underlying relevance criterion ξ and parameters θ. In the simplest case we can assume a Boolean relevance function, i.e. $s(u) = 1$ if the document pointed to by u is relevant and $s(u) = 0$ if the document is not interesting. More generally, we can think of $s(d)$ as a real-valued function, such as the conditional probability that the document belongs to a certain category given its contents. Note that in all of these cases the scoring function depends only on the URL and the criterion, but does not depend on the state of the crawler. Moreover, we will assume that the scoring function does not depend on time.

One general approach for building selective crawlers consists of changing the policy of insertions and extractions in the queue of discovered URLs. For simplicity, let us consider again the algorithm SIMPLE-CRAWLER on page 47. Suppose now that URLs in Q are sorted according to the value returned by $s(u)$. In this way we obtain a *best-first* search strategy (see, for example, Russell and Norvig 1995) such that URLs

Modeling the Internet and the Web   P. Baldi, P. Frasconi and P. Smyth
© 2003 P. Baldi, P. Frasconi and P. Smyth   ISBN: 0-470-84906-1

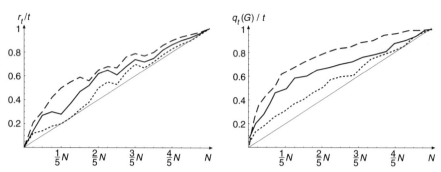

Figure 6.1 Efficiency curves obtained by Cho *et al.* (1998) for a selective crawler. Both were obtained on a set of 784 592 Stanford University URLs. Dotted lines correspond to a BFS crawler, solid lines to a crawler guided by the estimated number of backlinks, and dashed lines to a crawler guided by estimated PageRank. The diagonal lines are the reference performance of a random crawler. The target measure of relevance is the actual number of backlinks. In the right-hand plot, the importance target G is 100 backlinks.

having higher score are fetched first. If $s(u)$ provides a good model of the relevance of the document pointed to by the URL, then we expect that the search process will be guided toward the most relevant regions of the Web. From this perspective the problem is to define interesting scoring functions and to devise good algorithms to compute them.

In the following we consider in detail some specific examples of scoring functions.

Depth. As we have seen in Chapter 3, the distribution of the number of pages on different Web servers generally follows a power law. Thus, many sites have few pages and, at the same time, there is a small but significant fraction of sites containing a huge number of pages. A very simple strategy for maximizing coverage breadth consists of limiting the total number of documents downloaded from a single site, for example, by setting a threshold, or by keeping track of the depth in the site directory tree, or by limiting the acceptable length of the path from the site homepage to the document,

$$s_\delta^{(\text{depth})}(u) = \begin{cases} 1, & \text{if } |\text{root}(u) \rightsquigarrow u| < \delta, \\ 0, & \text{otherwise,} \end{cases} \tag{6.1}$$

where $\text{root}(u)$ is the root of the site containing u. The rationale behind this approach is that by maximizing breadth it will be easy for the end-user to eventually find the desired information. For example a page that is not indexed by the search engine may be easily reachable from other pages in the same site.

Popularity. It is often the case that we can define criteria according to which certain pages are more important than others. For example, search engine queries are answered by proposing to the user a sorted list of documents. Typically, users tend to inspect only the few first documents in the list (in Chapter 8 we will discuss empirical data that support this statement). Thus, if a document rarely appears near the top of any lists, it may be worthless to download it and its importance should be small. A simple way of assigning importance to documents that are more popular is to introduce a relevance function based on the number of backlinks

$$s_\tau^{(\text{backlinks})}(u) = \begin{cases} 1, & \text{if indegree}(u) > \tau, \\ 0, & \text{otherwise,} \end{cases} \tag{6.2}$$

where τ is an assigned threshold.

Clearly $s_\tau^{(\text{backlinks})}(u)$ can be computed exactly only if we have already crawled the entire Web. In practice we can use an approximated value from the partial subgraph obtained in a previous crawl, or we can incrementally update the score as soon as new parents of u are encountered during the crawl.

PageRank. As discussed in Section 5.4, PageRank is a measure of popularity that recursively assigns each link a weight that is proportional to the popularity of the source document. It may be seen as a refinement of the indegree measure defined above in Equation (6.2). Also PageRank needs to be estimated using partial knowledge of the graph that is refined during crawling.

The efficiency of a selective crawler can be measured by comparing it to an ideal crawler that fetches documents in order of relevance (whatever the relevance function). Suppose we are given a website with N pages and for $n = 1, \ldots, N$ let s_n denote the score of the page that is ranked in position n by the relevance function. Suppose at some point in time the crawler has fetched t pages and let r_t denote the number of fetched pages whose score is higher than s_t. In the ideal crawler we would have $r_t = t$ and the efficiency curve r_t/t would be constant and equal to one. In a random crawler, only a fraction t/N of the fetched pages will have a score higher than s_t, i.e. $r_t/2 = t/N$. Selective crawlers should obtain intermediate efficiency curves. Alternatively, one may assign a fixed 'importance target', G, and consider as relevant all the pages whose score is higher than G. For example, we might be interested in quickly fetching pages having at least 20 backlinks. The efficiency in this case may be measured as $q_t(G)/t$, where q_t is the number of fetched pages having a score higher than G. Cho *et al.* (1998) report several experimental results comparing the above strategies on a controlled data set of Stanford University pages. Figure 6.1 shows some of the results when the target measure of relevance is the number of backlinks. Interestingly, the estimated PageRank outperforms the estimated number of backlinks as a guiding criterion.

In a more recent study by Najork and Wiener (2001) it was found that traversing the Web in breadth-first order is a good strategy and allows us to download 'hot' pages first. The study was conducted on a large set of 328 million pages from 7 million distinct servers.

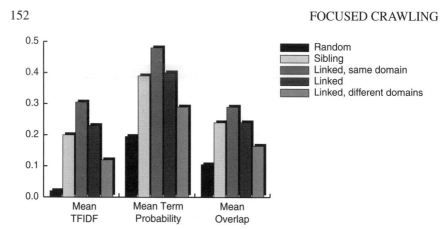

Figure 6.2 Results of the study carried out by Davison in 2000. (a) Three indicators that measure text similarity (see Section 4.3) are compared in five different linkage contexts. In this data set, about 56% of the linked documents belong to the same domain. It can be seen that the similarity between linked documents is significantly higher than the similarity between random documents. (b) Similarities measured between anchor text and text of documents in five different contexts.

6.2 Focused Crawling

In many cases a search engine does not need to be completely general purpose and cover every possible site on the Web, but instead can be focused on particular topics that are of interest to a specific community of users. In these cases, relevance can be defined with respect to the expected utility of the pages for users of the search engine. For example, a crawler that provides information to a vertical portal specialized in, say, music, could be quite safely programmed to ignore sites with content in different topics such as health or sports. A *focused crawler* is a refined selective crawler that searches for information related to certain topics rather than being driven by generic quality measures. The fraction of sites that specialize in a particular topic may be small enough to enable a focused crawler to download them almost exhaustively and in a relatively short time.

6.2.1 Focused crawling by relevance prediction

The URL ordering technique suggested for selective crawlers can be also extended to focused crawlers. In this case, it is necessary to determine whether a document is relevant or not to the topic of interest in order to define a scoring function for driving the search process. One way to do this is to use the text categorization techniques presented in Section 4.6. More precisely, if we denote by c the topic of interest, a score can be computed as the conditional probability that the document is relevant, given the text in the document:

$$s_\theta^{(\text{topic})}(u) = P(c \mid d(u), \theta). \tag{6.3}$$

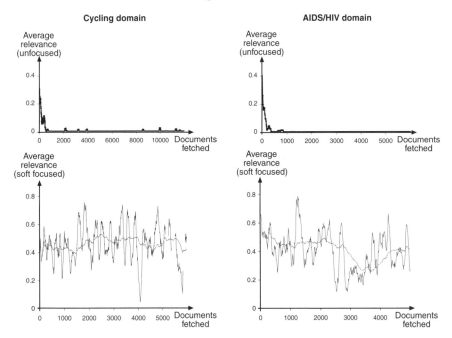

Figure 6.3 Efficiency of a BFS and a focused crawler compared in the study by Chakrabarti *et al.* (1999b). The plots show the average relevance of fetched documents versus the number of fetched documents. Results for two topics are shown (cycling on the left, AIDS/HIV on the right).

Here θ represents the adjustable parameters of the classifier. An obvious consideration is that the score cannot be computed exactly (we have not yet downloaded the document pointed to by u). There are a number of different strategies for approximating the topic score.

Parent based. In this case we compute the score for a fetched document and extend the computed value to all the URLs in that document. More precisely, if v is a parent of u, we approximate the score of u as

$$s_\theta^{(\text{topic})}(u) \simeq P(c \mid d(v), \theta). \tag{6.4}$$

The rationale is a general principle of 'topic locality'. If a page deals with, say, music, it may be reasonable to believe that most of the outlinks of that page will deal with music as well. In a systematic study based on the analysis of about 200 000 documents, Davison (2000b) found that topic locality is measurably present in the Web, under different text similarity measures (see Figure 6.2a).

Chakrabarti *et al.* (1999b) use a hierarchical classifier and suggest two implementations of the parent-based scoring approach. In *hard focusing*, they check if at least one node in the category path of $d(v)$ is associated with a topic of interest; if not

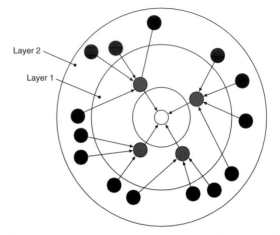

Figure 6.4   Example of a two-layered context graph. The central white node
is a target document. Adapted from Diligenti *et al.* (2000).

the outlinks of v are simply discarded (not even inserted in the crawling queue Q).
In *soft focusing*, relevance is computed according to Equation (6.4), but if the same
URL is discovered from multiple parents, a revision strategy is needed in order to
update $s_\theta^{(\text{topic})}(u)$ when new evidence is collected. Chakrabarti *et al.* (1999b) found
no significant difference between the two approaches.

Anchor based.   Instead of the entire parent document, $d(v)$, we can just use the text
   $d(v, u)$ in the anchor(s) where the link to u is referred to, as the anchor text is often
   very informative about the contents of the document pointed to by the corresponding
   URL. This 'semantic linkage' was also quantified by Davison (2000b), who showed
   that the anchor text is most similar to the page it references (see Figure 6.2b).

To better illustrate the behavior of a focused crawler on real data, consider the
efficiency diagrams in Figure 6.3 that summarize some results obtained by Chakrabarti
et al. (1999b). An unfocused crawler starting from a seed set of pages that are relevant
to a given topic will soon begin to explore irrelevant regions of the Web. As shown in
the top diagrams, the average relevance of the fetched pages dramatically decreases
as crawling goes on. Using a focused crawler (in this case, relevance is predicted
by a Naive Bayes classifier trained on examples of relevant documents) allows us to
maintain an almost steady level of average relevance.
       There are several alternatives to a focused crawler based on a single best-first queue,
as detailed in the following.

6.2.2   Context graphs

Diligenti *et al.* (2000) suggested a strategy that takes advantage of knowledge of
the Internet topology to train a machine-learning system to predict 'how far' some

relevant information can be expected to be found starting from a given page. Intuitively, suppose the crawler is programmed to gather homepages of academic courses in artificial intelligence. The backlinks of these pages are likely to lead to professors' home pages or to the teaching sections of the department site. Going one step further, backlinks of backlinks are likely to lead into higher level sections of department sites (such as those containing lists of the faculty). More precisely, the *context graph* of a node u (see Figure 6.4) is the layered graph formed inductively as follows. Layer 0 contains node u. Layer i contains all the parents of all the nodes in layer $i - 1$. No edges jump across layers. Starting from a given set of relevant pages, Diligenti *et al.* (2000) used context graphs to construct a data set of documents whose distance from the relevant target was known (backlinks were obtained by querying general purpose search engines). After training, the machine-learning system predicts the layer a new document belongs to, which indicates how many links need to be followed before relevant information will be reached, or it returns 'other' to indicate that the document and its near descendants are all irrelevant. Denoting by n the depth of the considered context graph, the crawler uses n best-first queues, one for each layer, plus one extra queue for documents of class 'other'. This latter queue is initialized with the seeds. In the main loop, the crawler extracts URLs from the first nonempty queue and in this manner favors those that are more likely to rapidly lead to relevant information.

6.2.3 Reinforcement learning

Reinforcement learning (see, for example, Sutton and Barto 1998) is a framework for deciding in an optimal way what actions an agent operating in a discrete environment should take in order to maximize its expected future reward. Compared to supervised learning, reinforcement learning is characterized by the absence of external supervision. The agent learns from the received rewards (or punishments). More formally, the discrete-time environment is characterized by a finite set of states S. At time t the environment perceived by the agent is in state $s_t \in S$ and, correspondingly, the agent has access to a finite set of actions $A(s_t)$. After performing action $a_t \in A(s_t)$ the state transitions to s_{t+1} and the agent receives an immediate reward r_{t+1}. A policy π describes the joint probability distribution on states and actions, i.e. $\pi(s, a)$ is the probability of taking action a when in state s. An optimal policy should maximize the expected rewards received by the agent over time. In order to specify what an optimal policy is, we first define the value of state s under a policy π as

$$V^\pi(s) = E_\pi \left[\sum_{k=0}^{\infty} \gamma^k r_{t+k+1} \mid s_t = s \right]. \tag{6.5}$$

Similarly, we can define the action-value function as

$$Q^\pi(s, a) = E_\pi \left[\sum_{k=0}^{\infty} \gamma^k r_{t+k+1} \mid s_t = s, a_t = a \right]. \tag{6.6}$$

An optimal policy π^* maximizes the value function over all the states: $V^*(s) \geqslant V^\pi(s)$ for all $s \in S$. According to the Bellman optimality principle, underlying the foundations of dynamic programming (Bellman 1957), a sequence of optimal decisions has the property that, regardless of the action taken at the initial time, the subsequent sequence of decisions must be optimal with respect of the outcome of the first action. This translates into

$$V^*(s) = \max_{a \in A(s)} E[r_{t+1} + \gamma V^*(s_{t+1}) \mid s_t = s, a_t = a], \tag{6.7}$$

which allows us to determine the optimal policy once V^* is known for all s. An optimal policy π^* also maximizes $Q^\pi(s, a)$ for all $s \in S, a \in A(s)$:

$$Q^*(s, a) = E[r_{t+1} + \gamma \max_b Q^*(s_{t+1}, b) \mid s_t = s, a_t = a]. \tag{6.8}$$

The advantage of the state-action representation is that once Q^* is known for all s and a, all the agent needs to do in order to maximize the expected future reward is to choose the action that maximizes $Q^*(s, a)$.

In the context of focused Web search, immediate rewards are obtained (downloading relevant documents) and the policy learned by reinforcement can be used to guide the agent (the crawler) toward high long-term cumulative rewards. As we have seen in Section 2.5.3, the internal state of a crawler is basically described by the sets of fetched and discovered URLs. Actions correspond to fetching a particular URL that belongs to the queue of discovered URLs. Even if we simplify the representation by removing from the Web all the off-topic documents, it is clear that the sets of states and actions are overwhelming for a standard implementation of reinforcement learning.

LASER (Boyan *et al.* 1996) was one of the first proposals to combine reinforcement learning ideas with Web search engines. The aim in LASER is to answer queries rather than to crawl the Web. The system begins by computing a relevance score $r_0(u) = \text{TFIDF}(d(u), q)$ and then propagates it in the Web graph using the recurrence

$$r_{t+1}(u) = r_0(u) + \gamma \sum_{v \in \text{pa}[u]} \frac{r_t(v)}{|\text{pa}[u]|^\ni}, \tag{6.9}$$

which is iterated until convergence for each document in the collection, where γ and $\ni$ are free parameters. After convergence, documents at distance k from u provide a contribution proportional to γ^k times their relevance to the relevance of u.

McCallum *et al.* (2000c) used reinforcement to search computer science papers in academic websites. In order to simplify the problem of learning $Q^*(s, a)$ for an enormous number of states and actions they propose the following two assumptions:

- the state is independent of the relevant documents that have been fetched so far;

- actions can only be distinguished by means of the words in the neighborhood of the hyperlink that correspond to each action (e.g. the anchor text).

In this way, learning Q reduces to learning a mapping from text (e.g. bag of words representing anchors) to a real number (the expected future reward).

6.2.4 Related intelligent Web agents

The Fish algorithm by De Bra and Post (1994) uses a population of agents that autonomously spider the Web. Like fishes in an information sea, agents accumulate energy as they collect relevant documents for a given query while they consume energy for using network resources. Since agents need energy to survive, those which end up exploring irrelevant portions of the Web perish, implementing a selection mechanism. The Shark algorithm by Hersovici *et al.* (1998) improves the Fish search by introducing a real-valued relevance score that also depends on anchor text and a discounted fraction of the score that was given to ancestor pages. One of the main limitations of these approaches is the lack of adaptation.

WebWatcher (Armstrong *et al.* 1995) is an interactive recommendation system that assists a user during browsing. It introduced the use of machine learning for predicting the best hyperlink to follow according to a given user goal. A similar approach based on heuristic searching was proposed in Lieberman (1995). Arachnid (Menczer 1997) is based on a distributed population of adaptive agents that search information related to user-provided keywords. Similar in spirit to the Fish algorithm, Arachnid agents receive energy by relevance feedback provided by the user. An important feature of this approach is that hyperlinks are selected by a neural network associated with the agent, whose weights are adjusted by reinforcement learning. An extension of this system is described in Menczer and Belew (2000).

A more recent framework that generalizes focused crawling is *intelligent crawling* (Aggarwal *et al.* 2001). In this case, the crawler does not need a collection of topical examples for training. Users describe their information needs by means of *predicates*, i.e. general specifications that generalize keyword-based queries to also include document-to-document similarity queries, or characterizations of a topic as obtained from a text classifier. The intelligent crawler proposed in Aggarwal *et al.* (2001) is capable of auto-focusing using documents that satisfy the query predicate.

CiteSeer (`http://citeseer.nj.nec.com/cs`) is a search engine focused on computer science literature (Bollacker *et al.* 1998; Lawrence *et al.* 1999). It fetches PostScript or PDF papers from paper repositories (for example, a researcher's Web pages) and autonomously builds a citation index. As a major difference with respect to the system described by McCallum *et al.* (2000c), documents are located by querying traditional Web search engines and by collecting submitted URLs of pages containing links to research articles. The system performs several information extraction operation from the documents and, in particular, it extracts citations and references in the body of a paper in order to build a scientific literature web. Having represented the collection of papers as a web, CiteSeer can also apply link analysis algorithm and compute for example authority and hubness weights of each paper. Link analysis in this case can also be used to determine which documents are related to any given one.

Interestingly, since CiteSeer creates a Web page for each online article, it effectively maps the online subset of the computer science literature web to a subset of the World Wide Web. A recent study has shown that papers that are available online tend to receive a significantly higher number of citations (Lawrence 2001).

The DEADLINER system described in Kruger *et al.* (2000) is a search engine specialized in conference and workshop announcements. One of the input components of the system is a context-graph focused crawler (see Section 6.2.2) that gathers potentially related Web documents. An SVM text classifier is subsequently used to refine the set of documents retrieved by the focused crawler.

6.3   Distributed Crawling

A single crawling process, even if multithreading is used, will typically be insufficient for large-scale engines that need to fetch large amounts of data rapidly. When using a single centralized crawler, all the fetched data must pass through a single physical link. This is often problematic, regardless of the bandwidth available at the crawling center, because it will be unlikely to have comparable connection speeds from different geographical regions. Distributing the crawling activity via multiple processes can be seen as a form of 'divide and conquer' that can help build a scalable system. Splitting the load decreases hardware requirements and at the same time increases the overall download speed and reliability if separate processes access the Internet through different physical links. The advantage is particularly evident if separate crawlers run in separate data centers that are located in different countries or continents. However, while the physical links reflect geographical neighborhoods, we know that the edges of the Web graph are instead associated with 'communities' that can and often do cross geographical borders. Hence, running separate and independent crawling processes can result in a significant overlap among the collections of fetched documents. The performance of a parallelization approach can be measured in terms of

- communication overhead – the fraction of bandwidth spent to coordinate the activity of the separate processes, with respect to the bandwidth usefully spent for document fetching;

- overlap – the fraction of duplicate documents downloaded by all the processes;

- coverage – the fraction of documents reachable from the seeds that are actually downloaded; and

- quality – e.g. some of the scoring functions defined in Section 6.1 depend on the link structure and this information can be partially lost using separate crawling processes.

The literature on this important topic is not abundant. In a recent study, Cho and Garcia-Molina (2002) have defined a framework based on several dimensions that characterize the interaction among a set of crawlers.

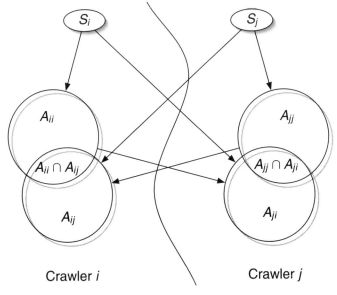

Figure 6.5 Two crawlers statically coordinated.

Coordination refers to the way different processes agree about the subset of pages each of them should be responsible for. If two crawling processes i and j are completely *independent* (not coordinated), then the degree of overlap can only be controlled by having different seeds S_i and S_j. If we assume the validity of topological models such as those presented in Section 3, then we can expect that the overlap will eventually become significant unless a partition of the seed set is properly chosen. On the other hand, making a good choice is a challenge, since the partition that minimizes overlap may be difficult to compute, for example, because current models of the Web are not accurate enough. In addition it may be suboptimal with respect to other desiderata that motivated distributed crawling in the first place, such as distributing the load and scaling-up.

A pool of crawlers can be coordinated by partitioning the Web into a number of subgraphs and letting each crawler be mainly responsible for fetching documents from its own subgraph. If the partition is decided before crawling begins and not changed thereafter, we refer to this as *static coordination*. This option has the great advantage of simplicity and implies little communication overhead. Alternatively, if the partition is modified during the crawling process, the coordination is said to be *dynamic*. In the static approach the crawling processes, once started, can be seen as agents that operate in a relatively autonomous way. In contrast, in the dynamic case each process is subject to a reassignment policy that must be controlled by an external supervisor.

Confinement specifies, assuming statically coordinated crawlers, how strictly each crawler should operate within its own partition. Consider two processes, i and

j, and let A_{ij} denote the set of documents belonging to partition i that can be reached from the seeds S_j (see Figure 6.5). The question is what should happen when crawler i pops from its queue 'foreign' URLs pointing to nodes in a different partition. Cho and Garcia-Molina (2002) suggest three possible modes: *firewall*, *crossover*, and *exchange*. In firewall mode, each process remains strictly within its partition and never follows interpartition links. In crossover mode, a process can follow interpartition links when its queue does not contain any more URLs in its own partition. In exchange mode, a process never follows interpartition links, but it can periodically open a communication channel to dispatch the foreign URLs it has encountered to the processes that operate in those partitions. To see how these modes affect performance measures, consider Figure 6.5 again. The firewall mode has, by definition, zero overlap but can be expected to have poor coverage, since documents in $A_{ij} \setminus A_{ii}$ are never fetched (for all i and j). Cross-over mode may achieve good coverage but can have potentially high overlap. For example, documents in $A_{ii} \cap A_{ij}$ can be fetched by both process i and j. The exchange mode has no overlap and can achieve perfect coverage. However, while the first two modes do not require extra bandwidth, in exchange mode there will be a communication overhead.

Partitioning defines the strategy employed to split URLs into non-overlapping subsets that are then assigned to each process. A straightforward approach is to compute a hash function of the IP address in the URL, i.e. if $n \in \{0, \ldots, 2^{32} - 1\}$ is the integer corresponding to the IP address and m the number of processes, documents such that $n \mod m = i$ are assigned to process i. In practice, a more sophisticated solution would take into account the geographical dislocation of networks, which can be inferred from the IP address by querying Whois databases such as the Réseaux IP Européens (RIPE) or the American Registry for Internet Numbers (ARIN).

6.4   Web Dynamics

In the final section of this chapter we address the question of how information on the Web changes over time. Knowledge of this 'rate of change' is crucial, as it allows us to estimate how often a search engine should visit each portion of the Web in order to maintain a fresh index.

In Chapter 3 we discussed general 'aging' properties of Web graphs. A precise notion of recency in this context has been proposed by Brewington and Cybenko (2000). The index entry for a certain document, indexed at time t_0, is said to be β-current at time t if the document has not changed in the time interval between t_0 and $t - \beta$. Basically β is a 'grace period': if we pretend that the user query was made β time units ago rather than now, then the information in the search engine would be up to date. A search engine for a given collection of documents is said to be (α, β)-current if the probability that a document is β-current is at least α. According to this definition, we can ask interesting questions like 'how many documents per day should a search engine refresh in order to guarantee it will remain (0.9,1 week)-current?' Answering this question requires that we develop a probabilistic model of Web dynamics. The model will be complicated because of two concomitant factors: pages change over time, and the Web itself grows and evolves in time (see Chapter 3).

6.4.1 Lifetime and aging of documents

To begin with consider a single document. The model we are going to develop is based on the same ideas underpinning reliability theory in industrial engineering (see, for example, Barlow and Proshan 1975). Let T be a continuous random variable representing the *lifetime* of a component in a piece of machinery or equipment. Assuming the component was initially installed or replaced at time zero, lifetime is the time when the component breaks down (dies). Let $F(t)$ be the cumulative distribution function (cdf) of lifetime. The reliability (or survivorship function) is defined as

$$S(t) \doteq 1 - F(t) = P(T > t), \tag{6.10}$$

i.e. the probability that the component will be functioning at time t. A variable closely related to lifetime is the *age* of a component, i.e. the time elapsed since the last replacement (see Figure 6.6 to better understand the relationship between lifetime and age). Its cdf, defined as $G(t) = P(\text{age} < t)$, is obtained by integrating the survivorship function and normalizing:

$$G(t) = \frac{\int_0^t S(\tau) \, d\tau}{\int_0^\infty S(\tau) \, d\tau}. \tag{6.11}$$

Note that $G(t)$ is the expected fraction of components that are still operating at time t. The age probability density function (pdf) $g(t)$ is thus proportional to the survivorship function. Returning from the reliability metaphor to Web documents, $S(t)$ is the probability that a document that was last changed at time zero will remain unmodified at time t, while $G(t)$ is the expected fraction of documents that are older than t. The probability that the document will be modified before an additional time h has passed is expressed by the conditional probability $P(t < T \leqslant t + s \mid T > t)$. The change rate $\lambda(t)$ (also known as the hazard rate in reliability theory, or mortality force in demography) is then obtained by dividing by h and taking the limit for small h,

$$\lambda(t) \doteq \lim_{h \to 0} \frac{1}{s} P(t < T \leqslant t + h \mid T > t)$$
$$= \lim_{h \to 0} \frac{1}{S(t)} \frac{1}{h} \int_t^{t+h} f(\tau) \, d\tau = \frac{f(t)}{S(t)}, \tag{6.12}$$

where $f(t)$ denotes the lifetime pdf. Combining (6.10) and (6.12) we have the ordinary differential equation (see, for example, Apostol 1969)

$$F'(t) = \lambda(t)(1 - F(t)), \tag{6.13}$$

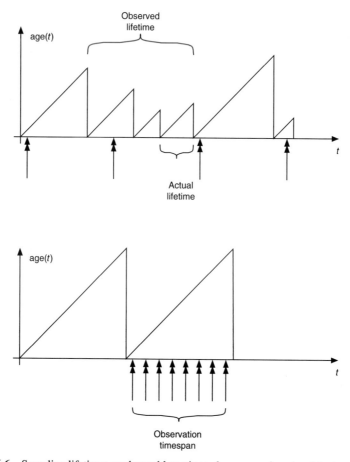

Figure 6.6 Sampling lifetimes can be problematic as changes can be missed for two reasons. Top, two consecutive changes are missed and the observed lifetime is overestimated. Bottom, the observation time-span must be large enough to catch changes that occur in a long range. Sampling instants are marked by double arrowheads.

with $F(0) = 0$. We will assume that changes happen randomly and independently. According to a Poisson process, the probability of a change event at any given time is independent of when the last change happened (see Appendix A). For a constant change rate $\lambda(t) = \lambda$, the solution of Equation (6.13) is

$$F(t) = 1 - e^{-\lambda t}, \qquad f(t) = \lambda e^{-\lambda t}.$$

Brewington and Cybenko (2000) observed that the model could be particularly valuable for analyzing Web documents. In practice, however, the estimation of $f(t)$ is problematic for any method based on sampling. If a document is observed at two different instants t_1 and t_2, we can check for differences in the document but we cannot know how many times the document was changed in the interval $[t_1, t_2]$, a phe-

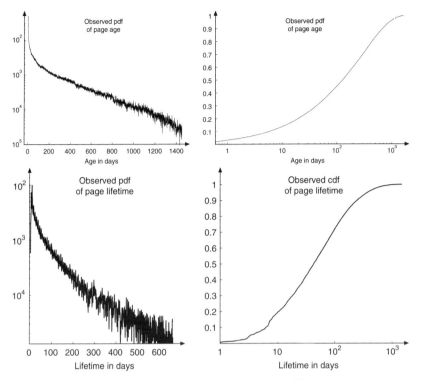

Figure 6.7 Empirical distributions of page age (top) and page lifetime (bottom) on a set of 7 million Web pages. Adapted from Brewington and Cybenko (2000).

nomenon known as *aliasing* (see Figure 6.6). On the other hand, the age of a document may be readily available, if the Web server correctly returns the `Last-Modified` timestamp in the HTTP header (see Section 2.3.4 for a sample script that obtains this information from a Web server). Sampling document ages is not subject to aliasing and lifetime can be obtained indirectly via Equation (6.11). In particular, if the change rate is constant, it is easy to see that the denominator in Equation (6.11) is a/λ and thus from Equation (6.12) we obtain

$$g(t) = f(t) = \lambda e^{-\lambda t}. \tag{6.14}$$

In other words, assuming a constant change rate, it is possible to estimate lifetime from observed age.

This simple model, however, does not capture the essential property that the Web is growing with time. Brewington and Cybenko (2000) collected a large data set of roughly 7 million Web pages, observed between 1999 and 2000 in a time period of seven months, while operating a service called 'The Informant'.[1] The resulting distributions of age and lifetime are reported in Figure 6.7.

---

[1] Originally `http://informant.dartmouth.edu`, now `http://www.tracerlock.com`.

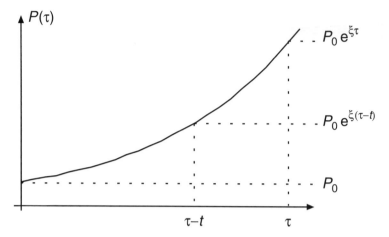

Figure 6.8 Assuming an exponentially growing Web whose pages are never changed, it follows that the age distribution is also exponential at any time τ.

What is immediately evident is that most of the collected Web documents are 'young', but what is interesting is why this is the case. Suppose that growth is modeled by an exponential distribution, namely that the size at time τ is $P(\tau) = P_0 e^{\xi \tau}$, where P_0 is the size of the initial population and ξ the growth rate. If documents were created and never edited, their age would be simply the time since their creation. The probability that a random document at time τ has an age less than t is

$$G_g(t, \tau) = \frac{\text{new docs in } (\tau - t, \tau)}{\text{all docs}} = \frac{e^{\xi \tau} - e^{\xi (\tau - t)}}{e^{\xi \tau} - 1} \qquad (6.15)$$

(see Figure 6.8) and thus the resulting age density at time τ is

$$g_g(t, \tau) = \frac{\xi e^{\xi (\tau - t)}}{e^{\xi \tau} - 1} [H(t) - H(t - \tau)], \qquad (6.16)$$

where $H(t)$ is the Heaviside step function, i.e. $H(t) = 0$ for $t < 0$ and $H(t) = 1$ otherwise.

In other words, this trivial growth model yields an exponential age distribution, like the model in Section 6.4.1 that assumed a static Web with documents refreshed at a constant rate. Clearly, the effects of both Web growth and document refreshing should be taken into account in order to obtain a realistic model. Brewington and Cybenko (2000) experimented with a hybrid model (not reported here) that combines an exponential growth model and exponential change rate. Fitting this model with ages obtained from timestamps, they estimated $\xi = 0.001\,76$ (in units of days^{-1}). This estimate corresponds to the Web size doubling in about 394 days. In contrast, if we estimated ξ from the lower bounds of 320 million pages in December 1997 (Lawrence and Giles 1998b) and 800 million pages in February 1999 (Lawrence and Giles 1999), roughly 426 days later, we would obtain $\xi = 0.022$, or a doubling time of 315 days. Despite the differences in these two estimates, they are of the same order of magnitude and give us some idea of the growth rate of the Web.

Another important problem when using empirically measured ages is that servers often do not return meaningful timestamps. This is particularly true in the case of highly dynamic Web pages. For example, it is not always possible to assess from the timestamp whether a document was just edited or was generated on-the-fly by the server. These considerations suggest that the estimation of change rates should be carried out using lifetimes rather than ages. The main difficulty is how to deal with the problem of potentially poorly chosen timespans, as exemplified in Figure 6.6. Brewington and Cybenko (2000) suggest a model that explicitly takes into account the probability of observing a change, given the change rate and the timespan. Their model is based on the following assumptions.

- Document changes are events controlled by an underlying Poisson process, where the probability of observing a change at any given time does not depend on previous changes. Given the timespan τ and the change rate λ, the probability that we observe one change (given that it actually was made) is therefore

$$P(c \mid \lambda, \tau) = 1 - e^{-\lambda \tau}. \tag{6.17}$$

It should be observed that, in their study, Brewington and Cybenko (2000) found that pages are changed with different probabilities at different hours of the day or during different days of the week (most changes being concentrated during office hours). Nonetheless, the validity of a memoryless Poisson model is assumed.

- Mean lifetimes are Weibull distributed (see Appendix A), i.e. denoting the mean lifetime by $t = 1/\lambda$, the pdf of t is

$$w(t) = \frac{\sigma}{\delta} \left(\frac{t}{\delta} \right)^{\sigma - 1} e^{(t/\delta)^{\sigma}}, \tag{6.18}$$

where δ is a scale parameter and σ is a shape parameter.

- Change rates and timespans are independent and thus

$$P(c \mid \lambda) = \int_0^{\infty} P(c, \tau \mid \lambda) \, d\tau = \int_0^{\infty} P(\tau) P(c \mid \tau, \lambda) \, d\tau.$$

The resulting lifetime distribution is

$$f(t) = \int_0^{\infty} \lambda e^{-\lambda t} \hat{w}(1/\lambda) \, d(1/\lambda), \tag{6.19}$$

where $\hat{w}(1/\lambda)$ is an estimate of the mean lifetime. Brewington and Cybenko (2000) used the data shown in Figure 6.7 to estimate the Weibull distribution parameters, obtaining $\sigma = 1.4$ and $\delta = 152.2$. The estimated mean lifetime distribution is plotted in Figure 6.9.

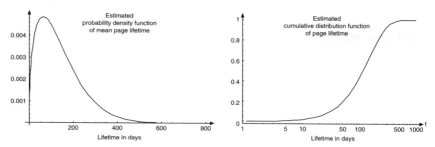

Figure 6.9 Estimated density and distribution of mean lifetime resulting from the study of Brewington and Cybenko (2000).

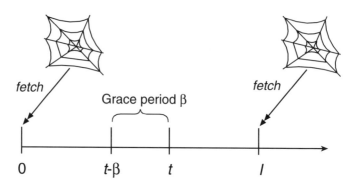

Figure 6.10 A document is β-current at time t if no changes have occurred before the grace period that extends backward in time until $t - \beta$.

These results allow us to estimate how often a crawler should refresh the index of a search engine to guarantee that it will remain (α, β)-current. Let us consider first a single document and, for simplicity, let $t = 0$ be the time when the crawler last fetched the document. Also, let I be the time interval between two consecutive visits (see Figure 6.10). The probability that for a particular time t the document is unmodified in $[0, t - \beta]$ is $e^{-\lambda(t-\beta)}$ for $t \in [\beta, I)$ and 1 for $t \in (0, \beta)$. Thus, the probability that a specific document is β-current at time t is

$$\int_0^\beta \frac{1}{I}\, dt + \int_\beta^I \frac{1}{I} e^{-\lambda(t-\beta)}\, dt = \frac{\beta}{I} + \frac{1 - e^{-\lambda(I-\beta)}}{\lambda I}, \qquad (6.20)$$

but since each document has a change rate λ, whose reciprocal is Weibull distributed, the probability that the collection of documents is β-current is

$$\alpha = \int_0^\infty w(t)\left[\frac{\beta}{I} + \frac{1 - e^{-(I-\beta/t)}}{t/I} \right] dt. \qquad (6.21)$$

This allows us to determine the minimum refresh interval I to guarantee (α, β)-currency once the parameters of the Weibull distribution for the mean change rate are known. Assuming a Web size of 800 million pages, Brewington and Cybenko (2000) determined that a reindexing period of about 18 days was required to guarantee that 95% of the repository was current up to one week ago.

6.4.2 Other measures of recency

Freshness and *index age* are two alternative and somewhat simpler measures of recency (Cho and Garcia-Molina 2000b). The freshness $\phi(t)$ at time t of a given document is a binary function that indicates whether the document is up-to-date in the index at time t. The expected freshness is therefore the probability that the document did not change in $(0, t]$, i.e. $E[\phi(t)] = e^{-\lambda t}$ (see Equation (6.17)). Note that freshness essentially corresponds to the concept of β-currency for $\beta = 0$. Hence, if d is refreshed regularly each I time units, the average freshness is

$$\bar{\phi} = \frac{1 - e^{-\lambda I}}{\lambda I}$$

as follows from Equation (6.20) with $\beta = 0$.

The *index age* of a document is the age of the document if the local copy is outdated, or zero if the local copy is fresh. Thus if the document was modified at time $s \in (0, t]$, its index age is $t - s$. From Equation (6.14) it follows that the expected age at time t is

$$E[a(t)] = \int_0^t (t - s)\lambda e^{-\lambda s}\, ds = t - \frac{1 - e^{-\lambda t}}{\lambda},$$

whose average in $(0, I]$ is

$$\bar{a} = \frac{1 - e^{-\lambda I}}{\lambda^2 I} - \frac{1}{\lambda} + \frac{I}{2}.$$

When a collection of documents is considered, the above quantities can be averaged over the collection. Because of the linearity of integral, we can average both time punctual values and time averages.

6.4.3 Recency and synchronization policies

The bandwidth requirements that can be inferred from the analysis reported in Section 6.4.1 could be somewhat pessimistic. Clearly, not all the sites change their pages at the same rate. Cho and Garcia-Molina (2000b) conducted another study in 1999, monitoring changes of about 720 000 popular[2] Web pages. Results are reported in Figure 6.11. In terms of the overall 'popular' Web, these diagrams are in good qualitative accordance with the findings of Brewington and Cybenko (2000), indicating that a vast fraction of this Web is dynamic. However, it is interesting to note that a very simple Web partitioning, based on four top-level domains, yields dramatically different results (Cho and Garcia-Molina 2000b). In particular, 'dot com' websites

---

[2] According to PageRank (see Section 5.4), so this forms a biased representative of the Web population.

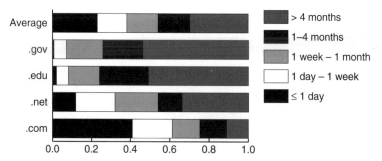

Figure 6.11   Average change interval found in a study conducted by Cho and Garcia-Molina (2000a). 270 popular sites were monitored for changes from 17 February to 24 June 1999. A sample of 3000 pages was collected from each site by visiting in breadth first order from the homepage.

are much more dynamic compared to educational or governmental sites. While this is not surprising, it suggests that a resource allocation policy that does not take into account site (or even document) *specific* dynamics may waste bandwidth re-fetching old information that, however, is recent in the index. From a different viewpoint, this nonuniformity suggests that, for a given bandwidth, the recency of the index can be improved if the refresh rate is differentiated for each document.

To understand how to design an optimal synchronization policy we will make several simplifying assumptions. Suppose there are N documents of interest and suppose we can estimate the change rate $\lambda_i, i = i, \ldots, N$, of each document. Suppose also that it will be practical to program a crawler that regularly fetches each document i with a refresh interval I_i. Suppose also that the time required to fetch each document is constant. The fact that we have limited bandwidth should be reflected in a constraint involving I_i. If B is the available bandwidth, expressed as the number of documents that can be fetched in a time unit, this constraint is

$$\sum_{i=1}^{N} \frac{1}{I_i} \leqslant N. \tag{6.22}$$

The problem of *optimal resource allocation* consists of selecting the refresh intervals I_i so that a recency measure of the resulting index (e.g. freshness or index age) will be maximized (Cho and Garcia-Molina 2000b). For example, we may want to maximize freshness

$$(I_1^*, \ldots, I_N^*) = \arg\max_{I_i, \ldots, I_N} \sum_{i=1}^{N} \bar{\phi}(\lambda_i, \phi_i) \tag{6.23}$$

subject to (6.22). We might be tempted to arrange a policy that assigns to each document a refresh interval I_i that is proportional to the change rate λ_i. However, this intuitive approach is suboptimal, and can be proven to be even worse than assigning the same interval to each document. The optimal intervals can be easily derived as the solution of a constrained optimization problem such as the one described by Equation (6.23).

If very large collections of documents need to be monitored, bandwidth limitations may even prevent us from frequently monitoring document changes. Cho and Ntoulas (2002) have recently proposed a sampling approach where a small fraction of pages from a given site is checked for changes and the results are used to estimate the change rate of the entire site.

WebFountain (Edwards *et al.* 2001) is a fully distributed and *incremental* crawler with no central control or centralized queue of URLs. 'Incremental' in this case means that the repository entry of a given document is updated as soon as it is fetched from the Web and the crawling process is never regarded as complete. The goal is to keep the repository as fresh and as complete as possible. The model in this case is not based on assumptions on the distribution of document change rates. Changes are simply detected when a document is re-fetched and documents are grouped into a set of buckets, each containing documents having similar rates of change. The trade-off between re-fetching (to improve freshness) and exploring (to improve coverage) is controlled in this case by maintaining separate queues for 'old' and 'new' URLs. The optimal ratio of between the number of old and new URLs is determined as the solution of a constrained optimization problem (see Edwards *et al.* (2001) for details).

Wolf *et al.* (2002) also formulate crawling as an optimization problem and suggest a pragmatic metric, the *embarrassment level*, that focuses on the expected number of times a search engine client is returned a stale URL.

Exercises

Exercise 6.1. Write a simplified crawling program that organizes the list of URLs to be fetched as a priority queue. Order the priority queue according to the expected indegree of the page pointed to by each URL and compare your results to a best-first search algorithm, using the actual indegree as a target measure of relevance.

Exercise 6.2. Extend the crawler developed in Exercise 6.1 to search for documents related to a specific topic of interest. Collect a set of documents of interest to be used as training examples and use one of text categorization tools studied in Chapter 4 to guide the crawler. Compare results obtained using the parent-based and the anchor-based strategies.

Exercise 6.3. Write a program that recursively scans your hard disk and estimates the lifetime of your files.

Exercise 6.4. Suppose the page mean lifetime is Weibull distributed with parameters δ and σ. What are the average mean lifetime, the most likely mean lifetime, and the median mean lifetime? What would be reasonable values for these quantities starting from an estimated distribution like the one shown in Figure 6.9?

Exercise 6.5. Explain what we mean when we say that a collection of stored documents is (α, β)-current.

Exercise 6.6. Suppose you want to build a (0.8,1-week)-current search engine for a collection of 2 billion documents whose average size is 10 Kb. Suppose the mean change is Weibull distributed with $\sigma = 1$ and $\delta = 100$ days. What is the required bandwidth (suppose the time necessary to build the index is negligible)?

7

Modeling and Understanding Human Behavior on the Web

7.1 Introduction

Up to this point in the book we have focused on how the Web works and its general characteristics, such as the inner workings of search engines and the basic properties of the connectivity of the Web. This provides us with a rich source of information on how the Web is constructed and interconnected, illustrating its inner workings from an engineering viewpoint. However, there is a critically important aspect of the picture that we have not yet discussed, namely how humans interact with the Web. In this chapter we will examine in some detail this aspect of human interaction across a broad spectrum of activities, such as how we use the Web to browse, navigate, and issue search queries.

The Web can be viewed as an enormous distributed laboratory for studying human behavior in a virtual 'digital environment'. From a data analysis viewpoint, the Web provides rich opportunities to gather observational data on a large-scale and to use such data to construct, test, refute, and adapt models of how humans behave in the Web environment. For example, if we were to record all search queries issued over a 12-month period at a large search-engine website, we could then use this data for

(1) exploration, e.g. we can generate summary statistics on how site-visitors are issuing queries, such as how many queries per session, or the distribution of the number of terms across queries;

(2) modeling, e.g. we can investigate whether there is any dependence between the content of a query and the time of day or day of week when the query is issued;

(3) prediction, e.g. we can construct a model to predict which Web pages are likely to be most relevant for a query issued by a site-visitor for whom we have past historical query and navigation data.

In this chapter we will explore how we can go from exploratory summary statistics (such as raw counts, percentages, and histograms) to more sophisticated predictive

Modeling the Internet and the Web   P. Baldi, P. Frasconi and P. Smyth
© 2003 P. Baldi, P. Frasconi and P. Smyth   ISBN: 0-470-84906-1

models. These types of models are of broad interest across a wide variety of fields: from the social scientist who wishes to better understand the social implications of Web usage, to the human factors specialist who wishes to design better software tools for information access on the Web, to the network engineer who seeks a better understanding of the human mechanisms that contribute to aggregate network traffic patterns, to the e-commerce sales manager who wants to better predict which factors influence the Web shopping behavior of a site-visitor.

We begin with an overview of Web data measurement issues. While the Web allows us to collect vast amounts of data, there are nonetheless some important factors that limit the quality of this data, depending, for instance, on how and where the data are collected. In Section 7.2 we discuss in particular those factors that can have a direct impact on any inferences we might make from the available data. We follow this in Section 7.3 by looking at a variety of empirical studies that investigate how we navigate and browse Web pages. Section 7.4 then discusses a number of different probabilistic mechanisms (such as Markov models) that have been proposed for modeling Web browsing. Section 7.5 concludes the chapter by looking at how we issue queries to search engines, discussing data from various empirical studies as well as a number of probabilistic models that capture various aspects of how we use Web search engines.

7.2   Web Data and Measurement Issues

7.2.1   Background

As with any data analysis endeavor, before we even look at our data it is critically important to understand how the data are collected. Web data are typically collected in a completely automated fashion via software logging tools. The basic features of Web server logging were discussed in Chapter 2. Automated logging is quite useful in the sense that the data are acquired continuously without any manual supervision being required on our part to collect the data (in comparison, for example, with the effort required to collect census data, or to collect patient data in a medical study).

However, the downside of automated logging is that we can never be completely certain that the collected data faithfully represents what we believe it to be. For example, the server log files for page requests at a website will contain page requests to the Web server from both human users and automated 'robots' (e.g. crawler programs as discussed in Chapter 2). In any analysis of the data it will be important to separate these two different sources of page requests (if possible), since otherwise the robot data may significantly skew the statistics for human users. For example, robots typically explore the entire website in a very systematic manner and, thus, tend to issue far more page requests (and often at a much quicker rate) than a typical human user. Tan and Kumar (2002) discuss general techniques for automatically detecting sessions generated by robots.

For example, Figure 7.1 shows three plots of the number of page requests per hour logged at the www.ics.uci.edu website during the first week of April 2002.

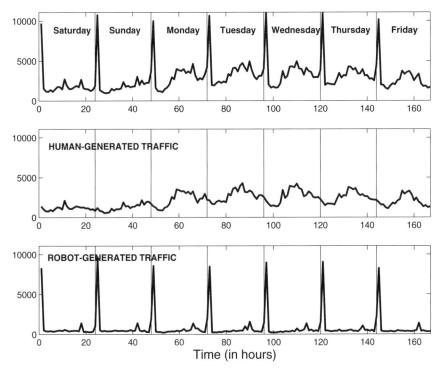

Figure 7.1 Number of page requests per hour as a function of time from page requests recorded in the www.ics.uci.edu Web server logs during the first week of April 2002. The solid vertical lines indicate boundaries between the different days (at midnight). The top plot is total traffic, the center plot is human-generated traffic, and the lower plot is robot-generated traffic.

Robot page requests were identified by classifying page requests using a variety of heuristics – for example, many robots self-identify themselves in the server logs. The top plot is the total number of page requests per hour, the center plot is the estimated number generated by humans per hour, and the lower plot is the estimated number of page requests generated by robots per hour. The top plot is the sum of the center and lower plots.

We can clearly see that the human-generated traffic is quite different from the robot-generated traffic. The robot traffic consists of two components: periodic spikes in traffic combined with a lower-level relatively constant stream of requests. The more constant component of the robot traffic can be assumed to be from robots from 'good citizen' search engines and cache sites, which attempt to distribute over time their crawler requests to any single site so as not to cause any significant bursts of activity (and potential deterioration in quality of service) at the Web server. The spiky traffic on the other hand can be assumed to be coming from a less well-intentioned source (or sources), whose crawler periodically floods websites with page requests and can potentially overload a server. In Figure 7.1 the dominant burst of page requests is arriving at the same time each night, just after midnight.

The human-generated traffic has quite a different characteristic. There is a noisy, but quite discernible, daily pattern of traffic, particularly during Monday to Friday. Human-generated traffic tends to peak around midday, and tends to be at its lowest between 2 am and 6 am (not surprisingly). This suggests that much of the traffic to the site is being generated 'in synchrony' with the daily Monday to Friday work schedule of university staff and students.

The main point here is to illustrate that Web server log data need careful interpretation. The presence of robot data in this example is a significant contributor to overall traffic patterns on the website. For example, if we were to mistakenly assume that the data in the top plot in Figure 7.1 were all human-generated, this assumption might lead us to rather different (and false) hypotheses about time-dependent patterns in human traffic to the site.

7.2.2   Server-side data

Data logging at Web servers

As discussed in Chapter 2 a primary function of a Web server (such as that running at www.ics.uci.edu) is to accept requests for particular pages at that website from individual clients (Web browser software running on machines anywhere on the Web) and to transmit the content of that page to the requester if it is a legal page request (e.g. after it checks if the page actually exists). In addition to serving Web pages, Web server software is typically configured to archive all page requests in a log file, recording the URL of the file requested, the time and date of the request, the IP address from which the request originated, information about the browser requesting the file, and an error code if the requested file could not be returned (e.g. if there is no such file). The URL of the *referrer page* may also be recorded if applicable; if a page request for page B is generated by a user clicking on a link on page A, then A is the referrer page for page B.

Server-side log files can potentially provide a wealth of information about how a website is being used, but require considerable care in interpretation. A detailed discussion of these issues can be found in Cooley *et al.* (1999), Mena (1999), and Shahabi *et al.* (2001).

Page requests, caching, and proxy servers

As described in Chapter 2, when a user requests a particular Web page using their browser, a message is sent via the HTTP protocol to the Web server, typically via the GET method. In theory, each such request is received by the Web server and both the request and the action of the server are logged in the server access log file with a timestamp. The server also records the IP address of the computer from which the user's request originated. The server can return the page if the URL is recognized, it can return an error message if the requested URL cannot be found on the website, or

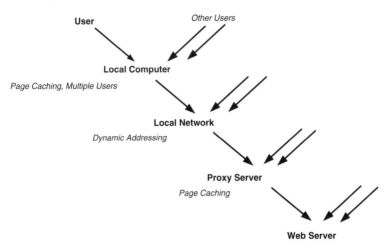

Figure 7.2 A graphical summary of how page requests from an individual user can be masked at various stages between the user's local computer and the Web server, via page caching at the browser level, multiple different users on the local computer, dynamic network address assignment at the local network level, and more page caching by proxy servers.

it can simply reply with a message that it is too busy to process the request at that time.

Thus, in theory, the Web server sees every individual request for pages on the site. In practice the situation is a little more complex, as illustrated graphically in Figure 7.2. Browser software, for example, often stores in local memory the HTML content of pages that were requested earlier by a user. This is known as caching and is a generally useful technique for improving response time for pages that are requested more than once over a period of time. For example, if the content of a particular Web page does not change frequently, and a user uses the back button on their browser to revisit the page during a session, then it makes sense for the browser to redisplay a stored or cached copy of the page rather than requesting it again from the Web server. Thus, although the user viewed the page a second time, this page viewing is not recorded in the Web server log because of local caching.

Caching can also occur at intermediate points between the user's local computer and the Web server. For example, an Internet service provider (ISP) might cache Web pages that are particularly popular and store them locally. In this manner, requests from users for these pages will be intercepted by the ISP cache and the cached version of the page is displayed on the user's browser. Once again the page request by the user for the cached page will not be recorded in the Web server logs.

Proxy servers can be thought of as a generalization of caching. A proxy server is typically designed to serve the Web-surfing needs of a large set of users, such as the customers of an ISP or the employees of a large organization. The proxy exists on the network at a point between each individual's computer and the Web at large. It acts as a type of intermediary for providing Web content to users. For example, the contents of entire websites that are frequently accessed might be downloaded by the proxy server during low-traffic hours, in order to reduce network traffic at peak hours. Various security mechanisms and filtering of content can also be implemented on the proxy server.

From a data analysis viewpoint the overall effect of caching and proxy servers is to render Web server logs somewhat less than ideal in terms of being a complete and faithful representation of individual page views.

In practice, in analyzing Web data, one can use various heuristics to try to infer the true actions of the user given the relatively noisy and partial information available in the Web server logs at the server. For example, one can try to combine pairs of referrer-requested pages with knowledge of the link structure of the website, to try to infer the actual path taken by an individual through a website. For example, if a user were to request the sequence page A, page B, page C, page B, page F, the Web server log would likely only record the sequence ABCF, since page B would likely be cached and not requested from the server the second time it was viewed. The logged referrer page for F would be page B, and using the timing information for the requested pages (say, for example that page F was requested relatively quickly after page C) one could then infer that the actual sequence of requests was ABCBF. In simple cases these heuristics can be quite useful (see, for example, Anderson *et al.* (2001) for a definition of *Web trails* in this context). However, in the general case, one can rarely be certain about what the user actually viewed. Statistical methods are likely to be quite useful for handling such uncertainty. At the present time, however, relatively simple heuristics, such as only using actual logged page requests in subsequent data analysis, and not making any assumptions about missed requests, are common practice in Web usage analysis.

Identifying individual users from Web server logs

In analyzing server log data over a specific period of time we would like to be able to associate specific page requests with specific individual users. We can certainly group requests from the same IP address. However, IP addresses do not necessarily have a one-to-one mapping to individual users (see for example Figure 7.2). One IP address might correspond to multiple different users, such as different individuals logged in to a multi-user system each running different browsers simultaneously, or to a single-user machine used by different people over time (e.g. in an academic laboratory). Even if there is only one user using a particular machine, the IP address may change over time, since it might be dynamically assigned by the individual's local network and, thus, we cannot link different page requests from this individual over time.

One approach to get around some of these problems is to use 'cookie' information. A cookie is small text file that is temporarily created on a user's hard disk by a Web server. This text file allows the Web server to uniquely identify and track requests from that individual user on that machine over time. When a user returns to a site they visited previously (e.g. a day earlier) the information in the cookie file can be accessed by the Web server and used to identify the user as the same user who visited on the previous day. Consequently, actions by the same user, such as page requests and product purchases, during different sessions can be linked together. Even if a user's IP address is dynamically re-assigned between sessions, the cookie information remains the same.

Thus, we can replace the IP address in the server log file with a more reliable identifier as provided by the cookie file. Commercial websites use cookies extensively and they can be quite effective in improving the quality of matching page requests to individual users, or at least matching to specific login accounts on specific machines. Although browser software allows individual users to disable the storing of cookies on their machines, many individual users either choose not to do so voluntarily (or are unaware of the fact that cookies can be disabled). Many commercial websites require that cookies are enabled in order for a user to access their site. It is estimated that in general well over 90% of Web surfers have the cookie feature permanently enabled on their browsers, and cookies generally tend to be quite effective for this reason. Of course, cookies in effect rely on the implicit cooperation of the user and, thus, raise the issue of privacy. For this reason some non-commercial sites (such as those at university campuses) often choose not to activate 'cookie creation' on their Web servers in the belief that cookies can be viewed as somewhat of an invasion of an individual's right to privacy.

Another option for user identification is to require users to voluntarily identify themselves via a login or registration process, e.g. for websites such as newspapers that charge fees for accessing content. As with cookies, this can provide much higher reliability in terms of being able to accurately associate particular requests with particular individuals, but again comes with the cost of imposing additional requirements on the user, which in turn may discourage potential visitors to the site.

7.2.3   Client-side data

Collecting data at the client side, for example, by recording a user's browser actions with software on the user's computer, solves many of the problems associated with server-side data logging. Advantages include

- direct recording of page requests, thus, eliminating 'masking' due to caching;

- recording of all browser-related actions by a user, including visits to multiple websites;

- more-reliable identification of individual users (e.g. by login ID for multiple users on a single computer).

The richness of client-side data means that it is the preferred mode of data collection for large-scale detailed studies of general navigation behavior on the Web. For example, companies such as comScore and Nielsen use client-side software to track home computer users' browser usage on an ongoing basis. Users are selected randomly using statistical experimental design methods to ensure a representative

Table 7.1   Estimates of monthly average user activity as reported from client-side data at www.netratings.com for September 2002.

	N. America	Japan	Ireland	France
Sessions per month	22	21	8	18
Unique sites visited	47	59	24	51
Total time (h)	11.6	11.6	3.6	8.5
Time per session (min)	31.6	33.3	26.7	28.9
Page-view duration (s)	55	35	44	42

cross-section. For example, as of October 2002, comScore's MediaMetrix software was actively monitoring browser actions of 60 000 Web users in North America. These data sets can provide relatively accurate estimates of which search engines and e-commerce sites are most popular, as well as providing the basis for studies that try to estimate the effect of user demographics on user behavior (we will see an example of such a study later, in Section 7.4.8).

Table 7.1 shows the estimated average monthly activity of a typical Web user in different countries. This type of client-side data is typically recorded on home computers and, thus, does not reflect a complete picture of Web browsing behavior, since it does not necessarily include the substantial volume of surfing data generated at business locations.

Statistics such as 'Time per session' and 'Page-view duration' in Table 7.1 can provide valuable information for user modeling in general. Server-side estimates of these same numbers would be very unreliable, if not impossible to obtain. Page-view duration for example, if estimated at the server-side, would be confounded by the fact that a particular Web server log only measures requests to that site. Thus, in between any two page requests to a particular site it is possible that the same user requested multiple other pages from other sites. Of course, even at the client-side, a statistic such as page-view duration does not directly measure how long a user views the page. Rather it merely measures the time duration between one page request and the next, and instead of viewing the page during that time the user might in fact be fetching a cup of coffee. One would in principle need to use video cameras and eye-tracking software to accurately monitor page viewing patterns at the client-side, but the expense and inconvenience of such setups have been a barrier to their use in any large-scale studies to date. However, as sensor technology continues to decline in size and cost it seems likely that user-monitoring systems (such as cameras that passively monitor and track a user's behavior) will become more prevalent.

Web surfing data can also be collected at other intermediate points in the network between an individual's computer and specific Web servers. For example, ISPs and proxy servers can log all outgoing page requests and browser actions from their customers. These data are often analyzed and used for a variety of commercial purposes. For example, by analyzing the sites that an ISP customer typically visits, a profile of that user can be constructed over time and used in real time for targeted advertising.

Handling massive Web server logs

Web server logs can be very large in size. Even a relatively small university department website can receive millions of page requests per month. Popular commercial websites, such as www.amazon.com or www.google.com, can receive tens of millions of page requests in a single day. If each page request requires roughly 100 bytes to store, then one month's worth of page-request data could be on the order of a terabyte or more. These data volumes pose significant practical challenges to conventional data analysis methods, since they clearly exceed the main memory capacities of current computers. Instead the server data are typically be stored in secondary memory on disk. Random access to records on a disk is typically orders of magnitude slower than random access to records in main memory (up to 10^5 times slower!). Even pipelined sequential scanning of data on disk is limited to about two Mb per second with current technology, so that to load one Gb of data from disk to main memory will take on the order of 8 min. These time-costs for data access place significant constraints on what types of data analyses we can carry out. In practice we must often resort to analyzing subsets of the data, e.g. selecting a single day's worth of data, perhaps filtering out events and fields that are not of direct interest, and focusing on the remainder of the data. In what follows we will ignore these issues of data access and in effect implicitly assume that the server log data of interest can be accessed in a time-efficient manner (e.g. assuming that we are in effect focusing on a subset of the log files and the data of interest reside in main memory). Techniques and algorithms for analyzing large data sets that reside outside of main memory are an interesting data mining topic in themselves, but somewhat outside the scope of this text. For further reading on such techniques and algorithms see Han and Kamber (2001).

7.3 Empirical Client-Side Studies of Browsing Behavior

A variety of empirical studies have examined in detail how groups of selected users use their Web browsers to navigate the Web. Data for these studies are typically collected at the client-side over a period of time so that, for example, reliable information on Web page revisitation patterns can be gathered.

Because the data sets are collected at the client-side, explicit permission needs to be requested from each user whose data are used in the study. For research studies conducted at universities (rather than large-scale commercial efforts) this often limits the number of individuals in each study to a relatively small number. Furthermore, there can be an implicit bias in the nature of the population being studied – individuals in an academic client-side study are typically all from the same university department and consist of computer science faculty, staff, and graduate students. Thus, some caution must be exercised in generalizing to the Web-using population at large from data that are gathered on small samples on individuals.

Client-side studies nonetheless provide relatively important clues about human behavior in the context of the Web. These clues can, for example, help design better network algorithms such as improved Web-caching and prediction strategies as well as support the development of better human interfaces for Web navigation such as more effective tools to allow a user to find pages previously visited.

7.3.1 Early studies from 1995 to 1997

Two of the earliest and most widely cited studies on client-side Web-browsing behavior are Catledge and Pitkow (1995) and Tauscher and Greenberg (1997). In both studies, data were collected by logging Web browser commands (page requests, use of the 'back' and 'forward' buttons, bookmarking, etc.) for a specific set of users over a period of time. In both studies the population consisted of faculty, staff, and students in a computer science department. Catledge and Pitkow collected data on 107 users over three weeks in 1994 and recorded 31 134 navigation commands, corresponding to roughly 14 page requests per user per day. Tauscher and Greenberg collected data on 23 subjects over a six-week period in 1995, with a total of about 19 000 navigation commands, and about 21 page requests per user per day.

Both studies found that clicking on hypertext anchors to navigate to another page was the most common Web-browser action, accounting for about 50% of all actions in each study. Using the 'back button' was the second most common action, 41% of all actions in the 1994 study and about 30% of actions in the 1995 study. The other categories of actions (such as bookmarking, navigating using a history list, etc.) were roughly evenly distributed among the remaining 10–20% of possible actions.

The Tauscher and Greenberg study focused in particular on how often pages are revisited by users. They found that the probability that a page is revisited (i.e. that the URL requested by a user is a URL that the user has requested in the past during the study) was about 0.58 across all users in the study. They also re-analyzed data from 55 users in the Catledge and Pitkow study and estimated the revisitation probability to be 0.61 from those data. Note that these estimates are likely to be underestimates of the actual revisitation rate, since any pages requested prior to the start of each study are not accounted for (and thus are not be counted as 'revisited' when requested during the study). Nonetheless, a lower bound of 61% indicates that the overall revisitation rate is quite high, suggesting that we are indeed creatures of habit. In practice, many of these revisits could be to pages whose contents change regularly (such as news pages) so that although the URL is the same, the content is different.

Tauscher and Greenberg also found a strong *recency* effect, namely that a page that is revisited is often a page that has been visited very recently rather than in the distant past. This is correlated with the empirical observation that users tend to use the 'back' button when navigating far more frequently than they use bookmarks or history lists.

It is interesting to note that repetition of user actions is not unique to Web-browsing and has been found to be a strong characteristic of user behavior in several other domains. Greenberg (1993) describes empirical data that estimated how many telephone numbers dialed are repeated (57%) and how many Unix command lines are repetitions of earlier commands (75%).

7.3.2   The Cockburn and McKenzie study from 2002

Although the 1994 and 1995 studies described above are widely cited in the research literature on empirical analysis of Web use, they are relatively out of date given the rate at which both the Web and our use of the Web has changed in the past few years. The more recent study of Cockburn and McKenzie (2002) provides a more up-to-date snapshot of browsing behavior. In their study, they analyzed the daily *history.dat* files produced by the Netscape browser for 17 users between October 1999 and January 2000. The population being studied consisted of faculty, staff, and graduate students at the University of Canterbury in New Zealand.

Over the 119 days of logging, 84 841 separate page requests were issued by the 17 users, which yields an approximate rate of 42 page requests per day per user. This is substantially higher than the roughly 14 and 21 page requests per day in the 1994 and 1995 studies, and is likely a confirmation of the fact that Web-browsers have become an increasingly integral part of our daily work patterns (again with the caveat that all of these studies are based on relatively small sample sizes in terms of the number of individuals involved).

Cockburn and McKenzie estimated that the revisitation rate was 0.81 in their study, averaged across users, ranging from a low of 0.61 for one user to a high of 0.92 for another user. These rates are significantly higher than the rates of about 0.6 obtained in the earlier studies. Two factors seem plausible as explanations for this increase. First, as mentioned earlier, the estimation of revisitation rates from a finite time-window of data is likely to be an underestimate of the true revisitation rate. Since the Cockburn and McKenzie time-window of four months was over three times longer than either of the other two studies, the longer time-window may have less of a 'downward' bias. The second and potentially more interesting factor is that our usage of the Web may have evolved from a more exploratory mode in 1994–1995, to a more utilitarian mode by 1999, a mode where we revisit certain sites such as general news, stock market information, and digital libraries on a regular basis.

Figure 7.3a shows a scatter plot of the number of distinct pages requested per user versus the total number of pages requested (over the 119 days). As we might expect there is a strong linear dependence: the more you surf the Web, the more new pages you tend to visit. It is also informative to look at the same data on a log–log plot (Figure 7.3b), where we now see that the marginal distributions of unique pages and total pages are somewhat less skewed and more symmetric on a log-scale. This suggests that if we wished to model the probability distribution of the average number of total requests per user (or the rate of requests per unit time) over a whole population, a log-scale distribution such as a log-normal might be a useful approach, since the data appear to have more symmetry and regularity on a log-scale (a log-normal distribution for a real-valued variable X is defined by assuming that log X has a normal distribution).

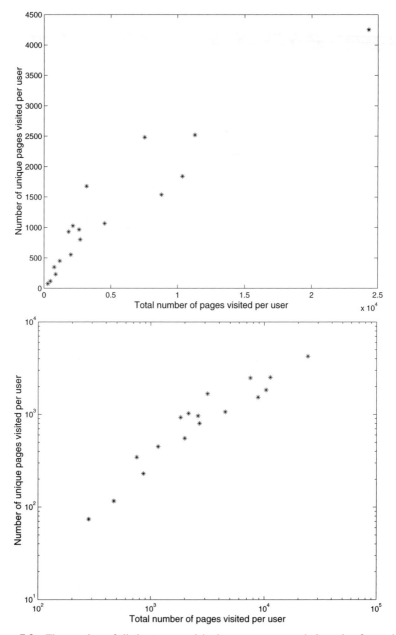

Figure 7.3   The number of distinct pages visited versus page vocabulary size for each of 17 users in the Cockburn and McKenzie (2002) study, on both (a) standard, and (b) log–log scales.

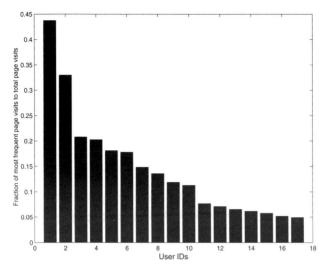

Figure 7.4 Bar chart of the ratio of the number of page requests for the most frequent page divided by the total number of page requests, for 17 users (sorted by ratio) in the Cockburn and McKenzie (2002) study.

Cockburn and McKenzie also found several other interesting and rather pronounced characteristics of client-side browsing. For example, the most frequently requested page for each user (e.g. their home page) can account for a relatively large fraction of all page requests for a given user. Figure 7.4 shows a bar chart of this fraction for each of the 17 users, ranging from about 50% of all requests for one user down to about 5%. This data also indicates that, while there is often considerable regularity in Web-browsing behavior for a particular individual, there can be considerable heterogeneity in behavior across different individuals.

Video-based analysis of Web usage

In addition to the studies mentioned above, other types of research studies have been used to improve our understanding of client-side Web behavior. For example, Byrne *et al.* (1999) analyzed video-taped recordings of eight different users as they used their Web browsers over a period of 15 min to about 1 h. The resulting audio descriptions by the users of what they were doing and the video recordings of their screens were then analyzed and categorized into a taxonomy of tasks. Among the empirical observations were the fact that users spent a considerable amount of time scrolling up and down Web pages (40 min out of 5 h). Another observation was that a considerable amount of time was spent by users waiting for pages to load, approximately 50 min in total, or 15% of the total time. This figure is of course highly dependent on bandwidth and other factors related to the network infrastructure between the user and the website being visited.

7.4   Probabilistic Models of Browsing Behavior

Moving beyond empirical studies, a natural next step is to build models that describe the browsing behavior of users on the Web. Modeling can both generate insight into how we use the Web as well as provide mechanisms for making predictions for a variety of applications such as pre-fetching of Web pages and personalization of Web content. In the next few sections we will examine some specific examples of various types of stochastic models that have been proposed for Web browsing data.

7.4.1   Markov models for page prediction

Consider building a model to describe how users navigate around a specific website. One general approach is to model their behavior by a finite-state Markov chain, where the states in the chain correspond to specific Web pages or general categories of Web pages. If we are only interested in the order in which users visit Web pages, and not in the time between page requests, then we can model each new page request as a transition between states. If each state corresponds to a Web page, it might be reasonable to assume that there are no 'self-transition' probabilities, i.e. that all transitions go from one page to a different page. However, if the states correspond to general categories of pages and contain multiple pages, then we can allow nonzero self-transition probabilities in the model to allow navigation between pages that are assigned to the same state in the model.

If we wish to explicitly include time in the model then we could (for example) use a discrete-time Markov chain (where the unit of time could be one second, for example) where we record at each 'tick' of the clock what state the user is in. Another option would be to use a continuous-time Markov process.

For simplicity we will focus on the order-dependent time-independent model where each transition corresponds to a page request. We will also assume that we can associate each page request with a specific individual and, thus, decompose a Web server log into individual user sessions.

A finite-state Markov chain is characterized by a set of M states (see Appendix A). Let s be a sequence of observed states of length L, e.g. such a sequence might be $ABBCAABBCCBBAA$ with three states A, B, and C. s_t denotes the state at position t in the sequence, $1 \leqslant t \leqslant L$. In general, for a sequence of fixed length L, the probability of generating a specific sequence s can be written as

$$P(s) = P(s_1) \prod_{t=2}^{L} P(s_t \mid s_{t-1}, \ldots, s_1).$$

Under a first-order Markov assumption, we assume that the probability of each state s_t given the preceding states $s_{t-1}, \ldots, s_1$ only depends on state s_{t-1}, i.e.

$$P(s) = P(s_1) \prod_{t=2}^{L} P(s_t \mid s_{t-1}). \tag{7.1}$$

This provides a simple generative model for producing sequential data, in effect a stochastic finite-state machine: we select an initial state s_1 from a distribution $P(s_1)$ then, given s_1, we choose state s_2 from the distribution $P(s_2 \mid s_1)$, and so on. Denoting $T_{ij} = P(s_t = j \mid s_{t-1} = i)$ as the probability of transitioning from state i to state j, where $\sum_{j=1}^{M} P(s_t = j \mid s_{t-1} = i) = 1$, we define T as an $M \times M$ *transition matrix* with entries T_{ij}, assuming a stationary Markov chain (see Appendix A).

This type of first-order Markov model makes a rather strong assumption about the nature of the data-generating process, namely that the next state to be visited is only a function of the current state and is independent of any previous states visited. The virtue of such a model is that it provides a relatively simple way to capture sequential dependence. Thus, although the true data-generating process for a particular problem may not necessarily be first-order Markov, such models can provide useful and parsimonious frameworks for analyzing and learning about sequential processes and have been applied to a very wide variety of problems ranging from modeling of English text to reliability analysis of complex systems (see, for example, Ross 2002).

Returning to the issue of modeling how a user navigates a website, consider a website with W Web pages, where we wish to describe the navigation patterns of users at this website. Since W may often be quite large (e.g. on the order of 10^5 or 10^6 for a large computer science department at a university), it may not be practical to represent each Web page with its own state, since a W-state model requires $O(W^2)$ transition probability parameters to specify the Markov transition matrix T. To alleviate this problem we can cluster or group the original W Web pages into a much smaller number M of clusters, each of which is assigned a state in the Markov model. The clustering into M states could be carried out in several different ways, e.g. by manual categorization into different categories based on content, by grouping pages based on directory structure on the Web server, or by automatically clustering Web pages using any of the clustering techniques described earlier in Chapters 1 and 4.

Transition probabilities $P(s_t = j \mid s_{t-1} = i)$ in such a model represent the probability that an individual user's next page request will be from category j, given that their most recent page request was in category i. We can add a special end-state to our model (call it E), which indicates the end of a session. For example, a commonly used heuristic is to declare the end of a session once a certain time duration, such as 20 min, has elapsed since the last page request from that user.

As an example, if we have three categories of pages plus an end state E the transition matrix for such a model can be written as

$$T = \begin{pmatrix} P(1 \mid 1) & P(2 \mid 1) & P(3 \mid 1) & P(E \mid 1) \\ P(1 \mid 2) & P(2 \mid 2) & P(3 \mid 2) & P(E \mid 2) \\ P(1 \mid 3) & P(2 \mid 3) & P(3 \mid 3) & P(E \mid 3) \\ P(1 \mid E) & P(2 \mid E) & P(3 \mid E) & 0 \end{pmatrix}. \tag{7.2}$$

This model can simulate a set of finite length sequences, where the boundaries between sequences are denoted by the 'invisible' symbol E. The state E has a self-transition probability of zero, to denote that after we get a single occurrence of E the system restarts and begins generating a new sequence, with probabilities $P(1 \mid E)$, $P(2 \mid E)$, and $P(3 \mid E)$ of starting in each of the three states. We could if we wish also define a fixed starting state S of a sequence, which is then followed by one of the other states – however, in effect we only need the end-state E here, since E is sufficient to denote the boundaries between the end of one sequence and the start of another.

Markov models have their limitations. The assumption of the first-order Markov model that we can predict what comes next given only the current state does not take into account 'long-term memory' aspects of Web browsing. For example, users can use the back button to return to the page from which they just came, imparting at least second-order memory to the process. For example, see Fagin *et al.* (2000) for an interesting proposal for a stochastic model that uses 'back buttons'.

We can try to capture more memory in the process by using a kth-order Markov chain, defined in general by the assumption that

$$P(s_t \mid s_{t-1}, \ldots, s_1) = P(s_t \mid s_{t-1}, \ldots, s_{t-k}), \qquad (7.3)$$

where k is some positive integer. Further model generalizations can be obtained by using various forms of stochastic grammars. However, fitting these more complex models to data can often require inordinate (and impractical) amounts of training data, and thus, we focus our discussion on the lowly first-order Markov model and later move on to various extensions of this model.

7.4.2   Fitting Markov models to observed page-request data

We can fit Markov models to observed data by collecting server-side page-request data from Web server logs. As discussed earlier in this chapter, server-side data have a number of limitations in terms of being an accurate representation of the actual page requests issued by a user. For example, caching by a user's browser software or by an intermediate proxy server can hide certain page requests from the server's viewpoint. There are several ways to try to deal with such problems (see, for example, Anderson *et al.* 2001; Cooley *et al.* 1999). We implicitly assume in what follows that these issues have been sufficiently resolved that we can proceed with fitting models to the session data that has been gathered.

Assume that we have collected such data in the form of N sessions, where the ith session s_i, $1 \leqslant i \leqslant N$, consists of a sequence of L_i page requests, categorized into $M - 1$ states and terminating in an 'invisible' symbol E. We denote the entire data set as $D = \{s_1, \ldots, s_N\}$.

Let θ denote the set of parameters for the Markov model that we wish to fit to the data. Using the representation in Equation (7.2), θ consists of the $M^2 - 1$ entries in the transition matrix T, and we can define θ_{ij} as the estimated probability of transitioning from state i to state j.

The *likelihood* function $\mathcal{L}(\theta)$ is defined as the probability of the data conditioned on a particular set of parameters θ for this model, i.e.

$$\mathcal{L}(\theta) = P(D \mid \theta) = \prod_{i=1}^{N} P(s_i \mid \theta). \qquad (7.4)$$

We assume on the right-hand side that page requests from different sessions are conditionally independent of each other given the model (a reasonable assumption in general, particularly if the sessions are being generated by different individuals). The statistical principle of *maximum likelihood* can now be used to find the parameters θ^{ML} that maximize the likelihood function $\mathcal{L}(\theta)$, i.e. the parameter values that maximize the probability of the observed data D conditioned on the model (see Chapter 1 for a general discussion of maximum likelihood).

To find the ML parameters requires that we determine the maximum of $\mathcal{L}(\theta)$ as a function of θ. Under our Markov model, it is straightforward to show that the likelihood above can be rewritten in the form

$$\mathcal{L}(\theta) = \prod_{i,j} \theta_{ij}^{n_{ij}}, \quad 1 \leqslant i, j \leqslant M, \qquad (7.5)$$

where n_{ij} is the number of times that we see a transition from state i to state j in the observed data D. Note that the initial starting states for each sequence are counted as transitions out of state E in our model.

We use the log-likelihood $l(\theta) = \log \mathcal{L}(\theta)$ for convenience:

$$l(\theta) = \sum_{ij} n_{ij} \log \theta_{ij}.$$

As shown earlier in Chapter 1, we can directly maximize this expression by taking partial derivatives with respect to each parameter θ_{ij} and incorporating (via Lagrange multipliers) the constraint that the sum of the transition probabilities out of any state must sum to one, i.e. $\sum_j \theta_{ij} = 1$. This leads straightforwardly to ML solutions

$$\theta_{ij}^{\mathrm{ML}} = \frac{n_{ij}}{n_i}$$

for the transition probabilities θ, where n_i is the total number of page requests for state i in the observed data. This is similar to the result obtained for the die model in Chapter 1. The ML solutions turn out to be the intuitive 'frequency-estimates' in this case: the estimated transition probability is the number of times a transition from i to j is observed divided by the number of observations of state i.

7.4.3 Bayesian parameter estimation for Markov models

In practice, if M is large (e.g. in the range of 10^2 or 10^3) we will end up estimating a very large number of transition probabilities, namely M^2. Even though our Web server logs may be quite large and, thus, D may contain potentially millions or even hundreds of millions of sequences, there is still a good chance that some of the observed transition counts n_{ij} may be zero in our data D, i.e. we will never have observed such a transition. For example, if we have a new e-commerce site we might wish to fit a Markov model to the data for prediction purposes but in the first few weeks of operation of the site the amount of data available for estimating the model might be quite limited.

As discussed in Chapter 1, a useful approach to get around this problem of limited data is to incorporate prior knowledge into the estimation process. Specifically, we can specify a prior probability distribution $P(\theta)$ on our set of parameters and then maximize $P(\theta \mid D)$, the posterior distribution on θ given the data, rather than $P(D \mid \theta)$.

The prior distribution $P(\theta)$ is supposed to reflect our prior belief about θ – it is a subjective probability. The posterior $P(\theta \mid D)$ reflects our posterior belief in the location of θ, but now informed by the data D. For the particular case of Markov transition matrices, it is common to put a distribution on each row of the transition matrix and to further assume that each of these priors are independent,

$$P(\theta) = \prod_i P(\{\theta_{i1}, \ldots, \theta_{iM}\}),$$

where $\sum_j \theta_{ij} = 1$. Consider the set of parameters $\{\theta_{i1}, \ldots, \theta_{iM}\}$ for the ith row in the transition matrix T. A useful general form for a prior distribution on these parameters is the Dirichlet distribution, defined in Appendix A as

$$P(\{\theta_{i1}, \ldots, \theta_{iM}\}) = \mathcal{D}_{\alpha q_i} = C \prod_{j=1}^{M} \theta_{ij}^{\alpha(q_{ij}-1)}, \tag{7.6}$$

where $\alpha, q_{ij} > 0$, $\sum_j q_{ij} = 1$, and C is a normalizing constant to ensure that the distribution integrates to unity (see Appendix A and Chapter 1 for details). The MP posterior parameter estimates are

$$\theta_{ij}^{\text{MP}} = \frac{n_{ij} + \alpha q_{ij}}{n_i + \alpha}. \tag{7.7}$$

If there is a zero count $n_{ij} = 0$ for some particular transition (i, j) then rather than having a parameter estimate of 0 for this transition (which is what ML would yield), the MP estimate will be $\alpha q_{ij}/(n_i + \alpha)$, allowing prior knowledge to be incorporated. If $n_{ij} > 0$, then we get a 'smooth' combination of the data-driven information (n_{ij}) and the prior information represented by α and q_{ij}.

For estimating transitions among Web pages the following is a relatively easy way to set the relative sizes of the priors.

(1) Let α be the 'weight' of the prior. One intuitive way to think of this is as the effective sample size required before the counts for a particular state n_i 'balance' the prior component in the denominator in Equation (7.7).

(2) Given a value for α, partition the states into two sets: set 1 contains all states that are directly linked to state i via a hyperlink, and set 2 contains all states that have no direct link from state i.

(3) Assign a total prior probability mass of ϵ, $0 \leqslant \epsilon \leqslant 1$, on transitions out of any state i into the states in set 2. We could further assume that transitions to all states in set 2 are equally likely *a priori*, so that $q_{ij} = \epsilon/K$ for states j in set 2 (those not linked by a direct hyperlink to any page in state i). This assumes that we have no specific information about which states are more likely in set 2. The total probability ϵ reflects our prior belief that a site visitor will transition to a Web page by a 'non-hyperlink' mechanism such as via a bookmark, by use of the history information in a browser, or by directly typing in the URL of the page. Observational data (e.g. Tauscher and Greenberg 1997) suggest that ϵ is typically quite small, i.e. users tend to follow direct hyperlinks when surfing.

(4) For the remaining transitions, those to states in set 1 (for which hyperlinks exist from state i), we have a total of probability mass of $1 - \epsilon$ to assign. We can either assign this uniformly, or we could weight it nonuniformly in various ways, such as (for example) weighting by the number of hyperlinks that exist between all of the Web pages in state i and those in set 1 (if the states represent sets of Web pages).

(5) The prior probabilities both into and out of the end-state E can be set based on our prior knowledge of how likely we think a user is to exit the site from any particular state, and our prior belief over which states users are likely to use when entering the site.

This assignment of a prior on transition probabilities is intended as a general guide of how Bayesian ideas can be utilized in building models for Web traffic, rather than a definitive prescription. For example, the model would need to be extended to account for other frequently used navigational mechanisms such as the back button. Nor does it account for dynamically generated Web pages (such as those at e-commerce sites), where the hyperlinks on a page vary over time dynamically in a non-predictable fashion.

The general idea of using a small probability ϵ for nonlinked Web pages is quite reminiscent of the methodology used to derive the PageRank algorithm earlier in Chapter 5 (we might call this the 'Google prior'). Sen and Hansen (2003) describe the use of Dirichlet priors for the problem of estimating Markov model parameters, with applications to characterizing and predicting Web navigation behavior.

7.4.4 Predicting page requests with Markov models

Markov models in a variety of flavors have been applied to the problem of predicting the next page, or a future page, that a user will request, given the sequence of pages that he or she has requested up to that point (Sarukkai 2000; Zukerman *et al.* 1999). Being able to make such predictions provides a basis for numerous different applications, such as pre-fetching and caching of Web pages and personalizing the content of a Web page by dynamically adding hyperlink short-cuts to other pages.

As mentioned earlier, for a typical website the number of pages on the site can be quite large. For example, for the UC Irvine website (Figure 7.1) it is estimated that there are on the order of 50 000 different pages on the site. Thus, it is typical to group pages into M categories and build predictive models for these groups with an M-state Markov chain. The groups can be defined, for example, in terms of page functionality (Li *et al.* 2002), page contents (Cadez *et al.* 2003), or the directory structure of the site (Anderson *et al.* 2001).

Simple first-order Markov models have generally been found to be inferior to other types of Markov models in terms of predictive performance on test data sets (Anderson *et al.* 2001; Cadez *et al.* 2003; Deshpande and Karypis 2001; Sen and Hansen 2003). The most obvious generalization of a first-order Markov model is the kth-order Markov model. A kth-order model is defined so that the prediction of state s_t depends on the previous k states in the model:

$$P(s_t \mid s_{t-1}, \ldots, s_1) = P(s_t \mid s_{t-1}, \ldots, s_{t-k}), \quad k \in \{1, 2, 3, \ldots\}. \tag{7.8}$$

This model requires the specification of $O(M^{k+1})$ transition probabilities, one for each of the possible combinations of M^k'histories' multiplied by M possible current states. An obvious problem with this model is that the number of parameters in the model increases combinatorially as a function of both k and M. Thus, for example, if we use simple frequency-based (maximum-likelihood) estimates of the transition probabilities we face the problem of having many subsequences of length $k + 1$ that will not have been seen in the historical data. For example, if $M = 20$ and $k = 3$, then we need to have at least $M^{k+1} = 20^4 = 160\,000$ symbols in the training data set. Even this number is optimistically low – some subsequences will be more frequent than others and in addition we will need to see multiple instances of each subsequence to get reliable estimates of the corresponding transition probabilities.

There are a number of ways to get around this problem. For example, Deshpande and Karypis (2001) describe a number of different schemes for pruning the state space of a set of kth-order Markov models such that histories with little predictive power are pruned from the model. Empirical results with page-request data from two different e-commerce sites showed that these pruning techniques provided systematic but modest improvements in predictive accuracy compared to using the full model. Techniques from language modeling are also worth mentioning in this context, since language modeling (e.g. at the word level) can involve both very large alphabets and dependencies beyond near-neighbors in the data. Of particular relevance here are empirical smoothing techniques that combine different models from order one to order k, with weighting coefficients among the k models that are (generally speaking) estimated to maximize the predictive accuracy of the combined model. See Chen and Goodman (1996) for an extensive discussion and empirical evaluation of various techniques and MacKay and Peto (1995a) for a description of a general Bayesian framework for this problem.

Cadez *et al.* (2003) and Sen and Hansen (2003) propose *mixtures of Markov chains*, where we replace the first-order Markov chain

$$P(s_t \mid s_{t-1}, \dots, s_1) = P(s_t \mid s_{t-1}) \tag{7.9}$$

with a mixture of first-order Markov chains

$$P(s_t \mid s_{t-1}, \dots, s_1) = \sum_{k=1}^{K} P(s_t \mid s_{t-1}, c = k) P(c = k), \tag{7.10}$$

where C is a discrete-valued hidden variable taking K values, $\sum_k P(c = k) = 1$, and $P(s_t \mid s_{t-1}, c = k)$ is the transition matrix for the kth mixture component. One interpretation of this model is that users consist of K different groups, each with different navigation behaviors described by K different Markov chains. Cadez *et al.* (2003) use this model to cluster sequences of page requests into K groups. The parameters of the models are learned using the EM algorithm (see Chapter 1). Sen and Hansen (2003) use the same type of model motivated by the problem of pre-fetching of Web pages to reduce user access times.

Consider the problem of predicting the next state, given some number of states t that have been observed for a given user, where $t \in \{1, 2, 3, \dots\}$. This is the 'one-step ahead prediction' problem. Let $s_{[1,t]} = \{s_1, \dots, s_t\}$ denote the sequence of t states that have been observed up to t. Under a standard first-order mixture model, the predictive distribution on the next state is

$$P(s_{t+1} \mid s_{[1,t]}) = P(s_{t+1} \mid s_t). \tag{7.11}$$

For a mixture of K Markov models, the predictive distribution is

$$
\begin{aligned}
P(s_{t+1} \mid s_{[1,t]}) &= \sum_{k=1}^{K} P(s_{t+1}, c = k \mid s_{[1,t]}) \\
&= \sum_{k=1}^{K} P(s_{t+1} \mid s_{[1,t]}, c = k) P(c = k \mid s_{[1,t]}) \\
&= \sum_{k=1}^{K} P(s_{t+1} \mid s_t, c = k) P(c = k \mid s_{[1,t]}).
\end{aligned}
\tag{7.12}
$$

The last line follows from the fact that, conditioned on component $c = k$, the next state s_{t+1} only depends on s_t. Thus, the Markov mixture model defines the probability of the next symbol as a weighted convex combination of the transition probabilities $P(s_{t+1} \mid s_t, c = k)$ from each of the component models. The weights are the membership probabilities $P(c = k \mid s_{[1,t]})$ based on the observed 'history' $s_{[1,t]}$:

$$P(c = k \mid s_{[1,t]}) = \frac{P(s_{[1,t]} \mid c = k) P(c = k)}{\sum_j P(s_{[1,t]} \mid c = j) P(c = j)}, \quad 1 \leqslant k \leqslant K, \tag{7.13}$$

where

$$P(s_{[1,t]} \mid c = k) = P(s_1 \mid c = k) \prod_{t=2}^{L} P(s_t \mid s_{t-1}, c = k). \qquad (7.14)$$

Intuitively, these membership weights 'evolve' as we see more data from the user, i.e. as t increases. In fact, it is easy to show that as $t \to \infty$ and if the model is correct, then one of the weights will converge to one and the others to zero. In other words, if a data sequence becomes long enough it will always be possible to perfectly identify which of the K components is generating the sequence. In practice, sequences representing page requests are often quite short. Furthermore, it is not realistic to assume that the observed data are really being generated by a mixture of K first-order Markov chains – the mixture model is a useful approximation to what is likely to be a more complex data-generating process. Thus, in practice, several of the weights $P(c = k \mid s_{[1,t]})$ will be nonzero and the corresponding models will all 'participate' in making predictions for s_{t+1} via Equation (7.12).

Comparing Equations (7.11) and (7.12) is informative – it is clear from the functional form of the equations that the mixture model provides a richer representation for sequence modeling than any single first-order Markov model. The mixture can in effect represent higher-order information for predictive purposes because the weights encode (in a constrained fashion) information from symbols that precede the current symbol.

A value for K in this type of mixture model can be chosen by evaluating the out-of-sample predictive performance of a number of models on unseen test data. For example, the one-step ahead prediction accuracy could be evaluated on a validation data set for different values of K and the most accurate model chosen.

Alternatively, the log probability score, $\sum_{t=1}^{L-1} \log P(s_{t+1} \mid s_{[1,t]})$, provides a somewhat different scoring metric, where a model is rewarded for assigning high predictive probability to the state that occurs next and, conversely, is penalized for assigning probability to the other states. This is suggestive of a 'prediction game', where predictive models compete against each other by 'spreading' a probability distribution over the M possible values of the unseen state s_{t+1}. If the model is too confident in its predictions and assigns a probability near unity to one of the states, then it must assign a very low probability to other states, and may be penalized substantially (via $\log P$) if one of these other states actually occurs at $t + 1$. On the other hand, if the model is very conservative and always 'hedges its bets' by assigning probability $1/M$ to each of the states, then it will never gain a high $\log P$ score. Good models provide a trade-off between these two extremes.

If we take the negative of the log probability score and normalize by the number of predictions made, $-1/L \sum_{t=1}^{L} \log P(s_{t+1} \mid s_{[1,t]})$, we get a form of entropy that is bounded below by zero. The model that always makes perfect predictions gets a score of $\log P = \log 1 = 0$ for each prediction and a predictive entropy of zero, while the model that always assigns probability $1/M$ to each possible outcome will have a predictive entropy of $\log M$. Cadez *et al.* (2003) describe the evaluation of predictive entropy for a variety of Markov and nonMarkov models using a test data set of roughly 100 000 page request sequences logged at www.msnbc.com with $M = 17$ page categories. The entropy ranged from about four bits (using log to the base two) for a single 'histogram' model down to about 2.2 bits for the best-performing mixture of Markov chains with about $K = 60$ components.

A number of other variations of Markov models have been proposed for page-request predictions, including variable-order Markov models (Sen and Hansen 2003) and position-dependent Markov models where the transition probabilities depend on the position of the symbol in the sequence (Anderson *et al.* 2001). This latter model can be viewed as a *nonstationary* Markov chain. The position-dependent model makes sense for websites that have a pronounced hierarchical structure, where (for example) a user descends through different levels of a tree-structured directory of Web pages. Several other variations of Markov models for prediction are explored in Zukerman *et al.* (1999), Anderson *et al.* (2001, 2002), and Sen and Hansen (2003).

7.4.5 Modeling runlengths within states

A notable characteristic of a Markov model is that the run-length distribution (the distribution on the length of 'runs' of the system in a given state) follows a geometric distribution. Specifically, letting r be the runlength for state i,

$$P(r) = T_{ii}^{r-1}(1 - T_{ii}), \quad r = 1, 2, \ldots,$$

since, after one enters state i, the probability of exiting to another state after exactly r subsequent state transitions requires $r - 1$ self-transitions with probability T_{ii} followed by a transition out of state i (with probability $1 - T_{ii}$). The mean of this geometric distribution is $1/1 - T_{ii}$, so that the mean runlength is inversely proportional to the exit probability.

This geometric form alerts us to a potential limitation of first-order Markov models. In practice the geometric model for run-lengths might be somewhat restrictive, particularly for certain types of states and websites. For example, for a news website or an e-commerce website, we might find that observed runlengths in a state such as 'weather' or 'business news' have a mode at $r = 2$ or more page requests (since users may be more likely to request multiple pages rather than just a single page) rather than the monotonically decreasing shape of a geometric distribution with a mode at $r = 1$.

We can relax the geometric distribution on runlengths implied by the Markov model by using a semi-Markov model. A semi-Markov model operates in a generative manner as follows: after we enter state i we draw a runlength r from a state-dependent distribution $P_i(r)$, and then after r time-steps stochastically transition to a different state according to a set of transition probabilities from state i to the other states. (The more usual variant of semi-Markov models uses continuous time, where on entering a state i a time duration τ is drawn from a distribution on time for that state, $\tau > 0$).

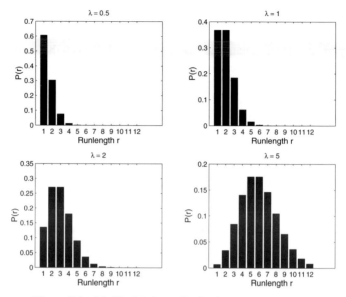

Figure 7.5   Modified Poisson distributions with parameters
(a) $\lambda = 0.5$, (b) $\lambda = 1$, (c) $\lambda = 2$, and (d) $\lambda = 5$.

For example, we could have $P_i(r)$ be a modified Poisson distributed random variable,

$$P_i(r) = \frac{\lambda^{r-1}}{(r-1)!}e^{-\lambda}, \tag{7.15}$$

where r is a positive integer, $\lambda > 0$ is a parameter of the Poisson model, and the distribution has mean $\lambda + 1$ and variance λ. The standard Poisson distribution (see Appendix A) has $r = 0, 1, 2, \ldots$, whereas here we use a slightly modified definition that has r starting at one, so that the mean is also shifted by one (in the standard model the mean and the variance are both λ (see Appendix A)).

The most likely run-length (the mode) is the largest integer k such that $k \leqslant \lambda + 1$. Thus, for example, if we choose $\lambda > 1$ for this model, the most likely runlength will be two or greater, providing a qualitatively different shape for our runlength distribution compared to the geometric model which has its mode at one (see Figure 7.5 for examples). For example, a Poisson distribution with $\lambda = 0.5$ might be appropriate for websites where visits tend to be relatively short and the most likely number of page requests is one, while a Poisson distribution with $\lambda = 5$ might be a better model for websites where most visitors request multiple pages.

7.4.6   Modeling session lengths

We now move from modeling run-lengths within states to modeling overall session lengths for individual users. In general, an M-state Markov model with an end-state

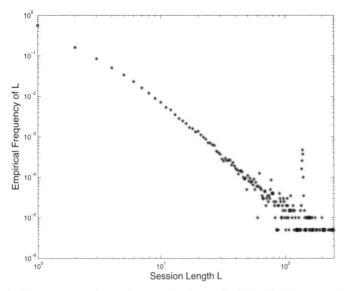

Figure 7.6 Histogram on a log-scale of session length for 200 000 different sessions at the
www.ics.uci.edu website in early 2002. Session boundaries were determined by 20 min
or longer gaps between page requests, and site visitors that were clearly identifiable as robots
were removed.

allows us to generate a variety of distributions on the lengths of strings (sessions), by
varying the parameters of the transition matrix T. While such a model may be quite
adequate in practice from a purely predictive viewpoint, in terms of understanding
the underlying data-generating process such a Markov model may not provide much
insight. With this in mind, there is motivation to explore general parametric models
for session lengths and this is the focus of this section.

Let L represent the session length, $L \in \{1, 2, 3, \ldots\}$. Figure 7.6 shows observed
session length data from the www.ics.uci.edu website, plotted on a log–log
scale. This figure is suggestive of a linear relationship between $\log P(L)$ and $\log L$,
which in turn suggests a power-law distribution

$$P(L \mid \gamma) = CL^{-\gamma} \tag{7.16}$$

as introduced in Chapter 1, where $\gamma > 1$ is the single parameter for this model, and
$C = 1/\sum L^{-\gamma}$ is a normalization constant, known as the Riemann zeta function,
which in general can only be evaluated in closed form for even integer values of γ.
This distribution is known in statistics as the zeta distribution, defined on positive
integers $L = 1, 2, 3, \ldots$.

We can use the technique of ML estimation once again to estimate the parameter
γ from observed data (e.g. from the data in Figure 7.6). Since we are only estimating
a single parameter and have a relatively large amount of data available, we expect
that ML and Bayesian estimation would yield roughly similar results – so we use the
simpler ML technique here. The likelihood function, $\mathcal{L}(\gamma)$ is simply $\prod_{i=1}^{N} P(L_i \mid \gamma)$,

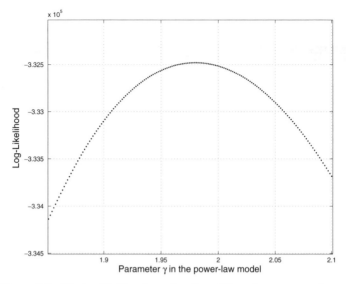

Figure 7.7 The log-likelihood as a function of the unknown parameter γ, under a power-law model, for the session-length data in Figure 7.6.

given an observed set of session lengths $\{L_1, \ldots, L_N\}$ that are assumed to be conditionally independent given the model, where $P(L_i \mid \gamma)$ is defined as in Equation (7.16). Differentiating this expression (or the log-likelihood) with respect to γ, and setting it to zero to find the maximum, does not yield a closed-form solution for the ML value for γ, since the term $C = 1/\sum L^{-\gamma}$ cannot be reduced to a closed-form function of γ.

The solution can, however, be obtained by standard numerical methods or, since we only have a single parameter to fit, we can just visually examine the log-likelihood as a function of γ and read-off an approximate value for γ^{ML}. Figure 7.7 shows the log-likelihood function, under the power-law distribution model, as a function of γ, for the empirical data shown in Figure 7.6, where we see that the maximum value occurs at $\gamma \approx 2$ ($\gamma = 1.98$ to be precise). Under the power-law model, maximum likelihood suggests that $P(L) \propto L^{-1.98}$. Figure 7.8 shows this model superposed on the empirical data.

Note that an alternative technique for fitting the parameter γ of a power-law distribution to empirical data is to use linear regression directly, i.e. to estimate the best-fitting slope for the line in Figure 7.6 with a least-squares penalty function. While this may seem like a reasonable approach, it can provide somewhat different results than maximum likelihood, since it in effect places equal weight on the error in predicting the count for each value of L, even though (for example) small values of L with large counts may account for a much larger fraction of the data than large values of L with very small counts. Maximum likelihood on the other hand seeks parameter values that maximize the probability of the data as a whole, rather than equally weighting the fit for different values of L. In many respects this is a more suitable criterion than

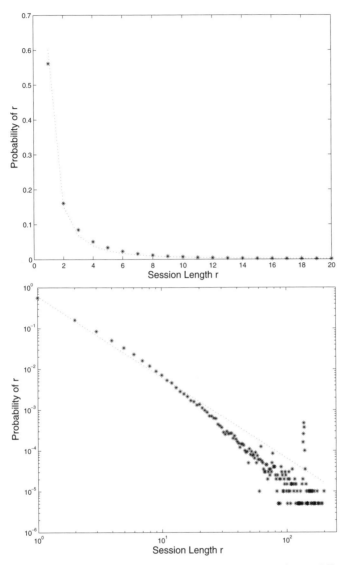

Figure 7.8 Comparing the fitted power-law model (dotted line) using an ML parameter estimate of $\gamma = 1.98$ with the observed empirical session lengths (dots), on both (a) standard scales, and (b) log–log scales.

the implicit equal-weighting criterion of least-squares. Thus, in assessing the fit in Figure 7.8 (on the log–log plot in particular) one should pay more attention to the quality of the fit for the smaller values of L (to the left), since this is the region that accounts for the bulk of the sessions. One could of course also use a weighted least squares penalty function, where the weight on the error for predicting the count at L

is multiplied by the size of the count at L, encouraging smaller prediction errors at the places where there are more data.

7.4.7   A decision-theoretic surfing model

Huberman *et al.* (1998) take a decision-theoretic approach to modeling navigation behavior at a website, and session lengths in particular. In their model, they consider the decision of a site-visitor visiting a particular page on whether to continue surfing to another page or to end the session. The utility to the user of viewing another page is modeled as the utility of the current page plus an additive Gaussian random term. Under this model, the decision on whether to continue or not depends on whether the expected cost of continuing is greater or less than the discounted expected utility of information to be gained from continuing. When the utility of the current page falls below some (negative) threshold, it is no longer optimal to continue. This leads to a stochastic random-walk type of model for user navigation. Under these assumptions on utilities and decision-making, Huberman *et al.* (1998) show that the distribution of session lengths will follow a particular functional form known as an inverse Gaussian distribution:

$$P(L) = \sqrt{\frac{\lambda}{2\pi L^3}} \exp\left(\frac{-\lambda(L - \mu)^2}{2\mu^2 L}\right),   \tag{7.17}$$

where $L \in 1, 2, 3, \ldots$ is the session length as before, $\mu > 0$ is the mean value and λ is a scale parameter (the variance of the distribution is inversely proportional to λ). The inverse-Gaussian distribution arises in statistical modeling of various physical problems and, strictly speaking, it defines a probability *density* function for all positive values of L, whereas here we have positive integers L for session lengths. Nonetheless, Huberman *et al.* argued that it provided a better fit to session-length data from two different websites than alternatives such as a simple geometric model or a power law.

We can use maximum likelihood once again to estimate the parameters of this inverse Gaussian model. It is straightforward to show that, given a data set $D = \{L_1, \ldots, L_n\}$ of n session lengths, the likelihood is maximized at the values

$$\mu^{\text{ML}} = \frac{1}{n} \sum_{i=1}^{n} L_i,   \tag{7.18}$$

i.e. the ML estimate of the mean under this model is the average session length, and

$$\lambda^{\text{ML}} = \left(\frac{1}{n} \sum_{i=1}^{n} \left(\frac{1}{L_i} - \frac{1}{\mu}\right)\right)^{-1}.   \tag{7.19}$$

In Figure 7.9 we show the fit of the inverse Gaussian distribution and three of the other models we have discussed earlier to the session length data from Figure 7.6. From this plot we see that the power-law fit matches the general shape of the empirical data distribution more closely than any of the other models. The Poisson model has the

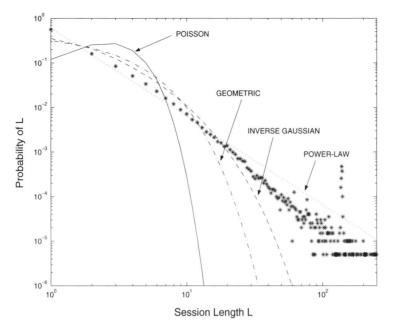

Figure 7.9 Comparing a variety of simple parametric session-length models with the observed empirical session lengths (dots), on a log–log scale, for the session-length data in Figure 7.6.

worst fit, with a mode at about $r = 3$. The geometric and inverse-Gaussian models both fail to capture the tail properties of the distribution, underestimating the probabilities of long session lengths. The power-law model is far from a perfect fit, but nonetheless it does capture the general characteristics of the session-length distribution with only a single parameter. Sen and Hansen (2002) discuss other approaches to modeling session lengths, such as higher-order Markov models.

7.4.8   Predicting page requests using additional variables

In the previous sections we looked at models that characterize the distribution of session lengths on a website and that try to predict the page a user will go to next based on pages visited in the past. It is natural to consider the incorporation of other predictor variables (sometimes referred to as *covariates*) into our characterization of navigation behavior. Relevant covariates might include

(a) the number of times a user has visited the site before (users with prior visits might tend to navigate the site more quickly and have shorter sessions);

(b) various demographic characteristics of users such as age, education level; and

(c) the bandwidth available to the user.

Bucklin and Sismeiro (2003) propose a covariate model of this form for site navigation behavior. As with the Huberman *et al.* model, they also consider a decision-theoretic framework for modeling the utility of continuing to a new page. However, in the Bucklin and Sismeiro model, this utility is modeled as a stochastic function of various covariates rather than just being a stochastic function of the utility of current and previous pages (the Huberman *et al.* approach). The covariates in the model include the number of previous site visits, the 'visit depth' (number of pages viewed on the site during this session up to this point), as well as various 'system' variables that take into account the complexity of a page, how long it takes to download, whether it contains dynamic content, and so forth. In the proposed model, both the utility of requesting an additional page and the logarithm of the page-view duration (conditioned on being at a particular page in a particular session), are modeled as linear functions of individual and page-specific covariates, plus an additive random noise term.

A useful general concept in the work of Bucklin and Sismeiro is *heterogeneity of behavior* across users. Rather than estimating a single set of model parameters that describe all site-users, a more general approach is to allow each user to have their own parameters. If we had a large amount of site-visit data for each individual user, then in principle we could estimate the parameters for each user independently using their own data. However, in practice we are usually in the opposite situation in terms of browsing behavior, i.e. we typically have very few data for many of the site visitors. In this situation, estimating model parameters independently for each user (using only their own data) would result in highly noisy parameter estimates for each individual. A general statistical approach to this problem is to couple together the parameter estimates in a stochastic manner, based on the simple observation that the behavior of different individuals will be dependent although different. In classical statistics this is known as a *random effects* model – individual behavior is modeled as a combination of a common systematic effect with an individual-specific 'random effect'.

In Bayesian statistics a somewhat more general approach is taken by specifying what is called a *hierarchical model*. In the hierarchical model we can imagine for simplicity that there are just two levels to the hierarchy: at the upper level we have a distribution that characterizes our belief in how the parameters of the model are distributed across the population as a whole; at the lower level we have a set of n distributions describing our belief in the distributions of parameters for each of n individuals. Both the population distribution and the n individual distributions can be learned from the observed data in a Bayesian manner. The upper level distribution in effect acts as a 'population prior' (learned from the data) that constrains the inferences we make on specific individuals. This population distribution typically acts to constrain the parameter estimates that we would get from looking at individual i's data alone, so that they are closer to those of the population as a whole. A proper description of Bayesian hierarchical models is beyond the scope of this text. A good introductory treatment can be found for example in Gelman *et al.* (1995).

For Web data analysis the use of hierarchical models (or random effects) is quite natural given that Web browsing behavior from populations of users can have both common patterns as well as significant individual differences. Bucklin and Sismeiro

Table 7.2 Toy example of query data from a search log.

User ID	Time stamp	Query
US12635123	1998 09 16 18 8102	World Cup soccer
YR19281101	1998 09 16 18 8312	Markov
YR19281101	1998 09 16 18 8410	Markov model
YR19281101	1998 09 16 18 8557	Markov state
FT29172531	1998 09 16 18 8876	Canary
FT29172531	1998 09 16 18 8992	Canary Islands

adopt a Bayesian hierarchical model approach in estimating the parameters of their covariate-dependent model. Their study was based on logged server-side data from 5000 visitors to a commercial website in the automotive industry, with 6630 sessions consisting of two or more page visits, and 40 560 page requests. These data were randomly sampled from a larger Web server log of 170 000 visitors in October 1999.

Using a Bayesian parameter estimation approach with the covariate linear models mentioned earlier, the authors were able to make a number of interesting inferences from the data about site-navigation behavior. For example, they found that there was significant variation in the parameters of the models fit at the individual level, supporting the hypothesis that significant heterogeneity exists among different individuals in terms of how they browse a website (see also Ansari and Mela 2003). They also found that repeat visits to a website have the effect of reducing the number of pages viewed but do not affect the length of time spent viewing a page. This can be thought of as a type of 'learning effect' over multiple visits – as users become more familiar with a site their sessions tend to get shorter in length.

Bucklin and Sismeiro also found evidence of a 'within-site lock-in' effect, as evidenced by a positive correlation in their model between visit depth and the mean length of time spent viewing a page – users who explore more pages tend to spend a longer time reading them. Such inferences from empirical data must be treated with some caution, given the relatively noisy nature of server data (as discussed earlier) and the fact that only a single website was used in the study. Nonetheless as both the quality and quantity of Web data continue to improve, these types of analytical studies are increasingly likely to provide substantial insights into site navigation behavior.

7.5 Modeling and Understanding Search Engine Querying

Up to this point in the chapter we have focused primarily on navigational behavior. An implicit 'driver' for much of our online navigation is the search for some specific information goal or set of goals. In the final section of this chapter we look in depth at one aspect of online search, namely, how users issue queries to search engines.

7.5.1   Empirical studies of search behavior

We can gain a broad understanding of the general characteristics of how search queries are issued by analyzing the search query logs of large commercial Web-based search engines. A number of different studies have followed this general approach, including Jansen *et al.* (1998), Silverstein *et al.* (1998), Lau and Horvitz (1999), Spink *et al.* (2002), and Xie and O'Hallaron (2002). These studies are typically based on the search query logs recorded by the search engine server and usually consist of three pieces of information per individual search query: a timestamp for when the query was received, the text string corresponding to the query, and a unique user ID (see, for example, Table 7.2).

The user ID is often a proxy used to replace the actual IP address from which the query is issued, for anonymity purposes. The term 'user' could be a little misleading since, for example, multiple individuals could be issuing queries from a single IP address. Xie and O'Hallaron (2002) used the user-agent field in the Web server log files for a particular search engine to check how many queries were coming from dial-up services such as America Online (AOL). They found that out of about 111 000 queries in total, roughly 2.7% were originating from AOL, a percentage which is relatively small but which nonetheless does indicate that data from dynamically assigned IP addresses can represent a nontrivial portion of log data and could in principle somewhat skew the overall statistics. Thus, as discussed earlier in the context of navigational studies, some caution in the interpretation of the empirical data is warranted since a set of queries from a particular 'user' might in reality be a set of individuals all sharing a particular IP address.

Similarly, since some search services answer search queries by querying other search engines and aggregating the results, a 'user' need not be a human searcher. The presence of non-human users (software programs such as robots) can significantly skew observed counts in the raw data, as illustrated earlier in Figure 7.1. For example, the Excite trace in Xie and O'Hallaron (2002) contained 130 200 queries from one particular IP address over 8 h – it is clearly not possible for a human to generate that many search queries in that amount of time. This type of outlier can make summary statistics such as means and standard deviations somewhat unreliable and emphasizes the need to examine the raw data, by plotting histograms for example, before any serious analysis or modeling.

The size and temporal extent of the search engine query logs analyzed in the different studies varies greatly.

- Lau and Horvitz (1999) analyze a sample of 4690 queries from a set of approximately one million queries collected by the Excite search engine in September 1997.

- Silverstein *et al.* (1998) analyze approximately one billion entries in the query log for the AltaVista search engine collected over six weeks in August and September 1998.

- Spink *et al.* (2002) describe a series of studies over multiple years involving the Excite search engine, with query logs collected at different time-points in September 1997, December 1999, and May 2001. The query logs analyzed at each time-point typically each contain about one million queries.

- Xie and O'Hallaron (2002) analyze 110 000 queries from the Vivisimo search query logs collected over a 35-day period in January and February of 2001, and 1.9 million queries collected over 8 h in a single day in December 1999 from the Excite query logs.

We will discuss below the main results from these four studies. Other empirical studies have also been reported – generally speaking, the inferred characteristics of query patterns across all of these studies are in broad agreement.

General characteristics of search query logs

The following general characteristics consistently appear across all of these studies.

- The average number of terms in a query is quite small, ranging from a low of 2.2 to a high of 2.6 across the studies.

- The most common number of terms in a query (the mode) in all four of the studies is two. In fact the distribution of the number of terms in a query is quite consistent across the different studies. Figure 7.10 plots the empirical histograms of the number of terms per query for both the Excite and Vivisimo data sets in the study by Xie and O'Hallaron. The histograms have been normalized by the total number of terms across all queries so that for each query length $L = 1, 2, 3, \ldots$ the height of each bar represents an empirical estimate of the probability $P(L)$.

  The modified Poisson model, defined in Equation (7.15) has also been fitted to each of the data sets as a simple hypothetical model for the query length data. Recall that the modified Poisson distribution has a single parameter λ which we estimate here as

  $$\frac{1}{N} \sum L_i - 1,$$

  where N is the total number of queries and L_i is the length of the ith query. The model provides a reasonable fit to both data sets – quite close to the observed values for the Vivisimo data, not quite as close for the Excite data. We can see that for both search engines the distribution of query lengths is qualitatively similar (i.e. mode at $L = 2$, significant mass for queries that are three or shorter, etc.). The tails of the empirical distributions and the Poisson models are in less agreement and diverge somewhat from each other if we plot the data on a log-scale on the Y-axis (not shown).

- The majority of users do not refine their query based on feedback from the results obtained with the initial query (e.g. by adding terms or modifying the syntax of the query). Silverstein *et al.* (1998) report that 77.6% of users issue

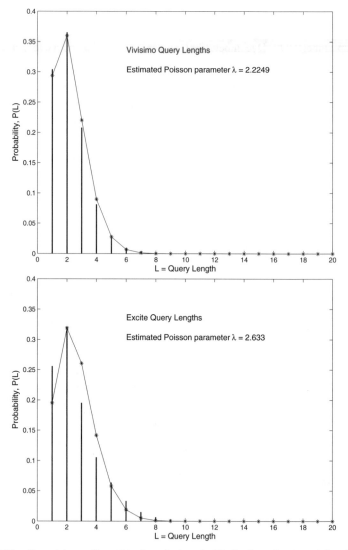

Figure 7.10   Comparison of empirical query length distributions (bars) and fitted Poisson models (dots and lines) for both Vivisimo query data (top) and Excite query data (bottom). The data were provided by Xinglian Yie and David O'Hallaron.

only a single query in a particular session, and Lau and Horvitz (1999) found similar results with only about 25% of the users in their study modifying queries.

• Spink *et al.* (2002) report that the number of users who viewed only a single page of Excite results (the top 10 websites as ranked by the search engine) increased from 29% in 1997 to 51% in 2001. In the AltaVista study of Silverstein *et al.* (1998) fully 85% of users were found to view only the first page of search results.

The difference in numbers between the two studies could reflect two different types of user populations (Excite users versus AltaVista users). It could also reflect a difference in methodology in terms of how 'number of pages viewed' was defined and measured in each study.

- There appears to have been a shift in the distribution of query topics over time. Both Lau and Horvitz (1999) and Spink *et al.* (2002) manually classified samples of several thousand queries into a predefined set of 15 information goals and 11 general categories, respectively. Both studies found that about one in six queries in the 1997 samples concerned adult content. In contrast, in the 2001 sample in the Spink study this figure had dropped to 1 in 12. The top two topics of 'commerce, travel, employment, or economy' and 'people, places or things' in the 2001 sample accounted for 45% of all queries. This is a considerable increase from the 20% share for the same two query topics in 1997, reflecting perhaps the increased use of the Web for business and commercial purposes. Conversely the top-ranked topic in 1997 with 20% of the queries, 'entertainment or recreation', had dropped to a 7% share by 2001. It should be noted that these topic-related figures are based on a particular sample of users (about 2500 per sample) at a particular search engine (Excite), and relied on a subjective process of manual assignment of query terms to categories.

Despite differences in sample populations, search engines, and dates of the studies, these four studies produced a generally consistent set of findings about user behavior in a search-engine context. For example, it is clear from the data that most users view relatively few pages per query and most users do not use advanced search features.

Power-law characteristics of repeated queries

Consider a set of observed queries $Q = \{q_1, \ldots, q_N\}$, and let $n(q_i)$, $1 \leqslant i \leqslant N$, be the number of observations (or count) of query q_i. If we sort the queries by their counts, we can define $C(r)$ as the count for the query of rank r, e.g. $C(1)$ is the number of times the most frequent query occurred, and so on. Equivalently we can define $f(r)$ as the empirical frequency of queries with rank r, calculated as $C(r)$, the number of times the query of rank r occurs, divided by the total number of queries.

Clearly $f(r)$ is a non-decreasing function of r; it varies inversely with r by definition. As discussed earlier in Chapter 1, for a variety of different empirical phenomena, $f(r)$ is often found to be well described by a power-law, i.e. $f(r) \propto r^{-\gamma}$.

Xie and O'Hallaron (2002) ranked queries by their frequency of occurrence for both the 110 000 queries from Vivisimo and the 1.9 million queries from Excite. Thus, the first query is the query that is most commonly issued, with a frequency count of approximately 100 in the Vivisimo data, and 10 000 in the Excite data, and so on. Plotting the log-frequency, $\log f(r)$, on the y-axis versus the log of the rank r on the x-axis reveals a near linear relationship in Figure 7.11, so that $\log f(r) \approx a \log r + b$. Once again we see power-law phenomena cropping up in another aspect of Web

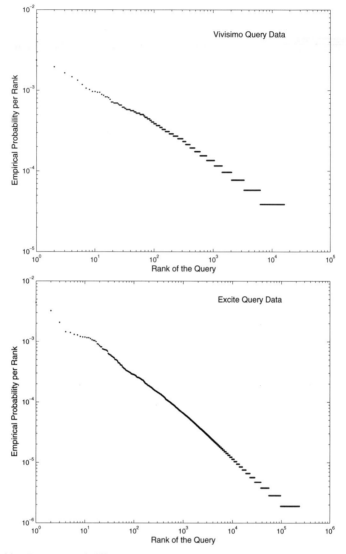

Figure 7.11 Frequency of different queries issued as a function of rank, for both Vivisimo query data (top) and Excite query data (bottom). The data were provided by Xinglian Yie and David O'Hallaron.

data analysis. Furthermore although the two curves are from two different search engines they are quite similar to each other. This again suggests that there are strong regularities in terms of patterns of behavior in how we search the Web, and these patterns appear to be relatively independent of the particular search engine being used.

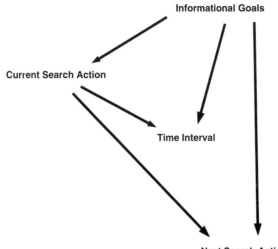

Figure 7.12 A graphical representation of the simple Bayesian network model used by Lau and Horvitz (1999) to model a user's search query actions over time.

7.5.2 Models for search strategies

In order to design search engines that can better serve the information needs of individual users it is important to understand the process by which a typical user navigates through search space when looking for information using a search engine. The four studies described earlier provide a general idea in terms of summary statistics of general characteristics of user behavior in an aggregate sense. Ideally, however, we would like to understand the search process in more detail. Lau and Horvitz (1999) describe a simple probabilistic model (a directed acyclic graph, or Bayesian network as defined in Chapter 1) for modeling how a user's search actions evolve over time. The four nodes in the network are:

(1) *current search action*, the first of two adjacent (in time) search actions, taking values *new query, reformulation, generalization, specialization, additional results* and *interruption*;

(2) *time interval*, a discrete-valued variable representing the elapsed time in seconds between the first user action and the second (if any), ranging in value from *10–20 s* to *more than 20 min*;

(3) *next search action*, taking the same values as *user search action*;

(4) *informational goals*, taking 15 different values such as *current events* and *education*.

The structure of the model proposed by Lau and Horvitz is illustrated in Figure 7.12 in the form of a Bayesian network. The conditional probability tables for each variable-value pair, given the variable-value combinations of all of its parents, were estimated by frequency counts (maximum likelihood) directly from the sample of 4690 queries in the study. This particular model imposes a specific (and somewhat natural) ordering on the variables (e.g. *informational goals* precedes all others) but otherwise does not make any conditional independence assumptions. More generally, if there were more variables in a model like this, a sparser graph (reflecting various conditional independence assumptions) would likely be used.

Although quite a simple model, Lau and Horvitz found that this network was able to produce interesting and potentially useful predictions. For example, according to the model, if a query specialization is followed by another action 10–20 s later then the user is most likely to be searching for entertainment-related information. The conditional probability for 'entertainment' given the user action and time-interval information is significantly higher than the marginal probability of 'entertainment'. Such correlations between sequences of actions, time between actions, and information goals, could in principle allow the search engine to make inferences about the likely 'search trajectory' of individual users based on their action history. In turn, these inferences could then be used to provide more relevant feedback to the user in real time and could also be used for marketing purposes such as real-time targeted advertising.

7.6   Exercises

Exercise 7.1. Obtain Web server logs from your organization's Web server (if you can) for one day's worth of data and carry out the following analyses:

- estimate the fraction of page requests that are coming from robots;

- segment the data into sessions, using a 20 minute time-out rule for declaring the end of sessions, and plot the empirical distribution of session lengths;

- fit at least two of the following models to the session-length data: geometric, Poisson, inverse Gaussian, and power-law. Plot and comment on the results.

Exercise 7.2. Derive the mean posterior estimate in Equation (7.7) from first principles.

Exercise 7.3. Prove that the ML estimate of the parameter λ for the Poisson distribution is the empirical mean (sample average) of the data.

Exercise 7.4. Construct three different session-length distributions by designing a transition matrix T for a four-state Markov chain (with one of the states being an end state), where each of the distributions should be quite different from the other two. Write a program to simulate data from each of the Markov models and construct histograms of the session lengths for each model from your simulations.

Exercise 7.5. Discuss how the Lau and Horvitz (1999) Bayesian network model for search queries and the Bucklin and Sismeiro (2003) model for site navigation might be combined to provide a more general model that encompasses both navigation and searching.

Exercise 7.6. We mentioned that for large M (e.g. a large number of pages M) estimating the $O(M^2)$ parameters of a first-order Markov chain may be quite impractical. Specify in detail two different algorithms for clustering M pages into K states, for the purposes of estimating a 'reduced category' K-state Markov chain for modeling navigation behavior.

Exercise 7.7. Discuss whether the first-order Markov models discussed in Section 7.4 can adequately model the process of users using the back button in their browsers when browsing. If you believe that the model is not adequate, discuss how one might develop a stochastic model that can handle this aspect of user behavior.

Exercise 7.8. Derive from first principles the ML estimates for the inverse Gaussian distribution in Equations (7.18) and (7.19).

Exercise 7.9. None of the models in Figure 7.9 fit the data perfectly well, and indeed to the right of the data (for session lengths greater than 10^2) there may be some outliers present. Comment on what aspects of real data sets these simple models may not be accounting for.

Exercise 7.10. In the techniques described in this chapter we did not discuss the issue of an individual's right to privacy. Research the issue of privacy in the context of Web data analysis and write an essay on the current state of affairs in your country in terms of privacy and the individual surfer. For example, if a surfer visits your website what does the law say about who owns the logged data? What are the laws (if any) that govern how Web servers can issue cookies? And so forth.

Exercise 7.11. Imagine that you are conducting a study on Web navigation behavior 10 years from now, with client-side data and demographic information for 100 000 individuals. Predict how you think both navigation and search behavior will have changed by then, compared to the data in the studies described in this chapter.

8

Commerce on the Web: Models and Applications

8.1 Introduction

In the late 1990s the Web quickly changed from being primarily an academic pursuit to an increasingly commercial one. This rapid commercialization of the Web, in areas such as e-commerce, subscription news services, and targeted advertising, has led to the infusion of the Web into modern daily life. While commercialization can have its negative aspects, on the plus side it presents significant new challenges and opportunities for academic researchers. Some examples of these are given in the following list.

- Can we design algorithms that can help recommend new products to site-visitors, based on their browsing behavior?

- Can we better understand what factors influence how customers make purchases on a website?

- Can we predict in real time who will make purchases given their observed navigation behavior?

In this chapter we will investigate these and related questions. Once again, as in Chapter 7, what is driving this research is the availability of vast amounts of raw data. For a standard bricks-and-mortar retail store, the only information about a customer's behavior within a store is usually in the form of scanner data or market basket data, namely, the list of items and prices that were scanned when the customer paid for their purchases. In contrast, for a store on the Web, a sequence of clickstream data is typically available for each customer prior to making a purchase, in addition to the purchase data (if the customer makes a purchase). This provides a much richer description of a customer's behavior over time. Naturally this clickstream data can itself be quite noisy, but nonetheless the volume of data available is often large enough (e.g. millions of customers per day visiting very large e-commerce sites such as www.amazon.com) that certain patterns clearly emerge even from noisy data.

Modeling the Internet and the Web   P. Baldi, P. Frasconi and P. Smyth
© 2003 P. Baldi, P. Frasconi and P. Smyth   ISBN: 0-470-84906-1

8.2   Customer Data on the Web

Data relevant to e-commerce can be obtained at the server side, at the client side, or at points in between (for example, by ISPs). Server-side data for e-commerce is similar in form to standard Web logs (i.e. a list of events that are timestamped, that record the page requested, the IP address requesting the page, and so forth), with the addition of purchase information if the customer made a purchase, as well as personal information on the customer such as address and credit card information. The purchase information is of course the primary event of interest to the retailer – we can, for example, try to estimate in real time whether a customer is likely to make a purchase or not as they begin to navigate the site. A predictive classification model for this purpose could be built from historical data on visitors to the site. We could also correlate personal information, such as city or country of origin, age, income (if known), with purchase behavior to try to get a better understanding of who is purchasing different types of products.

E-commerce logs are often more reliable by necessity than standard Web logs. For example, many e-commerce sites require that a site-visitor have cookies enabled, so that sessions can be reliably tracked and customers reliably identified. Many e-commerce sites personalize the information being presented to the site-visitor, and to do so they must identify who the customer is early on in the session and link them to a database of prior information. This identification can happen implicitly via cookies, or explicitly via customer login to the site. The overall effect is much more reliable identification of the site-visitor from session to session than with a typical non-commercial site.

Marketing researchers also find client-side data quite useful for modeling customer behavior. For example, how many websites (for a particular class of product) do customers typically visit before making a purchase? As discussed in Chapter 7, one source of such client-side data is companies such as Nielsen and comScore that provide industry data on Web usage. These companies select a representative random sample of Web users (e.g. perhaps 50 000 users in the United States), who agree to have their Web usage recorded over a period of time, usually by installing special software on their computers. This type of data, where the behavior of a set of selected individuals is tracked over time, is referred to as *panel data* in economics and marketing. We will look in detail at a sample of such data in Section 8.5 later in this chapter.

8.3   Automated Recommender Systems

A particular topic which has attracted considerable research attention in e-commerce in recent years is the problem of building automated recommender systems for Web applications. There are two main variants of this problem. In both variants we typically have a large number n of individuals and a large number m of items.

In the first variation of the problem, the observed data can be thought of as a very sparse binary $n \times m$ matrix V, where $v_{i,j} = 1$ indicates that individual i *purchased* item j, and the entry is assumed to be zero otherwise. See Table 8.1 for a toy example.

Table 8.1 A toy matrix R of binary votes, where items correspond to columns and users correspond to rows. If entry (i,j) is '1' it means that the ith user voted positively for the jth item. A blank entry means that no positive vote was recorded. Note the sparsity of the data.

	1	2	3	4	5	6	7	8	9	10	11
User 1	1				1						
User 2		1	1					1			
User 3		1									1
User 4	1							1			
User 5			1								
User 6					1						
User 7								1			
User 8										1	
User 9							1			1	
User 10	1								1		
User 11											
User 12	1			1				1		1	
User 13											1
User 14			1								
User 15						1	1				

In the second variation of the problem the items are not products that are purchased, but instead are items that are *rated* by the user, either explicitly (e.g. movies rated on a scale of one to five) or implicitly (e.g. documents in a digital library rated by how long the user is estimated to have spent viewing the document).

In either case the matrix V is usually very sparse, since the typical user will only have purchased or rated a very small fraction of the overall number m of items. For example, Popescul *et al.* (2001) report a study of recommender systems applied to the ResearchIndex online scientific literature database (Lawrence *et al.* 1999), with $n = 33\,050$ users accessing $m = 177\,232$ documents. Each user accessed on average only $0.01\% \approx 18$ documents in the database, so that 99.99% of the possible user–item pairs are in effect not present.

An interesting issue is the treatment of missing data in the matrix V for items j that were not voted on by user i, where missing is considered to be all the zeros. These missing data are not 'missing completely at random' but instead are likely to be affected by a form of negative selection bias, where users may be more likely to *not vote* on items that they do not like (Breese *et al.* 1998). In much of the work on recommender systems this bias is not explicitly dealt with. We also choose to conveniently ignore this issue in the discussion below, under the assumption that it may be somewhat of a second-order factor in modeling the data compared to the issue of how to handle the positive votes.

In our discussion we will generically refer to the user–item pairs $v_{i,j}$ as 'votes', keeping in mind that 'voting' might represent purchasing an item on a website, accessing a document on a website, or rating an item such as a piece of music or a movie on a website. All of these actions can be loosely interpreted as positive votes for the item in question, whether implicit or explicit.

The recommender problem is to be able to automatically recommend and rank a list of new items to a user based both on items this user has already voted on, as well as past voting patterns of other users. This can be viewed as a high-dimensional prediction problem where the training data from which to build a model is very sparse. Let a be the active user for whom we wish to make predictions and let $\mathcal{I}_a$ be the set of items that user a has already voted on (e.g. the items already in a's electronic basket, while purchasing books at an online bookstore). $\mathcal{I}$ is the total set of items. The prediction problem is to predict which of the items in the set $\mathcal{I} \setminus \mathcal{I}_a$ the user is most likely to provide a positive vote for, if they were asked to vote on that item.

Typically the number of items in the set $\mathcal{I} \setminus \mathcal{I}_a$ is quite large. For example, in the document database example earlier, the average user only voted on or accessed 18 documents, leaving 177 214 other documents that we need to rank in terms of how the user might like them. Thus, the computational aspects of recommender systems can be quite challenging, especially given that on a website the ranking needs to be computed in real time and for potentially thousands of different users who are visiting the site at any given time. Schafer *et al.* (2001) discuss various other important practical aspects of implementing recommender systems for real e-commerce applications.

8.3.1   Evaluating recommender systems

An important issue is that of evaluating the effectiveness of any particular recommender algorithm. Large e-commerce retailers conduct live experiments with real customers to test different algorithms, but academic researchers do not usually have the ability to attract large numbers of users to websites for experimental purposes. Unfortunately, from a research viewpoint the results of e-commerce experiments are rarely published for competitive reasons. For this reason recommender systems in research papers are evaluated based on historical data rather than actual recommendations to real customers. There is an important issue here: a system that recommends item X for individual a will be rewarded if in the test data individual a did indeed purchase that item. However, its entirely possible that the individual would have purchased the item anyway (whether recommended or not), or indeed the perverse situation might arise where the recommendation might actively discourage individual a from purchasing a product that they would have otherwise purchased!

This is an important caveat to keep in mind in terms of how recommender systems are evaluated, particularly when the 'votes' represent purchases. The point is aptly summarized by Zhang and Iyengar (2002):

> the true value of a recommender system can only be measured by controlled experiments with actual users. Such an experiment could measure the improvement achieved by a specific recommendation algorithm when compared to, say, recommending the most popular item. Experiments with historical data have been used to estimate the value of recommender algorithms in the absence of controlled live experiments ....

Many different ideas have been proposed as the basis for recommender algorithms, the main ones being nearest-neighbor and model-based collaborative filtering. We discuss both of these approaches as well as some related techniques in the next few sections.

8.3.2   Nearest-neighbor collaborative filtering

Basic principles

The term 'collaborative filtering' generally refers to the use of prior votes of individuals to predict new votes and make new recommendations. The nearest-neighbor approach to collaborative filtering (Resnick *et al.* 1994; Shardanand and Maes 1995) uses a memory-based approach for prediction: for user a, find the most similar users to a in the training data matrix V, and then use the votes of these neighbors to 'fill in' predictions for a on the items that a has not yet voted on. This is somewhat similar in concept to the K-nearest neighbor classifier described in Chapter 4.

A variety of different approaches have been proposed for (a) finding neighbors, (b) combining their histories (often using some form of weighting), and (c) making predictions. A typical scheme is as follows: let

$$v_i = \frac{1}{|\mathit{l}_i|} \sum_{j \in \mathit{l}_i} v_{i,j}, \quad j \in \mathit{l} \setminus \mathit{l}_a,$$

be the mean vote for user i, $1 \leqslant i \leqslant n$, where l_i is the set of items that user i has voted on (i.e. $v_{i,j} > 0$ for $j \in \mathit{l}_i$, and $v_{i,j} = 0$ otherwise). It is customary to calibrate user i's votes by subtracting the mean vote, to create an 'adjusted vote' matrix

$$v_{i,j}^* = v_{i,j} - v_i$$

so that if (for example) one user tends to vote high and another tends to vote low, this systematic 'mean difference' is removed.

We can then make predictions for the vote of new user a on items j that he or she has not voted on by using

$$\hat{v}_{a,j} = v_a + \frac{1}{C} \sum_{i=1}^{n} w_{a,i} v_{i,j}^*,$$

where $C = \sum_{i=1}^{n} |w_{a,i}|$ is a normalizing constant. The weights $w_{a,i}$ are defined by a function that estimates how similar user i is to user a. These weights are usually calculated based on the similarity of the votes that a has already made on items in l_a and the votes of other users on this same set of items. The overall effect of the prediction equation is to generate a predicted vote for items j, where the prediction is more heavily weighted toward votes of individuals who are have similar past voting histories to a. Once the predicted vote for each individual item in $\mathit{l} \setminus \mathit{l}_a$ is calculated, these predictions can then be ranked and the highest ranked items can be presented to user a.

Defining weights

One way to differentiate between different collaborative filtering algorithms is in terms of how the similarity weights are defined in the prediction equation above. One widely used weighting scheme (Resnick *et al.* 1994) is the correlation coefficient between users i and a, defined as

$$w_{a,i} = \frac{1}{C_2} \sum_j (v_{a,j} - v_a)(v_{i,j} - v_i),$$

where

$$C_2 = \left(\sum_j (v_{a,j} - v_a)^2 \sum_j (v_{i,j} - v_i)^2 \right)^{1/2}$$

is another normalizing constant, so that $-1 \leqslant w_{a,i} \leqslant 1$. The sums over items j in the expressions above are on items that both user a and i have voted on. If there are no common items in a and i's voting histories, then $w_{a,i} = 0$ by default. This alerts us to a potential limitation of nearest-neighbor methods – namely, when the intersection of $\mathcal{I}_a$ and $\mathcal{I}_i$ is small the weight will be based on relatively few matching items and may not provide reliable predictions. One technique to get around this problem is to introduce 'default votes' on certain items that neither a nor i have voted on (e.g. popular items). We can view this as filling in default values for missing data. From a statistical modeling perspective this may be somewhat suboptimal and a more principled alternative may be to estimate the missing values in a model-based fashion (Little and Rubin 1987). For example, we could use a Dirichlet prior to smooth out the votes in a manner analogous to the way we discussed the use of such a prior for generating reliable estimates of transition probabilities in Chapter 7. We will discuss model-based recommender algorithms later in this chapter.

Another example of a weighting scheme is to use vector similarity from information retrieval, the cosine of the angle formed by two vectors of word frequencies (see Chapter 4), where the vote vectors for individual i takes the place of word frequencies (Breese *et al.* 1998):

$$w_{a,i} = \sum_j \frac{v_{a,j}}{\sqrt{\sum_{k \in \mathcal{I}_a} v_{a,k}^2}} \frac{v_{i,j}}{\sqrt{\sum_{j \in \mathcal{I}_i} v_{i,k}^2}}.$$

The denominator has the effect of normalizing the contribution of user i relative to the total number of votes that user i has made. Thus, in an extreme example, if a has voted on only two items, but there is a user i who has voted on all items with the same value as user a on the two items in common, then the correlation weight as defined earlier would have value 1, even though user a and user i may really have very little in common. In contrast, the vector similarity function would have a weight $w_{a,i}$ that is much closer to zero than to one, since the contribution of user i to a's items would be downweighted by the fact that i has so many votes in general.

In addition to the weighting scheme, there are a significant number of other factors that can be 'tweaked' in terms of constructing an automated algorithm for predictions.

For example, for a user a with many votes, there may be a very large number of users i that will match with a but that have relatively small weights. Including these matches in the prediction equation may actually lead to worse predictions than simply leaving them out, and thus, one can impose a threshold on $w_{a,i}$ such that only users with similarity weights above the threshold are used in the prediction, or only select the top k most similar users, and so forth. The paper by Sarwar *et al.* (2000) discusses a variety of such extensions.

The curse of dimensionality

One general problem with similarity-based approaches is the sparseness of the data due to the high-dimensionality m of the space of items. Recall that typically only a tiny fraction of all user–item pairs are nonzero (Table 8.1). As an example of the effect that this can have, consider a user a who has a positive vote for a single book about mountain climbing in the Himalayas. There is a good chance that this person will be interested in other mountain-climbing books. With enough data from other users, we will eventually be able to *infer* a general preference by user a for mountain-climbing books in general. However, it may take a very large training data set of users to pick up these correlations by nearest-neighbor methods.

An alternative approach is to search for correlation structure among the items directly and to incorporate this into the prediction model. In essence, this idea is a model-based approach, where we try to model the structure of the item dependencies directly in some manner (and which we discuss in the next section in more detail). One such approach is dimensionality reduction via singular value decomposition (SVD) of the vote matrix V (Sarwar *et al.* 2000). In this approach we approximate the set of row-vectors of votes as a linear combination of r 'basis' columns, or equivalently, we approximate the matrix V with a rank-r approximation (see Appendix A). r is typically chosen to be much smaller than m (e.g. $r = 100$ while $m = 10\,000$). This general approach is also sometimes referred to as latent semantic indexing (LSI) when applied to text or recommender data (see Chapter 4).

The 'basis columns' can be estimated by standard numerical techniques for deriving SVD representations from sparse matrix data (Berry 1992; Letsche and Berry 1997) – these computations can be carried out offline. The general approach is to find the set of k columns such that the distance between the approximated versions of the rows and the original rows is minimized in a least-squares sense. The 'basis columns' can be thought of as defining a set of *meta-items*, each of which represent a set of items that are highly correlated with each other. In this way, we can perform the nearest-neighbor types of calculations for collaborative filtering in the r-dimensional space, and then 'project' out the predictions to all m items using the relevant linear algebra from the SVD representation.

This type of technique can be quite useful for alleviating the sparsity problem, since it can exploit item correlations directly. For example, Billsus and Pazzani (1998) used SVD to reduce item dimensionality before applying machine-learning algorithms to vote data, and found that this led to significantly higher predictive accuracies compared to simpler memory-based methods based on direct correlation of users' votes. An important practical issue, particularly for online recommender systems, is updating the basis vectors for an SVD representation as new data arrive. Ideally one would like to be able to do this calculation in an incremental manner given the current representation rather than recomputing a new SVD from scratch – incremental SVD computation is still somewhat of a research issue (Brand 2002).

Computational issues and clustering

In the memory-based approach to collaborative filtering, given a new user we must calculate the weights of potential neighbors in the training data V to make a prediction. For very large values of n and with many users on the site at any given time (e.g. large e-commerce websites) the time required to compute the weights for all users may be much greater than that required for a real-time response. As a result, nearest-neighbor approaches to collaborative filtering do not scale well to large n.

One approach to this problem is to cluster the historical data contained in V into K clusters, where K is much smaller than n (Breese *et al.* 1998; Ungar and Foster 1998). For prediction, a user is mapped into the cluster that they are most likely to belong to (an operation that scales as $O(K)$ rather than $O(n)$ per user) and properties of the cluster (e.g. the centroid or the mean prediction vector) are then used to make predictions. A potential weakness of clustering is that any individual user might really belong to multiple different clusters simultaneously. For example, in book buying, a user a might have interests in kayaking, Italian history, and blues music. While there might well be clusters representing each of these topics individually, it is unlikely that there would be a cluster that has all three topics within it. Thus, by forcing user a to be represented by a single cluster model, we will inevitably lose information about the heterogeneity of his or her interests. Breese *et al.* (1998) found that cluster-based collaborative filtering was outperformed by both nearest-neighbor and probabilistic model-based collaborative filtering, the second of which we discuss next.

8.3.3   Model-based collaborative filtering

General concepts

Model-based approaches to recommender systems make predictions by building (offline) an explicit model of the relationship among items, as inferred from historical voting patterns represented by V. This model is then used (online) in prediction mode for making predictions for a new user a. This is quite different in principle from the nearest-neighbor collaborative filtering approach discussed in the previous section, which relies on memory-based computations to compute predictions in real time, without ever explicitly constructing a global model.

Model-based approaches have an advantage in real-time computation requirements. Typically, once a model is constructed, the time to make a prediction is independent of n (the number of training examples used to construct the model) in contrast to memory-based approaches that scale directly with n.

Model-based approaches to collaborative filtering typically fall into a few general classes. The first general approach constructs a full joint probability distribution over the m items in $\mathcal{I}$, and then uses this joint distribution to make predictions online. The challenge here is to construct an accurate high-dimensional joint distribution from a very sparse data set. The second general approach builds m conditional models, where each model predicts the likelihood of an individual item given any combination of observed votes for the remaining $m - 1$, in contrast to a single joint model. Using this 'multiple conditional model' paradigm a variety of different predictive modeling techniques have been investigated, including dependency networks (Heckerman *et al.* 2000) and linear classifiers (Zhang and Iyengar 2002). A third general approach is to directly learn a model than can produce ranked lists, rather than models which produce scores that are then ranked (see, for example, Cohen *et al.* 1999).

We next examine some proposed methods for model-based recommender systems, specifically the joint density model proposed by Hofmann and Puzicha (1999) and the conditional density approach of Heckerman *et al.* (2000).

Joint density modeling

An unconstrained joint probability distribution for the full m-dimensional set of items requires the specification (or estimation) of 2^m independent parameters (for binary votes), which is clearly impossible for the ranges of m one typically encounters in recommender systems, e.g. $m \approx 10^3$ and upwards. Thus, a practical strategy is to seek more parsimonious representations for item dependencies.

One approach is to model the joint distribution as a finite mixture (a weighted combination) of simpler distributions, e.g.

$$P(v_1, \ldots, v_m) \approx \sum_{k=1}^{K} P(v_1, \ldots, v_m \mid c = k) P(c = k),$$

where the sum is over K mixture components, $P(c = k)$ is the probability that a randomly chosen row in the data set belongs to component k (where $\sum_k P(c = k) = 1$), and $P(v_1, \ldots, v_m \mid c = k)$ is a probability model for each component. If we think of this model in a generative fashion (for simulating the data) we can imagine for each user first selecting one of the K models with a K-sided die having 'face probabilities' $P(c = k)$, and then generating a set of votes for that user using the model $P(v_1, \ldots, v_m \mid c = k)$ given that the votes are coming from model k, $1 \leqslant k \leqslant K$ (see Chapter 1).

Of course to gain anything in terms of reducing the number of parameters needed in the overall model, we need to make the component models much simpler than a full unconstrained joint distribution on m items. A common choice for component models in mixtures for high-dimensional data is to assume that the votes for items are generated independently within a component

$$P(v_1, \ldots, v_m \mid c = k) \approx \prod_{j=1}^{m} P(v_j \mid c = k) = \prod_{j=1}^{m} \theta_{jk}^{v_j} (1 - \theta_{jk})^{1-v_j},$$

where $v_j \in 0, 1$ indicates whether the vote in the jth column is zero or one and $\theta_{jk} = P(v_j \mid c = k)$. This is known as a conditional independence or 'Naive Bayes' model in statistics and machine learning discussed earlier in Chapters 1 and 4. More generally one could use various forms of Bayesian networks or log-linear models for each component.

The number of parameters of the Naive Bayes model per component is linear in dimensionality m, so overall we only have $O(Km)$ parameters compared to $O(2^m)$ for a full joint distribution (for binary votes). The parameters of the model can be estimated from training data using a simple application of the expectation maximization algorithm that iteratively adjusts the parameters of the 'hidden' model to maximize the likelihood of the observed data (see, for example, the earlier discussion in Chapter 1 and more generally in Hand *et al.* (2001)).

The mixture of conditional independence models may be too simple to fully capture many aspects of the real data. For example, it ignores any dependencies between items within a cluster, i.e. all pairs $P(v_j, v_l \mid c = k)$ are modelled as $P(v_j \mid c = k)P(v_l \mid c = k)$. Nonetheless, it does capture *unconditional marginal dependence* of items, in that the model does allow $P(v_j, v_l)$ to be different from $P(v_j)P(v_l)$. Intuitively, we can imagine that for each component there are sets of items that are relatively likely to have 'on' votes, where $P(v_j = 1 \mid c = k)$ is much greater than $P(v_j = 1)$, and the rest of the items are more likely to have votes $v_j = 0$.

A limitation of this type of model for collaborative filtering is that it implies that each user can be described by a single component model – we are assuming that each user was 'generated' by one (and only one) of the K components. This is the same assumption made in the clustering approach discussed earlier, and indeed our mixture model is essentially a form of clustering with probabilistic semantics. Thus, as before with clustering, if a user's interests span multiple different types of topics (e.g. books on kayaking, Italian history, and blues music) there may be no single cluster that represents the combination of these three topics. However, there may well be mixture components that represent each of the individual topics, e.g. clusters of books for each of water sports, European history, and music.

Hofmann and Puzicha (1999) proposed an interesting extension of the conditional mixture model above that directly addresses the problem of multiple interests, based on a more general model proposed earlier by Hoffman (1999). Their model can be interpreted as allowing each user's interests to be generated by a superposition of K different underlying simpler component models. Thus, instead of assuming that each user's row vector of votes is generated by a single component model $P(v_1, \ldots, v_m \mid c = k)$, in the Hofmann and Puzicha model each row vector of votes can be generated by a combination of up to K of the component models. This is a potentially powerful idea in modeling high-dimensional data. To represent arbitrary combinations of K different interests we do not need 2^K different models but can instead combine K models appropriately.

Parameters of the Hofmann and Puzicha model can be fit to the observed data matrix V using the EM algorithm. In the original formulation of the model each user has his/her own set of parameters, leading to potential overfitting when using

EM in a standard ML framework. Another limitation of the original model is that it is not a complete probabilistic model in the usual sense: it only defines a distribution over users in the training data V and does not directly provide a mechanism for making predictions on a new user a who is not in the training data. More recent work by Blei *et al.* (2002a,b) extends the framework to address both of these problems, demonstrating a systematic improvement in prediction power over conditional independence mixture models, when the model is used as the basis for making recommendations on the widely used EachMovie ratings data set (available online at www.research.compaq.com/SRC/eachmovie/).

Conditional Distribution Models

An alternative to modeling the full joint density $P(v_1, \ldots, v_m)$ is to build m different conditional density models, $P(v_j \mid \mathscr{S} \setminus v_j)$, $1 \leqslant j \leqslant m$, where $\mathscr{S}$ is the full set of m random variables. The intuition behind this idea is that it may be more effective to directly predict the probability of each item j conditioned on the other items, rather than the less direct approach of modeling the full joint density and then using the joint density to infer predictions on how likely individual items are given information about other items.

One such approach proposed by Heckerman *et al.* (2000) is to use probabilistic decision trees (Buntine 1992) for each of the m conditional models. Trees are widely used in classification for high-dimensional problems, since they automatically perform variable selection and (in principle) only select the variables relevant to the prediction task at hand. They can also be used to approximate a high-dimensional conditional distribution for one variable v_j conditioned on many others. The tree is typically estimated from the data in a greedy fashion, by choosing a root node and recursively growing a binary tree beneath this node.

In Heckerman *et al.* (2000) each internal node corresponds to the addition of a single predictor binary variable v_k, where one branch corresponds to a particular value of v_k and the other branch corresponds to all other values of v_k. The node (variable plus split) that is greedily added to the current tree is the one that leads to the greatest increase in a Bayesian score. This score approximates the probability of a particular tree structure T given the data, $P(T \mid D) \propto P(D \mid T)P(T)$, where D is the data, $P(D \mid T)$ is the likelihood of the data under the current tree model T, and $P(T)$ is a suitably defined prior distribution over tree structures. If there are no variable-split combinations available that can be added to any node to increase the score then tree-growing halts. Conditional probability vectors for v_j are estimated at the leaves using only variable-value pairs in the branches from the root to each leaf. In this manner only variable-value pairs that are relevant to the prediction of v_j are included in the tree model, and thus, it is hoped that a relatively parsimonious tree structure can be found from the data that provides a useful approximation to the true conditional density.

The probability tree approach is relatively straightforward to apply to recommender data sets. m different probability trees are constructed (as described above) to predict the probability of each of the m different items in the data conditioned on the other items, using the vote matrix V. To make predictions for user a, if the votes are implicit (e.g. product purchases or Web page requests) we can use *all* of the other $m - 1$ implicit votes as inputs to predict the vote of interest. Predictions are made for each item that has an implicit vote of zero, i.e. that has not been purchased or visited, and then the resulting set of probabilities are ranked to provide a recommendation to the user a.

Heckerman *et al.* (2000) applied this technique to recommender data sets involving website visits (a visit to a particular page by a user is considered an implicit binary vote for that page, and items correspond to pages) and TV viewing records consisting of data collected by companies that estimate TV viewership – here the items correspond to programs watched by viewers. The tree models were compared to a method proposed earlier by Breese *et al.* (1998) that constructs a full Bayesian network as a model for the joint distribution $P(v_1, \ldots, v_m)$ and then uses the network to make predictions on each v_j that does not have a positive vote, using all the others. Breese *et al.* found that the Bayesian network model empirically outperformed memory-based and cluster-based collaborative filtering on different ratings data sets, and thus, it provided a useful comparative benchmark for the tree models.

The models were trained on training subsets and evaluated on disjoint out-of-sample test subsets. In the test data sets, each user's positive votes were randomly partitioned into two subsets of items: an *input set*, where the votes were assumed known and used as inputs to each model, and a *measurement set*, where the votes were assumed unknown and used to test the models' predictive power. Predictions were made for a variety of scenarios using this train/test arrangement. Under one scenario all but one of the positive votes for user a were placed in the input set of items and then used to predict the other item (this corresponds to knowing as much as possible about the user). Under the other scenarios only a fixed number $k \in 2, 5, 10$ of positive votes were put in the input set for prediction, so that smaller k values corresponded to less information about the user.

A particular type of objective function was constructed to estimate the utility of each ranked list of ratings to a user – a full description of this function is provided in Heckerman *et al.* (2000). For the Web and TV viewing data the probabilistic trees were generally slightly less accurate than Bayesian networks in terms of this objective function, but the differences were quite small.

Table 8.2 summarizes a variety of other aspects of the experiments, across the three data sets, for the specific scenario of 'all but one' prediction – results for the other scenarios were quite similar. The trees are significantly faster in terms of making predictions (e.g. 23.5 versus 3.9 per second on the second Web data set), which is an important feature for real-time prediction on websites. Presumably both algorithms were implemented efficiently so that these numbers provide fair comparisons between the two. The probabilistic trees also have advantages offline: they require less time and an order of magnitude less memory to train than the Bayesian network models. For both techniques it is quite impressive that models for high-dimensional data can be constructed so quickly, e.g. in about 100 s for the first Web data set based on 1000 dimensions (items) and 10 000 users.

Table 8.2 Summary of data sets and experimental results for 'all but one' prediction experiments from Heckerman *et al.* (2000). BN is the belief network model and PT is the probability tree model.

	Web data 1	Web data 2	TV data
Users in training data	10 000	32 711	1 637
Users in test data	5 000	32 711	1 637
Number of items	1 001	294	203
Mean positive votes per row	2.7	3.0	8.6
Predictions per second (BN)	7.1	3.9	23.5
Predictions per second (PT)	11.8	23.5	37.4
Training time [s] (BN)	105.8	144.6	7.7
Training time [s] (PT)	98.9	98.3	6.5
Training memory [MB] (BN)	43.0	42.4	3.3
Training memory [MB] (PT)	3.7	5.3	2.1

In practical settings, e.g. for a real-world e-commerce application, both the number of users and the number of items will often be significantly larger than those in the data sets described above (Schafer *et al.* 2001). Nonetheless, these experiments provide useful guidelines on some of the tradeoffs and options available when using model-based techniques for recommender systems.

8.3.4   Model-based combining of votes and content

Another variation on recommender systems is where we have additional content information about each item to be recommended. For example, for documents we usually know the terms that are in each document, and for movies we may have information about the type of movie, the actors, the rating of the movie, the director, and so forth. This type of content information can be used to estimate how similar certain items are to each other. For instance, we could represent the content information as a vector of values and use any of a variety of vector-based similarity measures. In principle, this type of content-based similarity alone could be used as the basis for a recommender system (Mooney and Roy 2000). For example, a 'user model' can be automatically constructed for each user based on the content descriptions of items purchased or rated, and then these user models are matched in terms of similarity to infer ratings and likely purchases for unseen items. Search engines can also be thought of as recommender systems that are purely based on content, where Web pages are recommended based on matches to a user query.

An advantage of content-based systems is that they can make recommendations for entirely new items for which there is no history, such as a new book or movie that no one has yet rated or purchased. Approaches based on historical patterns of ratings and purchases, such as collaborative filtering, cannot generalize in this manner to new items. On the other hand, systems that are only based on content are clearly ignoring potentially valuable information that lurks in historical rating and purchasing patterns of the community.

An example of a model-based approach that combines both content and collaborative filtering is that proposed in Popescul *et al.* (2001). Their approach is an extension of the hidden variable model of Hofmann and Puzicha (1999) discussed earlier in Section 8.3.3, where now the features of the items are included in the model as well votes from users. A specific model was developed for the case of users browsing documents at an online digital library (the NEC ResearchIndex document database), where items correspond to documents, 'content' consists of the words in each document, and a positive vote corresponds to user u requesting a particular document m. In this model, a joint density is constructed by assuming the existence of a hidden latent variable z that renders users u, documents m, and words w conditionally independent, i.e.

$$P(u, d, w) \approx \sum_{z} P(u \mid z) P(d \mid z) P(w \mid z) P(z). \tag{8.1}$$

As in the Hofmann and Puzicha (1999) approach, the hidden variable z represents different (hidden) topics for the documents, and multiple topics can be 'active' within a single document d or for a single user u. The inclusion of the term $P(w \mid z)$ allows the inclusion of content information in a natural manner. The EM algorithm can be used to estimate the conditional probability parameters that relate the hidden variable z to the observed data. Popescul *et al.* (2001) found that this particular model had problems with overfitting due to the sparsity of the data, even based on a relatively active set of 1000 users accessing 5000 documents, where the density of ones in the data matrix was 0.38% versus 0.01% for randomly selected users. To combat this overfitting, they also proposed a simpler model $P(u, w)$ based on content alone that in effect infers preferences in word-space, which is much more dense than document-space since there are far more words, and found empirically that this model produced better predictions than the original model. Thus, while relatively sophisticated models can be proposed for combining content and votes, fitting these models to sparse high-dimensional data remains a significant challenge.

8.4   Networks and Recommendations

8.4.1   Email-based product recommendations

While email-based recommending evokes images of being inundated with junk or 'spam' email messages, there are a number of situations where the concept works rather well in favor of the consumer. We are all familiar with the concept of 'word-of-mouth', where commercial products (such as movies) gain wide acceptance with very little explicit advertising. Instead, users of the product recommend the product to their circle of friends, who in turn recommend it to other friends, and so forth. For example, the Google search engine became (in 2002) the most widely used search engine on the Web, primarily based on 'word-of-mouth' recommendations and with very little direct advertising. From a marketing viewpoint this is an ideal way to gain market share without having to pay for expensive advertising campaigns.

Perhaps the most well-known and famous example of marketing by email is the case of Hotmail. The general idea was to provide free email service to users, with revenue for Hotmail being generated by selling various forms of advertising and additional fee-based services. The Hotmail free email service was introduced in July 1996 and had 20 000 subscribers by the end of the month. By September 1996 it had 100 000 subscribers, by January 1997 it had over one million subscribers, and 18 months after starting it had 12 million subscribers. By April 2002 the number of Hotmail subscribers (now part of Microsoft) was reported to be 110 million. What is particularly impressive about this rate of growth is the fact that Hotmail spent very little on conventional advertising during the initial growth period. Instead, they gained customers by embedding a link to the Hotmail sign-up page at the bottom of every email being sent out by their current subscribers. In a network-based system such as the Web this has a tremendous 'spreading activation' effect. In effect, compared to our discussion earlier in this chapter on collaborative filtering, the emails being sent by each individual can be thought of as an *implicit* recommendation of the Hotmail service, e.g. 'if my colleague is using this service perhaps I should look into it.'

Intuitively, we can imagine the Web as a very large dynamic network, with Hotmail emails emanating from an initially tiny fraction of nodes in the graph, e.g. the initial 20 000 Hotmail subscribers in July 1996. As Hotmail emails are sent from one individual to other individuals, each receiver will see the advertisement at the bottom of the email and some fraction (which need not be very large) will sign up for the service. At the next time-step (say the next day), there is now a larger set of Hotmail subscribers sending out emails, and thus, an even greater number of individuals receiving Hotmail emails. It is easy to see under fairly reasonable assumptions that large portions of the network will soon have received emails from someone who is a Hotmail subscriber, and some fraction of this large number will have signed up for the service.

This is precisely the type of 'spreading activation' mechanism that appears to have occurred with Hotmail. Even if only a very small fraction of individuals receiving Hotmail emails decided to sign up for the service (say 0.1% or less), because of the effective connectivity of email-based networks, and the number of individuals who send and receive email on a daily basis, there is an excellent chance that the product recommendations will 'spread like wildfire'.

The success of network-based recommendations relies of course on the assumption that the underlying product or service is something that is of a certain quality and of general benefit to consumers. For example, we can contrast the implicit recommendations from Hotmail subscribers with the now prevalent and unfortunate practice of 'spam' email, where emails that advertise products and services of dubious quality are broadcast to millions of email addresses. It is likely that most of these emails are not even read by the recipients and are almost never forwarded on to other colleagues.

Thus, there is clearly no network effect at play here but rather a brute force attempt to use the Web for direct advertising.

8.4.2 A diffusion model

Montgomery (2001) provides an interesting discussion of the Hotmail effect using *diffusion models* from the marketing literature – our discussion closely follows the treatment in this paper. These types of models were used by Bass (1969) to model the 'word of mouth' network effect in sales of consumer products such as televisions, refrigerators, and so forth. The model predicts how many individuals $k(t)$ at time t will adopt the product from a total population of N individuals.

The model is based on two simple assumptions.

(1) At time t, $N - k(t)$ individuals have not adopted the product. It is assumed that a constant proportion $\alpha \geqslant 0$ of these individuals will adopt the product at the next instant in time because of advertising. This part of the model represents the effect of direct advertising.

(2) The second part of the model represents the network effect. At time t there are $k(t)(N - k(t))$ potential links between pairs of individuals who have already adopted the product and those that have not. (For Hotmail this could be thought of as potential emails between pairs of individual email accounts.) It is assumed that the rate at which new individuals adopt the product is proportional to this number, with weighting coefficient $\beta \geqslant 0$.

From these two assumptions, the rate of change of $k(t)$ can be expressed as

$$\frac{\partial k(t)}{\partial t} = \alpha(N - k(t)) + \beta k(t)(N - k(t)), \tag{8.2}$$

which has a solution in the form

$$k(t) = N\left(\frac{1 - e^{-(\alpha+\beta N)t}}{1 + (\beta N/\alpha)e^{-(\alpha+\beta N)t}}\right) \tag{8.3}$$

that converges to N as $t \to \infty$.

Montgomery (2001) fit this model to the weekly Hotmail subscriber numbers for the first year of its operation (using data from Roberts and Mahesh (1999)), by estimating the parameters α, β, and N. The resulting estimates were $\alpha = 0.0012$, $\beta = 0.008$, and $N = 9.67$ (in millions of individuals) with t measured in weeks. It is noteworthy that the value for βN (which measures the network effect) is much larger than the coefficient α which measures the direct advertising effect.

This model of course has its limitations. It ignores effects such as individuals who stop using their Hotmail accounts (perhaps after initially trying it out), and the fact that the number of individuals using email in general is not constant but is growing as a function of time t. Nonetheless, the model provides quite a good fit to the observed weekly data (see Figure 8.1) as reported by Montgomery (2001), although it should be

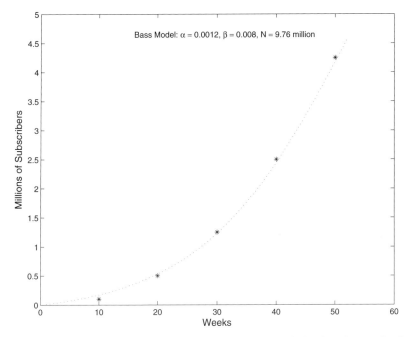

Figure 8.1 Fitted Bass distribution model for the first 52 weeks of adoption of Hotmail, where the curve represents the fitted model and the dots represent five of the 52 data points used to estimate the parameters of the model. (Adapted from Montgomery (2001).)

noted that these particular data could also be well-approximated by a simple quadratic or exponential for this initial growth period.

If we extrapolate the model to a later time-period we see that the diffusion model has a characteristic 'S'-shaped curve (Figure 8.2), that asymptotes at N, the estimated ultimate number of adopters. In Figure 8.2 we have added the number of Hotmail subscribers after 72 weeks (12 million) and after six years (110 million). We can see that the coefficients estimated using only the first 52 weeks of data do not extrapolate well. It is perhaps not surprising that a model fitted to the first year's worth of data do not extrapolate well to predicting what will happen five years later. Figure 8.2 also shows a different set of hand-chosen parameters for the diffusion model, where the asymptotic value of $N = 110$ is assumed known and the values of α and β are each reduced, resulting in a much better overall fit – of course in hindsight it is always somewhat easy to explain the data!

The diffusion model as presented above is certainly too simple to fully explain the success of Hotmail or other similar Web phenomena. Such models nonetheless serve the useful purpose of providing a starting point for generative modeling of network-based recommendations on the Web. In related work on this topic, Domingos and Richardson (2001) proposed a Markov random field model based on social networks for estimating an individual's *network value*, which depends on how much a particular individual can influence other individuals in the network. In a broader context, Daley

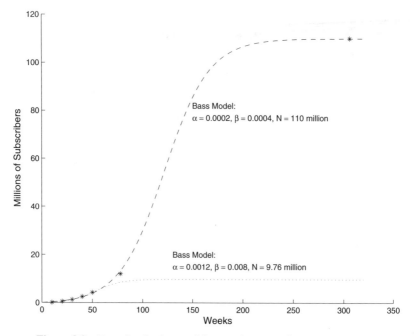

Figure 8.2   Bass distribution models over six years of Hotmail adoption.

and Gani (1999) provides a comprehensive overview of a variety of deterministic and stochastic models that have been investigated for characterizing the spread of epidemics and rumors in a population.

8.5   Web Path Analysis for Purchase Prediction

Analysis of individual navigation paths through e-commerce websites can provide valuable insights into customer behavior and provide clues about whether improvements in site design might be useful. From a marketing viewpoint it is often desirable to be able to predict early in a session whether an individual is likely to make a purchase or not. In this section we look at how an individual's browsing behavior can be related to purchase probability, using illustrative data from a study by Li *et al.* (2002).

When consumers search for products their behavior can be generally categorized into (1) a goal-directed search, or (2) an exploratory search (Janiszewski 1998). In a goal-directed search a consumer is typically looking for information about a specific product and is directly interested in purchasing that product. In exploratory search a consumer is 'just browsing' and may purchase any of a variety of products if he or she decides to do so on the spur of the moment, or may choose not to make any purchase at all. For consumers browsing an e-commerce website, this simple categorization seems quite reasonable. We might expect the goal-directed consumers

Table 8.3   Estimated transition probabilities among eight states from sessions where items were *purchased*. The value for row i, column j indicates the probability of going from a page in category i to a page in category j. The marginal probabilities (the probability of a page being in a particular category, as estimated from the data) are in the rightmost column.

	H	A	L	P	I	S	O	E	Marginal
H	0.21	0.03	0.36	0.00	0.30	0.07	0.00	0.03	0.01
A	0.01	0.67	0.01	0.00	0.04	0.25	0.01	0.01	0.13
L	0.01	0.00	0.63	0.21	0.08	0.06	0.00	0.01	0.27
P	0.01	0.01	0.32	0.44	0.15	0.06	0.00	0.01	0.16
I	0.02	0.05	0.11	0.12	0.56	0.06	0.03	0.05	0.23
S	0.01	0.19	0.05	0.02	0.13	0.45	0.14	0.01	0.15
O	0.00	0.01	0.00	0.00	0.71	0.12	0.00	0.17	0.03
E	0.50	0.00	0.00	0.00	0.50	0.00	0.00	0.00	0.02

to have different navigation patterns through a site than the consumers who are just browsing in an exploratory manner.

We next examine this hypothesis further using a data set collected by Li *et al.* (2002), consisting of the browsing patterns of 1160 different individuals who visited the online bookstore www.barnesandnoble.com between 1 April and 30 April 2002. The data were collected at the client side using the Comscore Media Metrix system that records the page requests and page-viewings of a randomly selected set of computer users. While relatively small in terms of the total number of individuals, this data set is nonetheless quite useful in terms of providing a general idea of the characteristics of online shopping behavior.

The 1160 individuals generated 1659 different sessions at the Barnes & Noble website. The sessions in total consisted of 9180 page requests and 14 512 page view-ings, where 'page viewings' count all pages viewed by the user and page requests only record requests that went to the server website (thus, pages that were viewed twice in a session and redisplayed by the caching software would be counted as page viewings, but not as page requests). The end of a session was declared if no page had been viewed for 20 min.

The mean session length (in terms of page viewings) was 8.75, the median was five, the standard deviation was 16.4, the minimum length was one, and the maximum was 570. 7% of sessions ended in purchases, a rate that is relatively high – purchase rates of between 1% and 2% have been reported in the media as being more typical for e-commerce sites in general (Tedeschi 2000).

Demographic data were also available for the 1160 individuals. They had an average age of 46, 53% were female, 77% were white, 40% had children, 29% were married, 82% had some college education, and 32% had an annual income in excess of $50 000. This suggests a relatively well-educated and affluent set of consumers.

One of the difficulties in analyzing Web navigation data is the very large number of possible pages that can be presented to a site visitor. In essence the number is unbounded, since in addition to there being a page for each product, the pages themselves can change dynamically (in terms of presentation, information, promotions, etc.) for different visitors and even for the same visitor at different times. To address this, Li *et al.* (2002) classified each page into one of seven different categories:

- home (H), the main home page and entry point for most visitors;

- account (A), containing account information for individuals;

- list (L), for pages with lists of items in certain categories or results of searches;

- product (P), with information on a specific product;

- information (I), with general information about shipping, order status, etc.;

- shopping cart (S), with current shopping cart information; and

- order (O), indicating confirmation of a purchase.

In addition an entry/exit category (E) was added to indicate the page prior to the first page on the site or the page after the last page visited on the site.

An example of an actual session where a purchase did not occur is IHHIIILIIE (taken from Li *et al.* (2002)). We see that the user visited several information pages, but did not visit any specific product page. An example of a session where a purchase did occur is IPPPPSASSSSSOIAAAHLLPLLLLLLE, where the user looks at specific product pages (P), visits the shopping cart (S), goes to the order confirmation page (O), then the account page (A), before finishing the session with some viewing of lists of information on category pages (L). For these two examples, the session with a purchase is much longer than that with no purchase. In fact this is true in general for this data set – the average length of sessions with purchases is 34.5 page views, versus only 6.8 page views for sessions with no purchases. While this is interesting retrospectively, it is not necessarily useful from a predictive viewpoint, since ideally we would like to be able to predict whether a visitor will make a purchase or not after only the first few page requests. One has to be careful here not to infer that there is necessarily a causal connection between the number of pages viewed and the probability of making a purchase. It might well be that any causal relationship (if it exists) is in the other direction, i.e. if a visitor is inclined to make a purchase then this might cause the visitor to view more pages than usual during this session.

From the data Li *et al.* (2002) estimated sample Markov transition matrices for the eight categories, both for sessions where purchases occurred, indicated by the presence of the state O for order confirmation in a session sequence, and for sessions where no purchases occurred. The estimated matrices are shown in Tables 8.3 and 8.4.

We note for example that the probability of visiting the home page (per pageview) is eight times higher at $P(H \mid \text{no purchase}) = 0.08$ for non-purchasers than it is for purchasers ($P(H \mid \text{purchase}) = 0.01$) indicating that non-purchasers spend much more time looking at general aspects of the website. We also see that the probability of exiting a session is also much higher for the non-purchasers than for the purchasers, $P(E) = 0.14$ versus $P(E) = 0.02$, respectively, which translates into much longer sessions on average for the purchasers, as noted earlier. Transition probabilities into

Table 8.4   Estimated transition probabilities among eight states from sessions where items were *not purchased*. The value for row *i*, column *j* indicates the probability of going from a page in category *i* to a page in category *j*. The marginal probabilities (the probability of a page being in a particular category, as estimated from the data) are in the rightmost column.

	H	A	L	P	I	S	O	E	Marginal
H	0.36	0.01	0.16	0.01	0.25	0.01	0.00	0.19	0.08
A	0.03	0.68	0.04	0.01	0.08	0.06	0.00	0.11	0.04
L	0.02	0.01	0.59	0.14	0.12	0.01	0.00	0.11	0.30
P	0.03	0.01	0.25	0.39	0.19	0.01	0.00	0.12	0.14
I	0.08	0.01	0.14	0.08	0.53	0.02	0.00	0.15	0.28
S	0.06	0.13	0.03	0.06	0.33	0.32	0.00	0.07	0.02
O	0.00	0.00	0.00	0.00	0.00	0.00	0.00	0.00	0.00
E	0.25	0.06	0.19	0.00	0.50	0.00	0.00	0.00	0.14

the Shopping Cart (S) and Account (A) states are much higher in the purchase group than in the non-purchase group, and of course the transition probabilities into the Order (O) state are all zero for the non-purchase group.

Li *et al.* fitted a series of relatively sophisticated statistical models to this data, the details of which are somewhat beyond the scope of this text. The most complex models included several components:

(1) latent variables that modelled the utility of individual *i* selecting a page from category *c* at time *t*;

(2) time-dependence via a hidden Markov model that allows switching between states over time;

(3) covariates based on both properties of the pages and individual demographics; and

(4) heterogeneity across different individuals using a hierarchical Bayes approach, as briefly discussed in Chapter 7.

The models were fitted using a widely used stochastic sampling technique for Bayesian estimation known as the Monte Carlo Markov chain. The different models were then evaluated in terms of their ability to provide out-of-sample predictions of both

(a) the next page category requested by an individual given the sequence of pages requested by that individual up to that time-point and given the individual's demographic data; and

(b) whether or not that person would make a purchase or not (given the same information).

In terms of predicting the next category, the results indicated that Markov models with memory were significantly more accurate than models that did not have memory, with the best memory-based models achieving 64% accuracy in predicting the next category (out of eight) versus only about 20% for non-memory models. It was also found that hidden Markov models with two states consistently had slightly better accuracy out-of-sample than models with only a single state, although it should be cautioned that this was based on only a single train/test split of the available data. The authors interpreted one of the states as being an 'exploratory browsing state' and the other as being 'purchase-oriented', lending some support to the hypothesis earlier about a simple two-state model for consumer behavior in an e-commerce context.

Demographics did not appear to have significant predictive power in the models, and browsing behavior seemed to be relatively independent of the duration of page views according to the model. Interestingly, the composition of the page appeared to have an influence on browsing behavior. Having prices on a page seemed to discourage visitors in the browsing-oriented state from continuing to browse, while the same information had a positive effect on visitors in the purchase-oriented state. Conversely, promotional items such as advertised discounts had the reverse effect, encouraging visitors in the browsing state to continue browsing, but discouraging users in the purchasing state from continuing in that state. Naturally, we need to keep in mind that all such interpretations of the data are being filtered through a particular model and as such may reflect artifacts of the model-fitting process and the random sample at hand rather than necessarily reflecting a true property of the processes generating the data. Nonetheless, these types of inferences are quite suggestive and lead naturally to hypotheses about consumer behavior that could be properly tested in further experimental studies.

In terms of predicting whether a visitor makes a purchase or not, the models fitted by Li *et al.* were somewhat less accurate than they were in predicting the next category. Recall that 7% of all sessions in this data set result in a purchase. The two-state model was used to predict whether a visitor would make a purchase or not, based on the first k page views, for site visitors whose sessions were not used in building the model (but whose demographic data were assumed known). After two page views, the model predicted that 12% of the true purchasers would purchase a product and only 5.3% of the true non-purchasers. After six page views, the model predicted that 13.1% of the true purchasers would purchase a product and made the same prediction for 2.9% of the true non-purchasers. Note that a perfect forecasting system would predict a purchase for 100% of the purchasers and 0% of the non-purchasers. More-accurate forecasts can be made as we receive more information – six page views are more accurate than two page views – but overall there is significant room for improvement, since some 87% of true purchasers went undetected by the model after six page views.

8.6 Exercises

Exercise 8.1. Consider a simple toy model that we could use to simulate a large sparse binary data set with n rows (users) and m columns (items). The model operates in the following way.

(1) Each user acts independently of all other users, using the same model (below).

(2) For a given user i, he or she considers each of the m items in turn, and independently makes a decision with probability p as to whether to purchase the item or not. If they decide to purchase the item then they enter a '1' for that item, otherwise a '0'.

(3) We set $p = k/d$, where k is a parameter of the model and represents the expected number of items purchased by each user. Typically k is much smaller than m, e.g. $k = 4, m = 1000$.

Calculate the probability, as a function of k, m, and n, that a new user acting under the same model will have no items at all in common with any of the n users in an $n \times m$ data matrix generated under the model above. Plot this probability for different values of k, m, and n, and comment on the results.

Exercise 8.2. An extension of the model in Exercise 8.1 allows each of the items to have a different probability p_j, $1 \leqslant j \leqslant m$, of being purchased. A further extension allows limited dependencies among items, e.g. the probability of a certain item j being purchased depends on whether another item i was purchased or not, represented by conditional probabilities $P(j \mid i)$. Clearly the most complex model we could consider in terms of item dependencies would be a full joint density on all m items.

How many parameters are needed to specify this full joint density? How many are required for the independence model in Exercise 8.1? Clearly the independence model is too simple, but the full joint density is impractical once m becomes large. Describe as many types of probabilistic models you can think of that are 'in between' the independence model and the full joint density model (e.g. a Markov model where each item depends on other item, in a specified order, would be one such model).

Exercise 8.3. A different probabilistic model than the independence model in Exercise 8.1 for generating sparse binary data operates as follows. Each user still generates data independently. To generate a row of data, the ith user now tosses an m-sided die some number of times. The number of tosses could also be drawn from a random distribution, or could be fixed for each user. The probability of the m-sided die coming up on side j is represented by p_j, $1 \leqslant j \leqslant m$, where $\sum p_j = 1$, and for simplicity we could assume that all p_j are equal, i.e. $p_j = 1/m$. If the die comes up on side j, the user puts a '1' in column j, otherwise a '0'. This type of data generation model is sometimes referred to as a *multinomial model* as discussed in Chapter 4.

Compare this model with the independence model in Exercise 8.1, in terms of the type of data that will be generated. For example, assume that in our model here that the die is tossed exactly k times for each user. How does the probability distribution of number of ones per row (we can think of this as the distribution of basket sizes) under the 'die model' compare to the probability distribution of basket sizes under the independence model in Exercise 8.1? (Hint: reviewing the material on Markov chains in Chapter 1 may help.)

You can also consider the problem of trying to model two items that have an exclusive-OR relationship: the user is likely to purchase one or the other, but not both together (e.g. Pepsi and Coke). Which model is more appropriate for capturing such effects and how would you specify such a relationship between items in such a model?

Exercise 8.4. Verify that the solution to the Bass diffusion equation as presented in Equation (8.3) is correct.

Exercise 8.5. Suggest an extension to the Bass diffusion equation that has a 'death' component to the model, i.e. after a certain period of time users grow tired of the product and stop using it. How might you capture this in an equation for the derivative of $k(t)$? Simulate data from your proposed model using selected parameter values to illustrate the 'death' effect and plot the results, i.e. the values of $k(t)$ over time.

Exercise 8.6. Consider the Markov chain represented by the transition probability matrix in Table 8.3. Assume that the probability that an individual starts at any of the seven categories (other than E) is proportional to the marginal probability for that category. The marginal probabilities are in the rightmost column in the Table. Derive in closed form the probability distribution on the lengths of sessions under this Markov model, where the length of a session is defined as all pages viewed before the category E is viewed. (Hint: you do not need to work with the full 8×8 transition matrix to find a solution. The material on Markov chains in Chapter 1 may help.)

We mentioned earlier in the chapter that the empirically observed mean session length for the 'purchase' group (as represented by the Markov chain in Table 8.3) was 34.5 page views. Compare this with the mean of the analytical distribution that you derived in the first part of this exercise and explain why the two numbers are not the same.

Appendix A

Mathematical Complements

In this appendix, we first begin with a short and rather informal introduction to some of the basic concepts and definitions of graph theory used primarily in Chapter 3 but also in the definition of Bayesian networks in Chapter 1. We then review a number of useful distributions that are used in various sections of this book, and provide a short introduction to Singular Value Decomposition (SVD), Markov chains, and the basic concepts of information theory. The list of distributions that are routinely encountered in probabilistic models and analysis of real world phenomena is not meant to be exhaustive in any way. In fact, a more detailed list can be found, for instance, in Gelman *et al.* (1995).

A.1 Graph Theory

A.1.1 Basic definitions

A graph $G = (V, E)$ represents a binary relationship and is defined by a set of vertices (or nodes) V and a set of edges E consisting of unordered pairs (v, w) of vertices. Depending on the problem, loops of the form (v, v) may or may not be allowed. Likewise, multiple edges between two nodes may or may not be allowed in different situations. With $|V| = n$ vertices numbered from 1 to n, the graph can be represented by an $n \times n$ 0-1 adjacency (or incidence) matrix A, where $a_{ij} = 1$ if and only if there is an edge between vertices i and j. If multiple edges between vertices are allowed, this can be reflected in the matrix A by allowing nonzero integer entries reflecting the multiplicity of the corresponding edge. The degree $d(v)$ of a vertex is the number of edges adjacent to that vertex. In a directed graph $G = (V, E)$ the edges are directed, or equivalently the corresponding pairs are ordered. The adjacency matrix can be defined in a similar way and, unlike what happens in the directed case, in the directed case the matrix is usually not symmetric. In a directed graph we can distinguish the indegree $d_{in}(v)$ and outdegree $d_{out}(v)$ of a vertex, by considering the number incoming and outgoing directed edges, respectively.

Modeling the Internet and the Web   P. Baldi, P. Frasconi and P. Smyth
© 2003 P. Baldi, P. Frasconi and P. Smyth   ISBN: 0-470-84906-1

A.1.2   Connectivity

A path of length k in a graph is an alternating sequence of edges and vertices $v_0, e_1, v_1, \ldots, e_k, v_k$ with each edge adjacent to the vertices immediately preceding and following it. A directed path is defined in the obvious way by adding the requirement that all edges be oriented in the same direction. The shortest directed or undirected path between any two vertices in a graph can be found using standard dynamic programming/breadth first methods (Cormen *et al.* 2001; Dijkstra 1959; Viterbi 1967). The distance between two vertices can be measured by the length of the shortest path. A loop is a path where $v_0 = v_k$ and a cycle is a directed loop. A directed acyclic graph, or DAG, is a directed graph which contains no directed cycles. A (strongly) connected component of a (directed) graph is a maximal set of vertices such that there is a (directed) path joining any two nodes. A graph is connected if it has only one connected component. The diameter of a component is the maximal length of the shortest path between any vertices in the component. The diameter of a graph is the largest diameter of its connected components. A clique of G is a fully connected subgraph of G (in some definitions, clique are also required to be *maximal* complete subgraphs). The mathematical definition of diameter corresponds to a 'worst-case' definition. For practical purposes, it is often more informative to know what is the average distance between any two vertices rather than the maximal distance. This is also called average diameter or average distance in the literature and throughout the book. Finally, it is easy to see how these connectivity notions can be modified and adapted as needed in the context of directed graphs.

A.1.3   Random graphs

In Chapter 3, we consider graphs that can be associated with the Internet and the Web, in particular the Web graph where the vertices are Web pages and the (directed) edges are hyperlinks, as well as several classes of other related graphs. To the extent that these graphs and their evolution have a stochastic component, it is useful to compare and contrast them with various models of random graphs. Several models of random graphs exist but, as a basic starting point, the reader should be familiar with two classical models – the uniform random graph models. The first model $G(n, e)$ (Bollobás 1985; Erdos and Rényi 1959, 1960) is associated with a uniform distribution over all graphs with n vertices and e edges. In the second model $G(n, p)$ (Bollobás 1985; Gilbert 1959), each possible edge occurs with a probability p, independently of all other edges. It is well known that for large n and $e = pn(n - 1)/2$ that these two models have very similar properties. In particular, the vertex degrees have approximately a Normal or Poisson distribution with mean $2e/n$. The classical theory of random graphs deals, in particular, with the asymptotic properties of such classes of random graphs, for instance, for large n as p is varied. For many properties, it has been observed and proved that there exist threshold functions (or phase transitions), i.e. values of $p = p(n)$ around which the asymptotic properties vary abruptly. For instance, it is well known that the connectivity of random uniform graphs undergoes a phase transition around $p = \log n/n$, so that the graph tends to be disconnected below the threshold value and has a 'giant' connected component above the threshold (see, for instance, Bollobás 1985). For reasons described in Chapter 3, the random uniform graphs sometimes are also called random exponential graphs, due to the exponentially decaying tail of their degree distribution.

A.2   Distributions

A.2.1   Expectation, variance, and covariance

For completeness, we provide a brief reminder on the definition of expectation, variance, and covariance. Let $P(X)$ be a probability density function for a real-valued variable X. The expected value of X is defined as

$$E[X] = \int_{-\infty}^{\infty} x P(x)\, dx$$

and the variance is defined as

$$\text{var}[X] = \int_{-\infty}^{\infty} (x - E[X])^2 P(x)\, dx = E[(X - E[X])^2] = E[X^2] - (E[X])^2.$$

The covariance of two random variables X_i and X_j is defined by

$$\text{cov}[X_i, X_j] = E[(X_i - E[X_i])(X_j - E[X_j])].$$

It should be clear how these expressions can also be applied to discrete random variables, or to vector random variables, on a component-by-component basis.

A.2.2   Discrete distributions

Binomial and multinomial distributions

The binomial distribution counts the number of heads or successes in a sequence of n independent and identically distributed (i.i.d.) coin flips (Bernouilli trials). If p is the probability of a single success, the probability of having k successes is given by

$$P(X = k \mid p, n) = \binom{n}{k} p^k (1 - p)^{n-k}. \tag{A.1}$$

The mean is $E[X] = np$ and the variance $\text{var}[X] = np(1 - p)$.

   The multinomial distribution is the generalization of the binomial distribution, when each trial has m different possible outcomes. Thus instead of a coin, we have a die with m faces, with probabilities $p_1, \ldots, p_m$. In this case, the probability of observing each one of the m categories $k_1, \ldots, k_m$ times in n trials, with $\sum_i k_i = n$, is given by

$$P(X_1 = k_1, \ldots, X_m = k_m) = \frac{n!}{k_1! \ldots k_m!} p_1^{k_1} \ldots p_m^{k_m}. \tag{A.2}$$

The mean of each component is $E[X_i] = np_i$, the variance $\text{var}[X_i] = np_i(1 - p_i)$, and the covariance $\text{cov}[X_i, X_j] = -p_i p_j$ for $i \neq j$.

Poisson distribution

The Poisson distribution corresponds to rare events. The Poisson distribution with parameter λ is given by

$$P(X = k \mid \lambda) = e^{-\lambda} \frac{\lambda^k}{k!}. \tag{A.3}$$

The mean and the variance are equal to λ. When p is small, the binomial distribution can be approximated by the Poisson distribution.

Geometric distribution

The geometric density with parameter p $(0 < p < 1)$ is described by

$$P(X = k) = p(1 - p)^k, \tag{A.4}$$

for $k = 0, 1, 2, \ldots$. In the framework of Bernouilli trials, this is the probability that heads appears for the first time on the kth trial. The geometric density has mean $E[X] = 1/p$ and variance $\text{var}[X] = (1 - p)/p^2$. It is a special case of the negative binomial density.

A.2.3   Continuous distributions

Gaussian distribution

The Gaussian or normal density is probably the most important continuous density. A one-dimensional Gaussian density has the form

$$\mathcal{N}(x \mid \mu, \sigma) = \frac{1}{\sqrt{2\pi}\sigma} e^{-(x-\mu)^2/2\sigma^2}, \tag{A.5}$$

where the parameters μ and σ are the mean and the standard deviation. The predominance and ubiquity of the Gaussian density has multiple origins. For instance, it can be shown that the sum of a large number of i.i.d. random variables approaches a Gaussian random variable. This is the well known Central Limit Theorem. The binomial distribution can be approximated by a Gaussian distribution when n is large. The Gaussian density is also its own conjugate in models parameterized only by the mean. Furthermore, when the only information available about a continuous density is its mean μ and its variance σ^2, then the Gaussian density $\mathcal{N}(\mu, \sigma)$ is the one achieving maximal entropy (Cover and Thomas 1991). Finally, if $\log X$ is normally distributed, then X is said to have a log-normal distribution. In particular if $\log X$ is normally distributed with mean μ and variance σ^2, then the density of X is

$$f(x \mid \mu, \sigma^2) = \frac{1}{\sqrt{2\pi}\sigma} \frac{1}{x} e^{-(\log x - \mu)^2/(2\sigma^2)} \tag{A.6}$$

for $x > 0$. The mean satisfies $E[X] = e^{\mu + (\sigma^2/2)}$ and the variance

$$\text{var}[X] = e^{2(\mu + \sigma^2)} - e^{2\mu + \sigma^2}.$$

Because the sum of two independent normal distributions is normal, the *product* of two independent log-normal distributions is log-normal.

Exponential distribution

The exponential density with parameter λ is given by

$$f(x \mid \lambda) = \lambda e^{-\lambda x} \tag{A.7}$$

for $x \geq 0$. The mean is $E[X] = 1/\lambda$ and the variance $\text{var}[X] = 1/\lambda^2$. The exponential distribution is a special case of Gamma distribution.

Gamma distribution

The gamma density (Feller 1971) with parameters α and λ is given by

$$\Gamma(x \mid \alpha, \lambda) = \frac{\lambda^\alpha}{\Gamma(\alpha)} x^{\alpha-1} e^{-\lambda x} \tag{A.8}$$

for $x > 0$, and zero otherwise. $\Gamma(\alpha)$ is the Gamma function $\Gamma(\alpha) = \int_0^\infty e^{-x} x^{\alpha-1}\, dx$. The mean is $E[X] = \alpha/\lambda$ and the variance $\text{var}[X] = \alpha/\lambda^2$. If X has a Gaussian distribution with mean zero and variance σ^2, then X^2 has a Gamma density with parameters $\lambda = 1/2$ and $\beta = 1/2\sigma^2$. The exponential density with parameter λ is the Gamma density with parameters $\alpha = 1$ and λ. The Chi-square density with k degrees of freedom, for instance, is the Gamma density with parameters $\alpha = k/2$ and $\lambda = 1/2$.

Dirichlet distribution

Finally, in the context of multinomial distributions that play an important role in this book, an important class of distributions are the Dirichlet distributions (Berger 1985). By definition, a Dirichlet density on the probability vector $\boldsymbol{p} = (p_1, \ldots, p_m)$, with parameters α and $\boldsymbol{q} = (q_1, \ldots, q_m)$, has the form

$$D_{\alpha q}(\boldsymbol{p}) = \frac{\Gamma(\alpha)}{\prod_i \Gamma(\alpha q_i)} \prod_{i=1}^m p_i^{\alpha q_i - 1} = \prod_{i=1}^m \frac{p_i^{\alpha q_i - 1}}{Z(i)}, \tag{A.9}$$

with $\alpha, p_i, q_i \geq 0$ and $\sum p_i = \sum q_i = 1$. For such a Dirichlet distribution, $E(p_i) = q_i$, $\text{var}[p_i] = q_i(1 - q_i)/(\alpha + 1)$, and $\text{cov}[p_i p_j] = -q_i q_j/(\alpha + 1)$. Thus $\boldsymbol{q}$ is the mean of the distribution, and α determines how peaked the distribution is around its mean. Alternatively, the distribution can be parameterized using $\boldsymbol{\alpha} = (\alpha_1, \ldots, \alpha_m)$ with $\alpha_i = \alpha q_i$. One reason Dirichlet distributions are important is that they are the natural conjugate prior for multinomial distributions, as seen in Chapter 1. In the two-dimensional case ($m = 2$), the Dirichlet distribution is also called the Beta distribution.

A.2.4   Weibull distribution

The Weibull distribution is used in Chapter 6 in connection with the life span of Web pages. More generally, it is used to model reliability and the distribution of life spans in a wide range of applications ranging from mechanical devices to biological organisms. The Weibull density function is defined by

$$f(x \mid \alpha, \beta) = \frac{\beta}{\alpha} \left(\frac{x}{\alpha} \right)^{\beta - 1} e^{-(x/\alpha)^{\beta}}, \tag{A.10}$$

where $\alpha > 0$ is a scale parameter, $\beta > 0$ is a shape parameter, and $x > 0$. The mean (also called MTBF, mean time between failures) is given by

$$E[X] = \alpha \Gamma \left(1 + \frac{1}{\beta} \right), \qquad \mathrm{var}[X] = \alpha \left(\Gamma \left(1 + \frac{2}{\beta} \right) - \Gamma \left(1 + \frac{1}{\beta} \right)^2 \right).$$

The cumulative distribution function is simply

$$F(x \mid \alpha, \beta) = 1 - e^{-(x/\alpha)^{\beta}}. \tag{A.11}$$

A.2.5   Exponential family

The *exponential family* (Brown 1986) is the most important family of probability distributions. It has a wide range of applications and unique computational properties: many fast algorithms for data analysis have some version of the exponential family at their core. Many general theorems in statistics can be proved for this particular family of parameterized distributions. The density in an exponential family with parameter $\boldsymbol{\theta}$ must have the form

$$f(x \mid \boldsymbol{\theta}) = h(x) c(\boldsymbol{\theta}) \exp \left(\sum_{i=1}^{m} w_i(\boldsymbol{\theta}) t_i(x) \right), \tag{A.12}$$

where $h(x) \geqslant 0$, $t_i(x)$ are real-valued functions of the observation x that do not depend on $\boldsymbol{\theta}$, and $c(\boldsymbol{\theta})$ and $w_i(\boldsymbol{\theta})$ are real-valued functions of the parameter vector $\boldsymbol{\theta}$ that do not depend on x.

Most common distributions belong to the exponential family, including the normal (with either mean or variance fixed), gamma (e.g. Chi square and exponential), Dirichlet (e.g. Beta), in the continuous case, and binomial and multinomial, geometric, negative binomial, and Poisson distributions in the discrete case. A characteristic of all exponential distributions is the exponential decay to zero for large values of x. This is in contrast with the polynomial decay of power-law distributions studied in Chapter 1 and often encountered in the other chapters of this book. Among the important general properties of the exponential family is the fact that a random sample from a distribution in the one-parameter exponential family always has a sufficient statistic S. Furthermore, the sufficient statistic itself has a distribution that belongs to the exponential family.

A.2.6   Extreme value distribution

Finally, the extreme value distribution is an important distribution when considering extreme effects and the distribution of the maximum of a sample of n points sampled from a density $g(x)$. If $G(x) = P(X \leqslant x) = \int_{-\infty}^{x} g(v)\, dv$ is the distribution function, then the distribution of the maximum in the sample is given by $F(x) = P(\max \leqslant x) = G(x)^n$ with corresponding density $f(x) = nG^{n-1}(x)g(x)$. The EVD (extreme valued density) is the limit of $f(x)$ when $n \to \infty$. The EVD distribution has many applications to the study of many extreme events ranging from catastrophes, to assessing the significance of a maximum score in data mining and information retrieval.

In the case where g is the exponential density with $g(x) = \lambda e^{-\lambda x}$ straightforward integration gives $G(x) = 1 - e^{-\lambda x}$. If we define $z = x - y$ with $n = e^{\lambda y}$ a simple calculation shows that

$$nG(x)^{n-1}g(x) \to \lambda e^{-\lambda z} \exp(-e^{-\lambda z}) \qquad (A.13)$$

when $n \to \infty$. This yields the distribution $P(\max \leqslant x) = \exp(-e^{-\lambda z})$, which is called the Gumbel distribution (Galambos 1987). A remarkable fact is that the Gumbel distribution is the EVD for many other densities besides the exponential density, including the case where g is Gaussian.

A.3   Singular Value Decomposition

It is a standard fact in linear algebra that any $m \times n$ real matrix X can be written as

$$X = USV^{\mathrm{T}}, \qquad (A.14)$$

where

- S is a diagonal $m \times n$ matrix having the form

$$S = \begin{cases} \begin{bmatrix} \Sigma \\ 0 \end{bmatrix} & \text{if } m \geqslant n, \\[4mm] \begin{bmatrix} \Sigma & 0 \end{bmatrix} & \text{if } m < n, \end{cases}$$

being $\Sigma = \mathrm{diag}(\sigma_1, \ldots, \sigma_{\min(m,n)})$, and σ_i are called the *singular values* of X;

- U is an $m \times m$ matrix such that $U^T U = I_m$ whose columns are called the left singular values of X;

- V is an $n \times n$ matrix such that $V^T V = I_n$ whose columns are called the right singular values of X.

Without loss of generality we may assume $\sigma_1 \geqslant \sigma_2 \geqslant \cdots \geqslant \sigma_{\min(m,n)}$. If

$$r = \text{rank}(X) < \min(m, n),$$

then only the first r singular values are nonzero.

SVD is often studied in conjunction with multivariate regression problems $y = f(x)$, $y \in \mathbb{R}$, $x \in \mathbb{R}^n$. In this case, assuming a linear model for f: $y = f(x) = x^T w$, the vector of parameters w can be estimated from a data set of m observations $\{(x(i), y(i)), \ i = 1, \ldots, m\}$ using the least squares method

$$\min_{w} \tfrac{1}{2} \| X w - y \|^2, \tag{A.15}$$

where $y = [y(1), \ldots, y(m)]^T$ and $X = [x(1), \ldots, x(m)]^T$.

The solution of (A.15) is

$$\hat{w} = (X^T X)^{-1} X^T y, \tag{A.16}$$

which can be rewritten as

$$\hat{w} = V S^{-1} U^T y \tag{A.17}$$

after plugging in Equation (A.14).

Another interesting interpretation of the SVD is in terms of *principal component analysis* (PCA) or Karhunen–Loève transform. If X represents a data set as in the least squares problem, then $X^T X / n$ is the sample covariance matrix and it is proportional to

$$X^T X = V S^T S V^T. \tag{A.18}$$

The eigenvectors v_i are known as principal components directions of X. The vector $u_1 \sigma_1$ is the principal component of X and satisfies

$$u_1 \sigma_1 = \arg \max_{\|w\|=1} E_x[(w^T x)^2], \tag{A.19}$$

i.e. it is the projection along the direction having maximum variance, which equals to σ_1^2 / n. Recursively, the kth principal component maximizes the projection variance subject to the constraint that is must be orthogonal to the previous $k - 1$ principal components. PCA is often used in statistics and machine learning for reducing the dimensionality of data. In this case, only the coordinates of data points with respect to the first $K < n$ principal components are conserved.

A.4   Markov Chains

A finite-state discrete-time Markov chain consists of a set of random variables $X = \{X_t, t \in \{0, 1, 2, \ldots\}\}$. The random variable at any time t, X_t, takes values in a finite set which we can index as $\{1, \ldots, M\}$ without loss of generality. The key property of a Markov chain is that the probability of the next state X_{t+1}, given the values of all previous states $X_0, \ldots, X_t$, can be written as

$$P(X_{t+1} = j \mid X_0 = i_0, \ldots, X_{t-1} = i_{n-1}, X_t = i) = P(X_{t+1} = j \mid X_t = i).$$

This is the standard 'first-order' Markov assumption: the probability of the future is independent of the past given the current state.

If the probabilities $P(X_{t+1} = j \mid X_t = i)$ are independent of time t, then we say that the Markov chain is *stationary* and denote the corresponding set of probabilities as P_{ij}, $1 \leqslant i, j \leqslant M$. The set of probabilities P_{ij} can be conveniently described by an $M \times M$ 'transition' matrix T, where the rows sum to unity, i.e.

$$\sum_j P_{ij} = 1.$$

For example, a three-state stationary Markov chain could have the transition matrix:

$$T = \begin{pmatrix} 0.7 & 0.2 & 0.1 \\ 0.1 & 0.8 & 0.1 \\ 0.2 & 0.3 & 0.5 \end{pmatrix}. \tag{A.20}$$

A.5   Information Theory

The concept of information (Aczel and Daroczy 1975; Chaitin 1987; Cover and Thomas 1991) is central to science, technology, and many other human endeavors, hence to the Internet and the World Wide Web. Information ought to play also a fundamental role in connection with the Web. Our starting point in this section is the definition provided by Claude Shannon over half a century ago (Shannon 1948a,b), which is at the center of modern telecommunication systems. As we shall see, however, the concept of information is complex and multi-faceted. Shannon's definition does not capture all aspects of information and, for Web applications in particular, there is room for refinements.

Before we review Shannon's definition of information, we must first review three mathematical concepts that are essential in information theory and also other fields such as statistical mechanics, and which are used in several chapters of this book. These are the entropy, the mutual information, and the relative entropy. These concepts are essential for the study of how information is transformed through a variety of operations such as information coding, transmission, and compression. The relative entropy is the most general concept, from which the other two can be derived. However, here we begin with the slightly simpler concept of entropy.

A.5.1 Mathematical background

Entropy

The three most basic concepts and measures of information are the entropy, the mutual information, and the relative entropy (Blahut 1987; Cover and Thomas 1991; Shannon 1948a,b). The entropy $\mathcal{H}(\boldsymbol{p})$ of a probability vector $\boldsymbol{p} = (p_1, \ldots, p_n)$ is defined by

$$\mathcal{H}(\boldsymbol{p}) = E(-\log \boldsymbol{p}) = -\sum_{i=1}^{n} p_i \log p_i. \tag{A.21}$$

The units used to measure entropy depend on the base used for the logarithms. When the base is two, the entropy is measured in bits. The entropy measures the prior uncertainty in the outcome of a random experiment described by $\boldsymbol{p}$, or the information gained when the outcome is observed. It is also the minimum average number of bits (when the logarithms are taken base 2) needed to transmit the outcome in the absence of noise. The corresponding concept in the case of a continuous random variable X with density $p(x)$ is called the differential entropy

$$\mathcal{H}(X) = -\int_{-\infty}^{+\infty} p(x) \log p(x) \, \mathrm{d}x \tag{A.22}$$

and likewise when x is vector valued.

Of all the densities with variance σ^2, the Gaussian $\mathcal{N}(\mu, \sigma)$ is the one with the largest differential entropy. The differential entropy of a Gaussian distribution with any mean and variance σ^2 is given by $[\log 2\pi e \sigma^2]/2$. More generally, in n dimensions, consider a random vector X, with vector mean μ, covariance matrix C, and density p. Then the differential entropy of p satisfies

$$\mathcal{H}(p) \leqslant \tfrac{1}{2} \log(2\pi e)^n |C| = \mathcal{H}(\mathcal{N}(\mu, C)) \tag{A.23}$$

with equality if and only if X is distributed according to $\mathcal{N}(\mu, C)$ almost everywhere. Here $|C|$ denotes the determinant of C.

The concept of entropy can be derived axiomatically. Indeed, consider a random variable X that can assume the values $x_1, \ldots, x_n$ with probabilities $p_1, \ldots, p_n$. The goal is to define a quantity $\mathcal{H}(\boldsymbol{p}) = \mathcal{H}(X) = \mathcal{H}(p_1, \ldots, p_n)$ that measures, in a unique way, the amount of uncertainty represented in this distribution. It is a remarkable fact that three commonsense axioms, really amounting to only one composition law, are sufficient to determine $\mathcal{H}$ uniquely, up to a constant factor corresponding to a choice of scale. The three axioms are as follows.

(1) $\mathcal{H}$ is a continuous function of the p_i.

(2) If all p_i are equal, then $\mathcal{H}(\boldsymbol{p}) = \mathcal{H}(n) = \mathcal{H}(1/n, \ldots, 1/n)$ is a monotonic increasing function of n.

(3) Composition law: partition all the events x_i into k disjoint sets. Let A_i represent the indices of the events associated with the ith set, so that $q_i = \sum_{j \in A_i} p_j$

represents the corresponding probability. Then

$$\mathcal{H}(\boldsymbol{p}) = \mathcal{H}(\boldsymbol{q}) + \sum_{i=1}^{k} q_i \, \mathcal{H}\left(\frac{\bar{\boldsymbol{p}}_i}{q_i}\right), \tag{A.24}$$

where $\bar{\boldsymbol{p}}_i$ denotes the set of probabilities p_j for $j \in A_i$. Thus, for example, the composition law states that, by grouping the first two events into one,

$$\mathcal{H}(\tfrac{1}{3}, \tfrac{1}{6}, \tfrac{1}{2}) = \mathcal{H}(\tfrac{1}{2}, \tfrac{1}{2}) + \tfrac{1}{2}\mathcal{H}(\tfrac{2}{3}, \tfrac{1}{3}) + \tfrac{1}{2}\mathcal{H}(1). \tag{A.25}$$

From the first condition, it is sufficient to determine $\mathcal{H}$ for all rational cases where $p_i = n_i/n$, $i = 1, \dots, n$. But from the second and third conditions,

$$\mathcal{H}\left(\sum_{i=1}^{n} n_i\right) = \mathcal{H}(p_1, \dots, p_n) + \sum_{i=1}^{n} p_i \mathcal{H}(n_i). \tag{A.26}$$

For example,

$$\mathcal{H}(9) = \mathcal{H}(\tfrac{3}{9}, \tfrac{4}{9}, \tfrac{2}{9}) + \tfrac{3}{9}\mathcal{H}(3) + \tfrac{4}{9}\mathcal{H}(4) + \tfrac{2}{9}\mathcal{H}(2). \tag{A.27}$$

In particular, by setting all n_i equal to m, from Equation (A.26) we get

$$\mathcal{H}(m) + \mathcal{H}(n) = \mathcal{H}(mn). \tag{A.28}$$

This yields the unique solution

$$\mathcal{H}(n) = C \log n, \tag{A.29}$$

with $C > 0$. By substituting in Equation (A.26), we finally have

$$\mathcal{H}(\boldsymbol{p}) = -C \sum_{i=1}^{n} p_i \log p_i. \tag{A.30}$$

The constant C determines the base of the logarithm. Base 2 logarithms lead to a measure of entropy and information in bits. For most mathematical calculations, however, we use natural logarithms so that $C = 1$.

It is not very difficult to verify that the entropy has the following properties:

- $\mathcal{H}(\boldsymbol{p}) \geqslant 0$;
- $\mathcal{H}(\boldsymbol{p} \mid \boldsymbol{q}) \leqslant \mathcal{H}(P)$ with equality if and only if $\boldsymbol{p}$ and $\boldsymbol{q}$ are independent;
- $\mathcal{H}(\boldsymbol{p}_1, \dots, \boldsymbol{p}_n) \leqslant \sum_{i=1}^{n} \mathcal{H}(\boldsymbol{p}_i)$ with equality if and only if $\boldsymbol{p}_i$ are independent;
- $\mathcal{H}(\boldsymbol{p})$ is convex ($\cap$) in $\boldsymbol{p}$;
- $\mathcal{H}(\boldsymbol{p}_1, \dots, \boldsymbol{p}_n) = \sum_{i=1}^{n} \mathcal{H}(\boldsymbol{p}_i \mid \boldsymbol{p}_{i-1}, \dots, \boldsymbol{p}_1)$;
- $\mathcal{H}(\boldsymbol{p}) \leqslant \mathcal{H}(n)$ with equality if and only if $\boldsymbol{p}$ is uniform.

Relative entropy

The relative entropy between two density vectors

$$p = (p_1, \ldots, p_n), \qquad q = (q_1, \ldots, q_n),$$

or the associated random variables X and Y, is defined by

$$\mathcal{H}(p, q) = \mathcal{H}(X, Y) = \sum_{i=1}^{n} p_i \log \frac{p_i}{q_i}. \qquad (A.31)$$

The relative entropy is also called cross-entropy, or Kullback–Leibler distance (Kullback 1968), or discrimination (see Shore and Johnson (1980), and references therein, for an axiomatic presentation of the relative entropy). It can be viewed as a measure of dissimilarity between p and q. The more dissimilar p and q are, the larger the relative entropy. The relative entropy can also be viewed as the amount of information that a measurement gives about the truth of a hypothesis compared with an alternative hypothesis. It is also the expected value of the log-likelihood ratio. Strictly speaking, the relative entropy is not symmetric and therefore is not a distance. It can be made symmetric by using the divergence $\mathcal{H}(p, q) + \mathcal{H}(q, p)$. But in most cases, the symmetric version is not needed. If $u = (1/n, \ldots, 1/n)$ denotes the uniform density, then $\mathcal{H}(p, u) = \log n - \mathcal{H}(p)$. In this sense, except for a constant, the entropy is equal to the relative entropy with respect to the uniform distribution.

By using the Jensen inequality (see below), it is easy to verify the following two important properties of relative entropy:

- $\mathcal{H}(p, q) \geqslant 0$ with equality if and only if $p = q$;

- $\mathcal{H}(p, q)$ is convex ($\cap$) in p and q.

The relative entropy is not symmetric but has well known theoretical advantages over other possible measures, including invariance with respect to reparameterizations.

Mutual information

The third concept for measuring information is the mutual information. Consider two density vectors p and q associated with a joint distribution r over the product space. The mutual information $\mathcal{I}(p, q)$ is the relative entropy between the joint distribution r and the product of the marginals p and q:

$$\mathcal{I}(p, q) = \mathcal{H}(r, pq). \qquad (A.32)$$

As such, it is always positive. When r is factorial, i.e. equal to the product of the marginals, the mutual information is zero. The mutual information is a special case of relative entropy. Likewise, the entropy (or self-entropy) is a special case of mutual information because $\mathcal{H}(p) = \mathcal{I}(p, p)$. Furthermore, the mutual information satisfies the following properties:

- $\mathcal{I}(p, q) = 0$ if and only if p and q are independent;

- $\mathcal{I}(p_1, \ldots, p_n, q) = \sum_{i=1}^{n} \mathcal{I}(p_i, q \mid p_1, \ldots, p_{i-1})$.

It is easy to understand mutual information in Bayesian terms: it represents the reduction in uncertainty of one variable when the other is observed, that is between the *prior and posterior distributions*. If we denote two random variables by X and Y, the uncertainty in X is measured by the entropy of its prior

$$\mathcal{H}(X) = \sum_x P(X = x) \log P(X = x).$$

Once we observe $Y = y$, the uncertainty in X is the entropy of the posterior distribution, $\mathcal{H}(X \mid Y = y) = \sum_x P(X = x \mid Y = y) \log P(X = x \mid Y = y)$. This is a random variable that depends on the observation y. Its average over the possible ys is called the *conditional entropy*:

$$\mathcal{H}(X \mid Y) = \sum_y P(y) \mathcal{H}(X \mid Y = y). \tag{A.33}$$

Therefore the difference between the entropy and the conditional entropy measures the average information that an observation of Y brings about X. It is straightforward to check that

$$\mathcal{I}(X, Y) = \mathcal{H}(X) - \mathcal{H}(X \mid Y) = \mathcal{H}(Y) - \mathcal{H}(Y \mid X)$$
$$= \mathcal{H}(X) + \mathcal{H}(Y) - \mathcal{H}(Z) = \mathcal{I}(Y, X), \tag{A.34}$$

where $\mathcal{H}(Z)$ is the entropy of the joint variable $Z = (X, Y)$. We leave the reader to draw the classical Venn diagram associated with these relations.

Jensen's inequality

A key theorem for the reader interested in proving some of the results in the previous sections is Jensen's inequality. If a function f is convex ($\cap$), and X is a random variable, then

$$E(f(X)) \leqslant f(E(X)). \tag{A.35}$$

Furthermore, if f is strictly convex, equality implies that X is constant. This inequality is intuitively obvious by thinking in terms of the center of gravity of a set of points on the curve f. The center of gravity of $f(x_1), \ldots, f(x_n)$ is below $f(x^*)$, where x^* is the center of gravity of $x_1, \ldots, x_n$. As a special important case, $E(\log X) \leqslant \log(E(X))$. This, for instance, immediately yields the properties of the relative entropy.

A.5.2   Information, surprise, and relevance

Shannon's information

While several approaches for quantifying information have been proposed, the most successful one so far has been Claude Shannon's definition introduced over half a century ago (Blahut 1987; Cover and Thomas 1991; McEliece 1977; Shannon 1948a,b). According to Shannon, the information contained in a data set D is given by $-\log P(D)$, and the average information over the family $\mathcal{D}$ of possible data sets is the entropy $\mathcal{H}(P(D))$.

Shannon's theory of information, although eminently successful for the development of modern computer and telecommunication technologies, does not capture subjective and semantic aspects of information that are not directly related to its transmission. As pointed out in the title of Shannon's seminal article, it is a theory of *communication*, in the sense of transmission rather than information. It concentrates on the problem of 'reproducing at one point either exactly or approximately a message selected at another point' regardless of the relevance of the message. But there is clearly more to information than data reproducibility and somehow information ought to depend also on the observer. Consider for instance the genomic DNA sequence of the AIDS virus. It is a string of about 10 000 letters over the four-letter DNA alphabet, of great significance to researchers in the biological or medical sciences, but utterly uninspiring to a layman. The limitations of Shannon's definition may in part explain why the theory has not been as successful as one would have hoped in other areas of science such as biology, psychology, economics, or the Web.

Shannon's theory fails to account how data can have different significance for different observers. This is rooted in the origin of the probabilities used in the definition of information.

These probabilities are defined according to an observer or a model M ('the Bell Labs engineer' (Jaynes 2003)) which Shannon does not describe explicitly, so that the information in a data set is rather the negative log-likelihood

$$I(D, M) = -\log P(D \mid M) \tag{A.36}$$

and the corresponding entropy is the average over all data sets

$$I(\mathcal{D}, M) = \mathcal{H}(P(D \mid M)) = -\int_{\mathcal{D}} P(D \mid M) \log P(D \mid M) \, dD. \tag{A.37}$$

There are situations, however, characterized by the presence of multiple models and/or observers and where the subjective/semantic dimensions of the data are more important than its transmission.

In a Web context, imagine surfing the Web in search of a car and stumbling on a picture of Marilyn Monroe. The Shannon information contained in the picture depends on the picture resolution, whether it is color or black and white, etc. In this situation, it is probably a secondary consideration. More important are the facts that the picture is unexpected, i.e. surprising, and irrelevant to the goal of finding a car. Thus there are at least three different aspects of 'information' contained in data: the transmission or Shannon's information, the surprise, and the relevance. We now provide a precise definition of surprise.

Surprise

The effect of the information contained in D is clearly to change the belief of the observer from $P(M)$ to $P(M \mid D)$. Thus, a complementary way of measuring information carried by the data D is to measure the distance between the prior and the posterior. To distinguish it from Shannon's communication information, we call this notion of information the surprise information or *surprise* (Baldi 2002)

$$\mathcal{S}(D, \mathcal{M}) = d[P(M), P(M \mid D)], \tag{A.38}$$

where d is a distance or similarity measure. It is natural to use the relative entropy so that surprise is given by

$$\mathcal{S}(D, \mathcal{M}) = \mathcal{H}(P(M), P(M \mid D)) = \int_{\mathcal{M}} P(M) \log \frac{P(M)}{P(M \mid D)} \, dM$$

$$= -\mathcal{H}(P(M)) - \int P(M) \log P(M \mid D) \, dM$$

$$= \log P(D) - \int_{\mathcal{M}} P(M) \log P(D \mid M) \, dM. \tag{A.39}$$

Alternatively, we can define the single model surprise by the log-odd ratio

$$\mathcal{S}(D, M) = \log \frac{P(M)}{P(M \mid D)} \tag{A.40}$$

and the surprise by its average

$$\mathcal{S}(D, \mathcal{M}) = \int_{\mathcal{M}} \mathcal{S}(D, M) P(M) \, dM, \tag{A.41}$$

taken with respect to the prior distribution over the model class. Unlike Shannon's entropy, which requires integration over the space of data, surprise is a dual notion that requires integration over the space of models.

Note that this definition addresses the 'TV snow' paradox: snow, the most boring of all television programs, carries the largest amount of Shannon information in terms of exact reproducibility. At the time of snow onset, the image distribution we expect and the image we perceive are very different and therefore the snow carries a great deal of both surprise and Shannon's information. Indeed snow may be a sign of storm, earthquake, toddler's curiosity, or military putsch. But after a few seconds, once our model of the image shifts toward a snow model of random pixels, television snow perfectly fits the prior and hence becomes boring. Since the prior and the posterior are virtually identical, snow frames carry zero surprise although carrying megabytes of Shannon's information.

Computing surprise

To measure surprise in the most simple settings, consider a data set consisting of N binary points. The simplest class $\mathcal{M}(x)$ of models contains a single parameter x, the probability of success. A natural prior distribution is the Dirichlet or Beta prior on x with parameters $(\alpha_1, \alpha2)$ and mean $\alpha_1/(\alpha_1 + \alpha_2)$. With a number n of successes, we have seen in Chapter 1 that the posterior is also a Dirichlet distribution with parameters $\alpha_1 + n$ and $\alpha_2 + (N - n)$. When $N \to \infty$, and $n = pN$ with $0 < p < 1$, it can be shown (Baldi 2002) that

$$\mathcal{S}(D, \mathcal{M}) \approx N\mathcal{H}(p, \alpha_1/(\alpha_1 + \alpha_2)), \tag{A.42}$$

where $\mathcal{H}$ represents the Kullback–Leibler divergence distance between the empirical distribution $(p, 1 - p)$ and the expectation of the prior $(\alpha_1/(\alpha_1 + \alpha_2), \alpha_2/(\alpha_1 + \alpha_2))$. Thus, asymptotically, surprise information grows linearly with the number of data points with a proportionality coefficient that depends on the discrepancy between the expectation of the prior and the observed distribution. Similar equations can be derived in the multinomial case and for simple continuous Gaussian models.

Surprise and learning

There is an immediate connection between surprise and computational learning theory (Vapnik 1995). If we imagine that data points from a training set are presented sequentially, we can consider that the posterior distribution after the Nth point becomes the prior for the next iteration (sequential Bayesian learning). In this case we can expect on average surprise to decrease after each iteration, since as a system learns what is relevant in a data set, new data points become less and less surprising. This can be quantified precisely, at least in simple cases.

For instance, in the discrete binary case above, where the prior and posterior at each step are Dirichlet distributions, it can be shown that the surprise $\mathcal{S}$ between step N and $N + 1$ satisfies

$$0 \leqslant \mathcal{S} \leqslant \frac{1 - p}{pN}. \tag{A.43}$$

Thus surprise decreases in time with the number of examples as $1/N$. Non-zero surprise is a prerequisite to learning, viewed as an adjustment of probability distributions, and repeated recalibration of surprise in training sets is a prerequisite to prevent boredom and sustain learning.

Relevance

Surprise is a measure of dissimilarity between the prior and posterior distributions and as such it lies close to the axiomatic foundation of Bayesian probability. Surprise is different from other definitions of information that have been proposed (Aczel and Daroczy 1975) as alternatives to Shannon's entropy. Most alternative definitions, such as Rényi's entropies, are actually algebraic variations on Shannon's definition rather than conceptually different approaches.

Scoring items by surprise provides a general principle for the rapid detection and ranking of unusual events and the construction of saliency maps, in *any* feature space, that can guide the deployment of attention (Itti and Koch 2001; Nothdurft 2000;

Olshausen *et al.* 1993) and other rapid filtering mechanisms in natural or synthetic information processing systems.

The notion of surprise, however, has its own limitations. In particular, it does not capture all the semantics/relevance aspects of data. When the degree of surprise of the data with respect to the model class becomes low, the data are no longer informative for the given model class. This, however, does not necessarily imply that one has a good model, since the model class itself could be unsatisfactory and in need of a complete overhaul. Conversely, highly surprising data could be a sign that learning is required, or that the data are irrelevant, as in the case of the TV snow or the Marilyn Monroe picture.

Thus, relevance, surprise, and Shannon's entropy are three different facets of information that can be present in different combinations. The notion of *relevance* in particular seems to be the least understood although there have been several attempts (Jumarie 1990; Tishby *et al.* 1999). A possible direction is to consider, in addition to the space of data and models, a third space $\mathcal{A}$ of actions or interpretations and define relevance as the relative entropy between the prior $P(A)$ and the posterior $P(A \mid D)$ distributions over $\mathcal{A}$. Whether this approach simply shifts the problem into the definition of the set $\mathcal{A}$ remains to be seen. In any event, the quest to understand the nature of information, and in particular of semantic relevance, goes well beyond the domain of Web applications and is far from being over.

Appendix B

List of Main Symbols and Abbreviations

General Guidelines

While we use the following guidelines, occasional exceptions are possible and clearly indicated in the text. In general, vectors and matrices are in bold face, with matrices represented by capital letters and vectors by lowercase letters. Unless otherwise specified, all vectors are column vectors by default. In general, Greek letters represent parameters.

$X = (x_{ij})$	matrix
$x = (x_i)$	vector
X^T	transpose of X
$\text{tr} X$	trace of X
$\|X\| = \det X$	determinant of X
D	data. In a typical unsupervised case the data are a matrix $D = (X)$ with one example per row. In a typical supervised case, the data are a double matrix $D = (X, Y)$, where Y denotes the targets, with one example per row.
$M\ (M(\theta))$	model (model with parameter θ)
θ	parameters for some model M
$\theta^*\ (\theta^{\text{ML}}, \theta^{\text{MP}}, \theta^{\text{MAP}})$	parameter estimates (using maximum likelihood, maximum *a posteriori* and mean posterior estimates)
$\mathcal{M}$	the universe of models under consideration
$\mathcal{D}$	the universe of possible data sets
n	number of training examples
m	number of dimensions

Modeling the Internet and the Web   P. Baldi, P. Frasconi and P. Smyth
© 2003 P. Baldi, P. Frasconi and P. Smyth   ISBN: 0-470-84906-1

x	generic input vector
x_j	jth component of x
x_i	ith training example, typically an m-dimensional row vector
x_{ij}	jth component of the ith training example
c_i	class label for training example x_i in supervised learning
K	number of classes, clusters, or mixture components. Classes are denoted by $1, 2, \ldots, K$.
T	time horizon or sequence length
subscript t	time index, in algorithmic iterations or in sequences

Probabilities

$P(\cdot)(Q, R \ldots)$	probability (probability density functions)
$E[\cdot]\,(E_Q[\cdot])$	expectation (expectation with respect to Q)
$\mathrm{var}[\cdot]$	variance
$\mathrm{cov}[\cdot]$	covariance
$P(x_1, \ldots, x_n)$	probability that $X_1 = x_1, \ldots, X_n = x_n$.
$P(X \mid Y)(E[X \mid Y])$	conditional probability (conditional expectation)
I	background information
$\mathcal{N}(\mu, \sigma)$, $\mathcal{N}(\mu, C)$, $\mathcal{N}(\mu, \sigma^2)$, $\mathcal{N}(x; \mu, \sigma^2)$	Normal (or Gaussian) density with mean μ and variance σ^2, or covariance matrix C
$\Gamma(w \mid \alpha, \lambda)$	Gamma density with parameters α and λ
$\mathcal{B}(n, p)$	Binomial distribution with n independent Bernouilli trials each with probability p of success
$\mathcal{P}(\lambda)$	Poisson distribution with parameter λ
$\mathcal{D}_{\alpha q}$, $\mathcal{D}_u$	Dirichlet distribution with parameters α and q, or u ($u_i = \alpha q_i$, $q_i \geqslant 0$, and $\sum_i q_i = 1$)

Functions

$\mathcal{E}$	energy, error, negative log-likelihood or log-posterior (depending on context)
$\mathcal{L}$	Lagrangian
$\mathcal{H}(p)$, $\mathcal{H}(X)$	entropy of the probability vector p, or the random variable X/differential entropy in continuous case
$\mathcal{H}(p, q)$, $\mathcal{H}(X, Y)$	relative entropy between the probability vectors p and q, or the random variables X and Y

$\mathcal{I}(\boldsymbol{p}, \boldsymbol{q}), \mathcal{I}(X, Y)$	mutual information between the probability vectors $\boldsymbol{p}$ and $\boldsymbol{q}$, or the random variables X and Y
$\mathcal{S}$	surprise
Z	partition function or normalizing factor (sometimes also C)
C	constant or normalizing factor
$\delta(x, y)$	Kronecker function equal to one if $x = y$, and zero otherwise
f, f'	generic function and derivative of f
$\Gamma(x)$	Gamma function
$B(\alpha, \boldsymbol{q})$	Beta function

We also use convex (∪) to denote upward convexity (positive second derivative), and convex (∩) to denote downward convexity (negative second derivative), rather than the more confusing 'convex' and 'concave' expressions.

Graphs and Sets

$G = (V, E)$	undirected graph with vertex set V and edge set E
$G = (V, \boldsymbol{E})$	directed graph with vertex set V and edge set E
u, v, w, etc.	vertices
T	tree
k_v	degree of vertex v (undirected case)
t_v	time at which vertex v is added
n	number of vertices
e	number of edges
d	average diameter
c	cliquishness index
$N(i)$	neighbors of vertex i
ch[v]	children of vertex v in a directed graph
pa[v]	parents of vertex v in a directed graph
$N(I)$	neighbors or boundary of a set I of vertices
$|S|$	size of set S

Abbreviations

DNS	domain name service
EM	expectation maximization
HMM	hidden Markov model

HTML	HyperText Markup Language
HTTP	Hypertext Transfer Protocol
i.i.d	independent, identically distributed
IP	Internet Protocol
LMS	least mean square
MAP	maximum *a posteriori*
MaxEnt	maximum entropy
MCMC	Markov chain Monte Carlo
ML	maximum likelihood
MP	mean posterior
pdf	probability density function
SGML	Standard Generalized Markup Language
SVM	support vector machine
TCP	Transmission Control Protocol
URL	uniform resource locator

References

Abello, J., Buchsbaum, A. and Westbrook, J. 1998 A functional approach to external graph algorithms. In *Proc. 6th Eur. Symp. on Algorithms*, pp. 332–343.

Achacoso, T. B. and Yamamoto, W. S. 1992 *Ay's Neuroanatomy of C. elegans for Computation*. Boca Raton, FL: CRC Press.

Aczel, J. and Daroczy, Z. 1975 *On measures of information and their characterizations*. New York: Academic Press.

Adamic, L., Lukose, R. M., Puniyani, A. R. and Huberman, B. A. 2001 Search in power-law networks. *Phys. Rev.* E **64**, 046135.

Aggarwal, C. C., Al-Garawi, F. and Yu, P. S. 2001 Intelligent crawling on the World Wide Web with arbitrary predicates. In *Proc. 10th Int. World Wide Web Conf.*, pp. 96–105.

Aiello, W., Chung, F. and Lu, L. 2001 A Random Graph Model for Power Law Graphs. *Experimental Math.* **10**, 53–66.

Aji, S. M. and McEliece, R. J. 2000 The generalized distributive law. *IEEE Trans. Inform. Theory* **46**, 325–343.

Albert, R. and Barabási, A.-L. 2000 Topology of evolving networks: local events and universality. *Phys. Rev. Lett.* **85**, 5234–5237.

Albert, R., Jeong, H. and Barabási, A.-L. 1999 Diameter of the World-Wide Web. *Nature* **401**, 130.

Albert, R., Jeong, H. and Barabási, A.-L. 2000 Error and attack tolerance of complex networks. *Nature* **406**, 378–382.

Allwein, E. L., Schapire, R. E. and Singer, Y. 2000 Reducing multiclass to binary: a unifying approach for margin classifiers. In *Proc. 17th Int. Conf. on Machine Learning*, pp. 9–16. San Francisco, CA: Morgan Kaufmann.

Amaral, L. A. N., Scala, A., Barthélémy, M. and Stanley, H. E. 2000 Classes of small-world networks. *Proc. Natl Acad. Sci.* **97**, 11 149–11 152.

Amento, B., Terveen, L. and Hill, W. 2000 Does authority mean quality? Predicting expert quality ratings of Web documents. In *Proc. 23rd Ann. Int. ACM SIGIR Conf. on Research and Development in Information Retrieval*, pp. 296–303. New York: ACM Press.

Anderson, C. R., Domingos, P. and Weld, D. 2001 Adaptive Web navigation for wireless devices. In *Proc. 17th Int. Joint Conf. on Artificial Intelligence*, pp. 879–884. San Francisco, CA: Morgan Kaufmann.

Anderson, C. R., Domingos, P. and Weld, D. 2002 Relational markov models and their application to adaptive Web navigation. In *Proc. 8th Int. Conf. on Knowledge Discovery and Data Mining*, pp. 143–152. New York: ACM Press.

Androutsopoulos, I., Koutsias, J., Chandrinos, K. and Spyropoulos, D. 2000 An experimental comparison of naive Bayesian and keyword-based anti-spam filtering with personal e-mail messages. In *Proc. 23rd ACM SIGIR Ann. Conf.*, pp. 160–167.

Ansari, A. and Mela, C. 2003 E-customization. *J. Market. Res.* (In the press.)

Apostol, T. M. 1969 *Calculus*, vols I and II. John Wiley & Sons, Ltd/Inc.

Appelt, D., Hobbs, J., Bear, J., Israel, D., Kameyama, M., Kehler, A., Martin, D., Meyers K. and Tyson, M. 1995 SRI International FASTUS system: MUC-6 test results and analysis. In *Proc. 6th Message Understanding Conf. (MUC-6)*, pp. 237–248. San Francisco, CA: Morgan Kaufmann.

Apté C., Damerau, F. and Weiss, S. M. 1994 Automated learning of decision rules for text categorization. (Special Issue on Text Categorization.) *ACM Trans. Informat. Syst.* **12**, 233–251.

Araújo M. D., Navarro, G. and Ziviani, N. 1997 Large text searching allowing errors. In *Proc. 4th South American Workshop on String Processing* (ed. R. Baeza-Yates), International Informatics Series, pp. 2–20. Ottawa: Carleton University Press.

Armstrong, R., Freitag, D., Joachims, T. and Mitchell, T. 1995 WebWatcher: a learning apprentice for the World Wide Web. In *Proc. 1995 AAAI Spring Symp. on Information Gathering from Heterogeneous, Distributed Environments*, pp. 6–12.

Aslam, J. A. and Montague, M. 2001 Models for metasearch. In *Proc. 24th Ann. Int. ACM SIGIR Conf. on Research and Development in Information Retrieval*, pp. 276–284. New York: ACM Press.

Baldi, P. 2002 A computational theory of surprise. In *Information, Coding, and Mathematics* (ed. M. Blaum, P. G. Farrell and H. C. A. van Tilborg), pp. 1–25. Boston, MA: Kluwer Academic.

Baldi, P. and Brunak, S. 2001 *Bioinformatics: The Machine Learning Approach*, 2nd edn. MIT Press, Cambridge, MA.

Baluja, S., Mittal, V. and Sukthankar, R. 2000 Applying machine learning for high performance named-entity extraction. *Computat. Intell.* **16**, 586–595.

Barabási, A.-L. and Albert, R. 1999 Emergence of scaling in random networks. *Science* **286**, 509–512.

Barabási, A.-L., Albert, R. and Jeong, H. 1999 Mean-field theory for scale-free random networks. *Physica* A **272**, 173–187.

Barabási, A.-L., Freeh, V. W., Jeong, H. and Brockman, J. B. 2001 Parasitic computing. *Nature* **412**, 894–897.

Barlow, R. and Proshan, F. 1975 *Statistical Theory of Reliability and Life Testing*. Austin, TX: Holt, Rinehart and Winston.

Barthélémy, M. and Amaral, L. A. N. 1999 Small-world networks: evidence for a crossover picture. *Phys. Rev. Lett.* **82**, 3180–3183.

Bass, F. M. 1969 A new product growth model for consumer durables. *Mngmt Sci.* **15**, 215–227.

Bellman, R. E. 1957 *Dynamic Programming*. Princeton, NJ: Princeton University Press.

Bengio, Y. and Frasconi, P. 1995 An input–output HMM architecture. *Adv. Neural Inf. Process Syst.* **7**, 427–434.

Berger, J. O. 1985 *Statistical Decision Theory and Bayesian Analysis*. Springer.

Bergman, M. 2000 The Deep Web: surfacing hidden value. *J. Electron. Publ.* **7**. (Available at http://www.completeplanet.com/Tutorials/DeepWeb/.)

Berners-Lee, T. 1994 Universal Resource Identifiers in WWW: A Unifying Syntax for the Expression of Names and Addresses of Objects on the Network as used in the World-Wide Web. RFC 1630. (Available at http://www.ietf.org/rfc/rfc1630.txt.)

Berners-Lee, T., Fielding, R. and Masinter, L. 1998 Uniform Resource Identifiers (URI): Generic Syntax. RFC 2396. (Available at http://www.ietf.org/rfc/rfc2396.txt.)

Berry, M. W. 1992 Large scale singular value computations. *J. Supercomput. Applic.* **6**, 13–49.

Berry, M. W. and Browne, M. 1999 *Understanding Search Engines: Mathematical Modeling and Text Retrieval*. Philadelphia, PA: Society for Industrial and Applied Mathematics.

Bharat, K. and Broder, A. 1998 A technique for measuring the relative size and overlap of public Web search engines. In *Proc. 7th Int. World Wide Web Conf., Brisbane, Australia*, pp. 379–388.

Bharat, K. and Henzinger, M. R. 1998 Improved algorithms for topic distillation in a hyperlinked environment. In *Proc. 21st Ann Int. ACM SIGIR Conf. on Research and Development in Information Retrieval*, pp. 104–111. New York: ACM Press.

Bianchini, M., Gori, M. and Scarselli, F. 2001 Inside Google's Web page scoring system. Technical report, Dipartimento di Ingegneria dell'Informazione, Università di Siena.

Bikel, D. M., Miller, S., Schwartz, R. and Weischedel, R. 1997 Nymble: a high-performance learning name-finder. In *Proceedings of ANLP-97*, pp. 194–201. (Available at http://citeseer.nj.nec.com/bikel97nymble.html.)

Billsus, D. and Pazzani, M. 1998 Learning collaborative information filters. In *Proc. Int. Conf. on Machine Learning*, pp. 46–54. San Francisco, CA: Morgan Kaufmann.

Blahut, R. E. 1987 *Principles and Practice of Information Theory*. Reading, MA: Addison-Wesley.

Blei, D., Ng, A. Y. and Jordan, M. I. 2002a Hierarchical Bayesian models for applications in information retrieval. In *Bayesian Statistics 7* (ed. J. M. Bernardo, M. Bayarri, J. O. Berger, A. P. Dawid, D. Heckerman, A. F. M. Smith and M. West). Oxford University Press.

Blei, D., Ng, A. Y. and Jordan, M. I. 2002b Latent Dirichlet allocation. In *Advances in Neural Information Processing Systems 14* (ed. T. Dietterich, S. Becker and Z. Ghahramani). San Francisco, CA: Morgan Kaufmann.

Blum, A. and Mitchell, T. 1998 Combining labeled and unlabeled data with co-training. In *Proc. 11th Ann. Conf. on Computational Learning Theory (COLT-98)*, pp. 92–100. New York: ACM Press.

Bollacker, K. D., Lawrence, S. and Giles, C. L. 1998 CiteSeer: an autonomous Web agent for automatic retrieval and identification of interesting publications. In *Proc. 2nd Int. Conf. on Autonomous Agents (Agents'98)* (ed. K. P. Sycara and M. Wooldridge), pp. 116–123. New York: ACM Press.

Bollobás, B. 1985 *Random Graphs*. London: Academic Press.

Bollobás, B. and de la Vega, W. F. 1982 The diameter of random regular graphs. *Combinatorica* **2**, 125–134.

Bollobás, B. and Riordan, O. 2003 The diameter of a scale-free random graph. *Combinatorica*. (In the press.)

Bollobás, B., Riordan, O., Spencer, J. and Tusnády G. 2001 The degree sequence of a scale-free random graph process. *Random. Struct. Alg.* **18**, 279–290.

Borodin, A., Roberts, G. O., Rosenthal, J. S. and Tsaparas, P. 2001 Finding authorities and hubs from link structures on the World Wide Web. In *Proc. 10th Int. Conf. on World Wide Web*, pp. 415–429.

Box, G. E. P. and Tiao, G. C. 1992 *Bayesian Inference In Statistical Analysis*. John Wiley & Sons, Ltd/Inc.

Boyan, J., Freitag, D. and Joachims, T. 1996 A machine learning architecture for optimizing Web search engines. In *Proc. AAAI Workshop on Internet-Based Information Systems*.

Brand, M. 2002 Incremental singular value decomposition of uncertain data with missing values. In *Proc. Eur. Conf. on Computer Vision (ECCV): Lecture Notes in Computer Science*, pp. 707–720. Springer.

Bray, T. 1996 Measuring the Web. In *Proc. 5th Int. Conf. on the World Wide Web, 6–10 May 1996, Paris, France. Comp. Networks* **28**, 993–1005.

Breese, J. S., Heckerman, D. and Kadie, C. 1998 Empirical analysis of predictive algorithms for collaborative filtering. In *Proc. 14th Conf. on Uncertainty in Artificial Intelligence*, pp. 43–52. San Francisco, CA: Morgan Kaufmann.

Brewington, B. and Cybenko, G. 2000 How dynamic is the Web? *Proc. 9th Int. World Wide Web Conf.* Geneva: International World Wide Web Conference Committee (IW3C2).

Brin, S. and Page, L. 1998 The anatomy of a large-scale hypertextual (Web) search engine. In *Proc. 7th Int. World Wide Web Conf. (WWW7). Comp. Networks* **30**, 107–117.

Broder, A., Kumar, R., Maghoul, F., Raghavan, P., Rajagopalan, S., Stata, R., Tomikns, A. and Wiener, J. 2000 Graph structure in the Web. In *Proc. 9th Int. World Wide Web Conf. (WWW9). Comp. Networks* **33**, 309–320.

Brown, L. D. 1986 *Fundamentals of Statistical Exponential Families*. Hayward, CA: Institute of Mathematical Statistics.

Bucklin, R. E. and Sismeiro, C. 2003 A model of Web site browsing behavior estimated on clickstream data. (In the press.)

Buntine, W. 1992 Learning classification trees. *Statist. Comp.* **2**, 63–73.

Buntine, W. 1996 A guide to the literature on learning probabilistic networks from data. *IEEE Trans. Knowl. Data Engng* **8**, 195–210.

Byrne, M. D., John, B. E., Wehrle, N. S. and Crow, D. C. 1999 The tangled Web we wove: a taskonomy of WWW use. In *Proc. CHI'99: Human Factors in Computing Systems*, pp. 544–551. New York: ACM Press.

Cadez, I. V., Heckerman, D., Smyth, P., Meek, C. and White, S. 2003 Model-based clustering and visualization of navigation patterns on a Web site. *Data Mining Knowl. Discov.* (In the press.)

Califf, M. E. and Mooney, R. J. 1998 Relational learning of pattern-match rules for information extraction. *Working Notes of AAAI Spring Symp. on Applying Machine Learning to Discourse Processing*, pp. 6–11. Menlo Park, CA: AAAI Press.

Callaway, D. S., Hopcroft, J. E., Kleinberg, J., Newman, M. E. J. and Strogatz, S. H. 2001 Are randomly grown graphs really random? *Phys. Rev.* E **64**, 041902.

Cardie, C. 1997 Empirical methods in information extraction. *AI Mag.* **18**, 65–80.

Carlson, J. M. and Doyle, J. 1999 Highly optimized tolerance: a mechanism for power laws in designed systems. *Phys. Rev.* E **60**, 1412–1427.

Castelli, V. and Cover, T. 1995 On the exponential value of labeled samples. *Pattern Recog. Lett.* **16**, 105–111.

Catledge, L. D. and Pitkow, J. 1995 Characterizing browsing strategies in the World-Wide Web. *Comp. Networks ISDN Syst.* **27**, 1065–1073.

Chaitin, G. J. 1987 *Algorithmic Information Theory*. Cambridge University Press.

Chakrabarti, S., Dom, B., Gibson, D., Kleinberg, J., Kumar, S. R., Raghavan, P., Rajagopalan S. and Tomkins, A. 1999a Mining the link structure of the World Wide Web. *IEEE Computer* **32**, 60–67.

Chakrabarti, S., Joshi, M. M., Punera, K. and Pennock, D. M. 2002 The structure of broad topics on the Web. In *Proc. 11th Int. Conf. on World Wide Web*, pp. 251–262. New York: ACM Press.

Chakrabarti, S., van den Berg, M. and Dom, B. 1999b Focused crawling: a new approach to topic-specific Web resource discovery. In *Proc. 8th Int. World Wide Web Conf., Toronto. Comp. Networks* **31**, 11–16.

Charniak, E. 1991 Bayesian networks without tears. *AI Mag.* **12**, 50–63.

Chen, S. F. and Goodman, J. 1996 An empirical study of smoothing techniques for language modeling. In *Proc. 34th Ann. Meeting of the Association for Computational Linguistics* (ed. A. Joshi and M. Palmer), pp. 310–318. San Francisco, CA: Morgan Kaufmann.

Chickering, D. M., Heckerman, D. and Meek, C. 1997 A Bayesian approach to learning Bayesian networks with local structure. *Uncertainty in Artificial Intelligence: Proc. 13th Conf. (UAI-1997)*, pp. 80–89. San Francisco, CA: Morgan Kaufmann.

Cho, J. and Garcia-Molina, H. 2000a Estimating frequency of change. Technical Report DBPUBS-4, Stanford University. (Available at http://dbpubs.stanford.edu/pub/2000-4.)

Cho, J. and Garcia-Molina, H. 2000b Synchronizing a database to improve freshness. In *Proc. 2000 ACM Int. Conf. on Management of Data (SIGMOD)*, pp. 117–128.

Cho, J. and Garcia-Molina, H. 2002 Parallel crawlers. In *Proc. 11th World Wide Web Conf. (WWW11), Honolulu, Hawaii.*

Cho, J. and Ntoulas, A. 2002 Effective change detection using sampling. In *Proc. 28th Int. Conf. on Very Large Databases (VLDB).*

Cho, J., Garcia-Molina, H. and Page, L. 1998 Efficient crawling through URL ordering. In *Proc. 7th Int. World Wide Web Conf. (WWW7). Comp. Networks* **30**, 161–172.

Chung, E. R. K., Graham, R. L. and Wilson, R. M. 1989 Quasi-random graphs. *Combinatorica* **9**, 345–362.

Chung, F. and Graham, R. 2002 Sparse quasi-random graphs. *Combinatorica* **22**, 217–244.

Chung, F. and Lu, L. 2001 The diameter of random sparse graphs. *Adv. Appl. Math.* **26**, 257–279.

Chung, F. and Lu, L. 2002 Connected components in random graphs with given expected degree sequences. Technical Report, Univeristy of California, San Diego.

Chung, F., Garrett, M., Graham, R. and Shallcross, D. 2001 Distance realization problems with applications to Internet tomography. *J. Comput. Syst. Sci.* **63**, 432–448.

Clarke, I., Sandberg, O., Wiley, B. and Hong, T. W. 2000 Freenet: a distributed anonymous information storage and retrieval system. In *Designing Privacy Enhancing Technologies: International Workshop on Design Issues in Anonymity and Unobservability, LNCS 2009* (ed. H. Federrath), pp. 311–320. Springer.

Clarke, I., Miller, S. G., Hong, T. W., Sandberg, O. and Wiley, B. 2002 Protecting free expression online with Freenet. *IEEE Internet Computing* **6**, 40–49.

Cockburn, A. and McKenzie, B. 2002 What do Web users do? An empirical analysis of Web use. *Int. J. Human Computer Studies* **54**, 903–922.

Cohen, W. W. 1995 Text categorization and relational learning. In *Proc. ICML-95, 12th Int. Conf. on Machine Learning* (ed. A. Prieditis and S. J. Russell), pp. 124–132. San Francisco, CA: Morgan Kaufmann.

Cohen, W. W. 1996 Learning rules that classify e-mail. In *AAAI Spring Symp. on Machine Learning in Information Access* (ed. M. Hearst and H. Hirsh). 1996 Spring Symposium Series. Menlo Park, CA: AAAI Press.

Cohen, W. W. and McCallum, A. 2002 Information Extraction from the World Wide Web. Tutorial presented at *15th Neural Information Processing Conf. (NIPS-15).*

Cohen, W. W., Schapire, R. E. and Singer, Y. 1999 Learning to order things. *J. Artif. Intell. Res.* **10**, 243–270.

Cohen, W. W., McCallum, A. and Quass, D. 2000 Learning to understand the Web. *IEEE Data Enging Bull.* **23**, 17–24.

Cohn, D. and Chang, H. 2000 Learning to probabilistically identify authoritative documents. In *Proc. 17th Int. Conf. on Machine Learning*, pp. 167–174. San Francisco, CA: Morgan Kaufmann.

Cohn, D. and Hofmann, T. 2001 The missing link: a probabilistic model of document content and hypertext connectivity. In *Advances in Neural Information Processing Systems* (ed. T. K. Leen, T. G. Dietterich and V. Tresp). Boston, MA: MIT Press.

Cooley, R., Mobasher, B. and Srivastava, J. 1999 Data preparation for mining World Wide Web browsing patterns. *Knowl. Informat. Syst.* **1**, 5–32.

Cooper, C. and Frieze, A. 2001 A general model of Web graphs. Technical Report.

Cooper, G. F. 1990 The computational complexity of probabilistic inference using Bayesian belief networks. *Art. Intell.* **42**, 393–405.

Cooper, W. S. 1991 Some inconsistencies and misnomers in probabilistic information retrieval. In *Proc. 14th Ann. Int. ACM SIGIR Conf. on Research and Development in Information Retrieval*, pp. 57–61. New York: ACM Press.

Cormen, T. H., Leiserson, C. E., Rivest, R. L. and Stein, C. 2001 *Introduction to Algorithms*, 2nd edn. Cambridge, MA: MIT Press.

Cortes, C. and Vapnik, V. N. 1995 Support vector networks. *Machine Learning* **20**, 1–25.

Cover, T. M. and Hart, P. E. 1967 Nearest neighbor pattern classification. *IEEE Trans. Inform. Theory* **13**, 21–27.

Cover, T. M. and Thomas, J. A. 1991 *Elements of Information Theory*. John Wiley & Sons, Ltd/Inc.

Cox, R. T. 1964 Probability, frequency and reasonable expectation. *Am. J. Phys.* **14**, 1–13.

Crammer, K. and Singer, Y. 2000 On the learnability and design of output codes for multiclass problems. *Computational Learning Theory*, pp. 35–46.

Craven, M., di Pasquo D., Freitag, D., McCallum A., Mitchell, T., Nigan, K. and Slattery, S. 2000 Learning to construct knowledge bases from the World Wide Web. *Artif. Intel.* **118**(1–2), 69–113.

Craven, M. and Slattery, S. 2001 Relational learning with statistical predicate invention: better models for hypertext. *Machine Learning* **43**(1/2), 97–119.

Cristianini, N. and Shawe-Taylor, J. 2000 *An Introduction to Support Vector Machines*. Cambridge University Press.

Daley, D. J. and Gani, J. 1999 *Epidemic Modeling: an Introduction*. Cambridge University Press.

Davison, B. D. 2000a Recognizing nepotistic links on the Web. *AAAI Workshop on Artificial Intelligence for Web Search*.

Davison, B. D. 2000b Topical locality in the Web. In *Proc. 23rd Ann. Int. ACM SIGIR Conf. on Research and Development in Information Retrieval*, pp. 272–279.

Dawid, A. P. 1992 Applications of a general propagation algorithm for probabilistic expert systems. *Stat. Comp.* **2**, 25–36.

Day, J. 1995 The (un)revised osi reference model. *ACM SIGCOMM Computer Communication Review* **25**, 39–55.

Day, J. and Zimmerman, H. 1983 The OSI reference model. In *Proc. IEEE* **71**, 1334–1340.

De Bra, P. and Post, R. 1994 Information retrieval in the World Wide Web: making client-based searching feasible. In *Proc. 1st Int. World Wide Web Conf.*

Dechter, R. 1999 Bucket elimination: a unifying framework for reasoning. *Artif. Intel.* **113**, 41–85.

Deering, S. and Hinden, R. 1998 Internet Protocol, Version 6 (IPv6) Specification. RFC 2460. (Available at http://www.ietf.org/rfc/rfc2460.txt.)

Del Bimbo A. 1999 *Visual Information Retrieval*. San Francisco, CA: Morgan Kaufmann.

Dempster, A. P., Laird, N. M. and Rubin, D. B. 1977 Maximum likelihood from incomplete data via the em algorithm. *J. R. Statist. Soc.* B **39**, 1–22.

Deshpande, M. and Karypis, G. 2001 Selective Markov models for predicting Web-page accesses. In *Proc. SIAM Conf. on Data Mining* SIAM Press.

Devroye, L., Györfi, L. and Lugosi, G. 1996 *A Probabilistic Theory of Pattern Recognition*. Springer.

Dhillon, I. S., Fan, J. and Guan, Y. 2001 Efficient clustering of very large document collections In *Data Mining for Scientific and Engineering Applications* (ed. R. Grossman, C. Kamath and R. Naburu). Kluwer Academic.

Dhillon, I. S. and Modha, D. S. 2001 Concept decompositions for large sparse text data using clustering. *Machine Learning* **42**, 143–175.

Dietterich, T. G. and Bakiri, G. 1995 Solving multiclass learning problems via error-correcting output codes. *J. Artificial Intelligence Research* **2**, 263–286.

Dijkstra, E. D. 1959 A note on two problem in connexion with graphs. *Numerische Mathematik* **1**, 269–271.

Diligenti, M., Coetzee, F., Lawrence, S., Giles, C. L. and Gori, M. 2000 Focused crawling using context graphs. In *VLDB 2000, Proc. 26th Int. Conf. on Very Large Data Bases, 10–14 September 2000, Cairo, Egypt* (ed. A. El Abbadi, M. L. Brodie, S. Chakravarthy, U. Dayal, N. Kamel, G. Schlageter and K. Y. Whang), pp. 527–534. Los Altos, CA: Morgan Kaufmann.

Dill, S., Kumar, S. R., McCurley, K. S., Rajagopalan, S., Sivakumar, D. and Tomkins, A. 2001 Self-similarity in the Web. In *Proc. 27th Very Large Databases Conf.*, pp. 69–78.

Domingos, P. and Pazzani, M. 1997 On the optimality of the simple Bayesian classifier under zero-one loss. *Machine Learning* **29**, 103–130.

Domingos, P. and Richardson, M. 2001 Mining the network value of customers. In *Proc. ACM 7th Int. Conf. on Knowledge Discovery and Data Mining*, pp. 57–66. New York: ACM Press.

Dreilinger, D. and Howe, A. E. 1997 Experiences with selecting search engines using metasearch. *ACM Trans. Informat. Syst.* **15**, 195–222.

Drucker, H., Vapnik, V. N. and Wu, D. 1999 Support vector machines for spam categorization. *IEEE Trans. Neural Networks* **10**, 1048–1054.

Duda, R. O. and Hart, P. E. 1973 *Pattern Classification and Scene Analysis.* John Wiley & Sons, Ltd/Inc.

Dumais, S., Platt, J., Heckerman, D. and Sahami, M. 1998 Inductive learning algorithms and representations for text categorization. In *Proc. 7th Int. Conf. on Information and Knowledge Management*, pp. 148–155. New York: ACM Press.

Jones, K. S. and Willett, P. (eds) 1997 *Readings in information retrieval.* San Mateo, CA: Morgan Kaufmann.

Edwards, J., McCurley, K. and Tomlin, J. 2001 An adaptive model for optimizing performance of an incremental Web crawler. In *Proc. 10th Int. World Wide Web Conf.*, pp. 106–113.

Elias, P. 1975 Universal codeword sets and representations of the integers. *IEEE Trans. Inform. Theory* **21**, 194–203.

Erdos, P. and Rényi, A. 1959 On random graphs. *Publ. Math. Debrecen* **6**, 290–291.

Erdos, P. and Rényi, A. 1960 On the evolution of random graphs. *Magy. Tud. Akad. Mat. Kut. Intez. Kozl.* **5**, 17–61.

Everitt B. S. 1984 *An Introduction to Latent Variable Models.* London: Chapman & Hall.

Evgeniou, T., Pontil, M. and Poggio, T. 2000 Regularization networks and support vector machines. *Adv. Comput. Math.* **13**, 1–50.

Fagin, R., Karlin, A., Kleinberg, J., Raghavan, P., Rajagopalan, S., Rubinfeld, R., Sudan., M. and Tomkins, A. 2000 Random walks with 'back buttons'. In *Proc. ACM Symp. on Theory of Computing*, pp. 484–493. New York: ACM Press.

Faloutsos, C. and Christodoulakis, S. 1984 Signature files: an access method for documents and its analytical performance evaluation. *ACM Trans. Informat. Syst.* **2**, 267–288.

Faloutsos, M., Faloutsos, P. and Faloutsos, C. 1999 On power-law relationships of the Internet topology. In *Proc. ACM SIGCOMM Conf., Cambridge, MA*, pp. 251–262.

Feller, W. 1971 *An Introduction to Probability Theory and Its Applications*, 2nd edn, vol. 2. John Wiley & Sons, Ltd/Inc.

Fermi, E. 1949 On the origin of the cosmic radiation. *Phys. Rev.* **75**, 1169–1174.

Fielding, R., Gettys, J., Mogul, J., Frystyk, H., Masinter, L., Leach, P. and Berners-Lee, T. 1999 Hypertext Transfer Protocol: HTTP/1.1. RFC 2616. (Available at http://www.ietf.org/rfc/rfc2616.txt.)

Fienberg, S. E., Johnson, M. A. and Junker, B. J. 1999 Classical multilevel and Bayesian approaches to population size estimation using multiple lists. *J. R. Statist. Soc.* A **162**, 383–406.

Flake, G. W., Lawrence, S. and Giles, C. L. 2000 Efficient identification of Web communities. In *Proc. 6th ACM SIGKDD Int. Conf. on Knowledge Discovery and Data Mining*, pp. 150–160. New York: ACM Press.

Flake, G. W., Lawrence, S., Giles, C. L. and Coetzee, F. 2002 Self-organization and identification of communities. *IEEE Computer* **35**, 66–71.

Fox, C. 1992 Lexical analysis and stoplists. In *Information Retrieval: Data Structures and Algorithms* (ed. W. B. Frakes and R. Baeza-Yates), ch. 7. Englewood Cliffs, NJ: Prentice Hall.

Fraley, C. and Raftery, A. E. 2002 Model-based clustering, discriminant analysis, and density estimation. *J. Am. Statist. Assoc.* **97**, 611–631.

Freitag, D. 1998 Information extraction from HTML: Application of a general machine learning approach. In *Proc. AAAI-98*, pp. 517–523. Menlo Park, CA: AAAI Press.

Freitag, D. and McCallum, A. 2000 Information extraction with HMM structures learned by stochastic optimization. *AAAI/IAAI*, pp. 584–589.

Freund, Y. and Schapire, R. E. 1996 Experiments with a new boosting algorithm. In *Proc. 13th Int. Conf. on Machine Learning*, pp. 148–146. San Francisco, CA: Morgan Kaufmann.

Frey, B. J. 1998 *Graphical Models for Machine Learning and Digital Communication*. MIT Press.

Friedman, N. and Goldszmidt, M. 1996 Learning Bayesian networks with local structure. In *Proc. 12th Conf. on Uncertainty in Artificial Intelligence, Portland, Oregon* (ed. E. Horwitz and F. Jensen), pp. 274–282. San Francisco, CA: Morgan Kaufmann.

Friedman, N., Getoor, L., Koller, D. and Pfeffer, A. 1999 Learning probabilistic relational models. In *Proc. 16th Int. Joint Conf. on Artificial Intelligence (IJCAI-99)* (ed. D. Thomas), vol. 2 , pp. 1300–1309. San Francisco, CA: Morgan Kaufmann.

Fuhr, N. 1992 Probabilistic models in information retrieval. *Comp. J.* **35**, 243–255.

Galambos, J. 1987 *The Asymptotic Theory of Extreme Order Statistics*, 2nd edn. Malabar, FL: Robert E. Krieger.

Garfield, E. 1955 Citation indexes for science: a new dimension in documentation through association of ideas. *Science* **122**, 108–111.

Garfield, E. 1972 Citation analysis as a tool in journal evaluation. *Science* **178**, 471–479.

Garner, R. 1967 *A Computer Oriented, Graph Theoretic Analysis of Citation Index Structures*. Philadelphia, PA: Drexel University Press.

Gelbukh, A. and Sidorov, G. 2001 Zipf and Heaps Laws' coefficients depend on language. In *Proc. 2001 Conf. on Intelligent Text Processing and Computational Linguistics*, pp. 332–335. Springer.

Gelman, A., Carlin, J. B., Stern, H. S. and Rubin, D. B. 1995 *Bayesian Data Analysis*. London: Chapman & Hall.

Ghahramani, Z. 1998 Learning dynamic Bayesian networks. In *Adaptive Processing of Sequences and Data Structures*. Lecture Notes in Artifical Intelligence (ed. M. Gori and C. L. Giles), pp. 168–197. Springer.

Ghahramani, Z. and Jordan, M. I. 1997 Factorial hidden Markov models. *Machine Learning* **29**, 245–273.

Ghani, R. 2000 Using error-correcting codes for text classification. In *Proc. 17th Int. Conf. on Machine Learning*, pp. 303–310. San Francisco, CA: Morgan Kaufmann.

Gibson, D., Kleinberg, J. and Raghavan, P. 1998 Inferring Web communities from link topology. In *Proc. 9th ACM Conf. on Hypertext and Hypermedia : Links, Objects, Time and Spacestructure in Hypermedia Systems*, pp. 225–234. New York: ACM Press.

Gilbert, E. N. 1959 Random graphs. *Ann. Math. Statist.* **30**, 1141–1144.

Gilbert, N. 1997 A simulation of the structure of academic science. *Sociological Research Online* **2**. (Available at http://www.socresonline.org.uk/socresonline/2/2/3.html.)

Gilks, W. R., Thomas, A. and Spiegelhalter, D. J. 1994 A language and program for complex Bayesian modelling. *The Statistician* **43**, 69–78.

Greenberg, S. 1993 *The Computer User as Toolsmith: The Use, Reuse, and Organization or Computer-Based Tools*. Cambridge University Press.

Guermeur, Y., Elisseeff, A. and Paugam-Mousy, H. 2000 A new multi-class SVM based on a uniform convergence result. In *Proc. IJCNN: Int. Joint Conf. on Neural Networks*, vol. 4, pp 4183–4188. Piscataway, NJ: IEEE Press.

Han, E. H., Karypis, G. and Kumar, V. 2001 Text categorization using weight-adjusted k-nearest neighbor classification. In *Proc. PAKDD-01, 5th Pacific–Asia Conferenece on Knowledge Discovery and Data Mining* (ed. D. Cheung, Q. Li and G. Williams). Lecture Notes in Computer Science Series, vol. 2035, pp. 53–65. Springer.

Han, J. and Kamber, M. 2001 *Data Mining: Concepts and Techniques*. San Francisco, CA: Morgan Kaufmann.

Hand, D., Mannila, H. and Smyth, P. 2001 *Principles of Data Mining*. Cambridge, MA: MIT Press.

Harman, D., Baeza-Yates, R., Fox, E. and Lee, W. 1992 Inverted files. In *Information Retrieval, Data Structures and Algorithms* (ed. W. B. Frakes and R. A. Baeza-Yates), pp. 28–43. Englewood Cliffs, NJ: Prentice Hall.

Hastie, T., Tibshirani, R. and Friedman, J. 2001 *Elements of Statistical Learning: Data Mining, Inference, and Prediction*. Springer.

Heckerman, D. 1997 Bayesian networks for data mining. *Data Mining Knowl. Discov.* **1**, 79–119.

Heckerman, D. 1998 A tutorial on learning with Bayesian networks. In *Learning in Graphical Models* (ed. M. Jordan). Kluwer.

Heckerman, D., Chickering, D. M., Meek, C., Rounthwaite, R. and Kadie, C. 2000 Dependency networks for inference, collaborative filtering, and data visualization. *J. Mach. Learn. Res.* **1**, 49–75.

Hersovici, M., Jacovi, M., Maarek, Y. S., Pelleg, D., Shtalhaim, M. and Ur, S. 1998 The shark-search algorithm – an application: tailored Web site mapping. In *Proc. 7th Int. World-Wide Web Conf. Comp. Networks* **30**, 317–326.

Heydon, A. and Najork, M. 1999 Mercator: a scalable, extensible Web crawler. *World Wide Web* **2**, 219–229. (Available at http://research.compaq.com/SRC/mercator/research.html.)

Heydon, A. and Najork, M. 2001 High-performance Web crawling. Technical Report SRC 173. Compaq Systems Research Center.

Hoffman, T. 1999 Probabilistic latent semantic indexing. In *Proc. 22nd Ann. Int. ACM SIGIR Conf. on Research and Development in Information Retrieval*, pp. 50–57. New York: ACM Press.

Hofmann, T. 2001 Unsupervised learning by probabilistic latent semantic analysis. *Machine Learning* **42**, 177–196.

Hofmann, T. and Puzicha, J. 1999 Latent class models for collaborative filtering. In *Proc. 16th Int. Joint Conf. on Artificial Intelligence*, pp. 688–693.

Hofmann, T., Puzicha, J. and Jordan, M. I. 1999 Learning from dyadic data. In *Advances in Neural Information Processing Systems 11: Proc. 1998 Conf.* (ed. M. S. Kearns, S. A. Solla and D. Cohen), pp. 466–472. Cambridge, MA: MIT Press.

Huberman, B. A. and Adamic, L. A. 1999 Growth dynamics of the World Wide Web. *Nature* **401**, 131.

Huberman, B. A., Pirolli, P. L. T., Pitkow, J. E. and Lukose, R. M. 1998 Strong regularities in World Wide Web surfing. *Science* **280**, 95–97.

Hunter, J. and Shotland, R. 1974 Treating data collected by the small world method as a Markov process. *Social Forces* **52**, 321.

ISO 1986 *Information Processing, Text and Office Systems, Standard Generalized Markup Language (SGML)*, ISO 8879, 1st edn. Geneva, Switzerland: International Organization for Standardization.

Itti, L. and Koch, C. 2001 Computational modelling of visual attention. *Nature Rev. Neurosci.* **2**, 194–203.

Jaakkola, T. S. and Jordan, I. 1997 Recursive algorithms for approximating probabilities in graphical models. In *Advances in Neural Information Processing Systems* (ed. M. C. Mozer, M. I. Jordan and T. Petsche), vol. 9, pp. 487–493. Cambridge, MA: MIT Press.

Jaeger, M. 1997 Relational Bayesian networks In *Proc. 13th Conf. on Uncertainty in Artificial Intelligence (UAI-97)* (ed. D. Geiger and P. P. Shenoy), pp. 266–273. San Francisco, CA: Morgan Kaufmann.

Janiszewski, C. 1998 The influence of display characteristics on visual exploratory behavior. *J. Consumer Res.* **25**, 290–301.

Jansen, B. J., Spink, A., Bateman, J. and Saracevic, T. 1998 Real-life information retrieval: a study of user queries on the Web. *SIGIR Forum* **32**, 5–17.

Jaynes, E. T. 1986 Bayesian methods: general background. In *Maximum Entropy and Bayesian Methods in Statistics* (ed. J. H. Justice), pp. 1–25. Cambridge University Press.

Jaynes, E. T. 2003 *Probability Theory: The Logic of Science*. Cambridge University Press.

Jensen, F. V. 1996 *An Introduction to Bayesian Networks*. Springer.

Jensen, F. V., Lauritzen, S. L. and Olesen, K. G. 1990 Bayesian updating in causal probabilistic networks by local computations. *Comput. Statist. Q.* **4**, 269–282.

Jeong, H., Tomber, B., Albert, R., Oltvai, Z. and Barabási, A.-L. 2000 The large-scale organization of metabolic networks. *Nature* **407**, 651–654.

Joachims, T. 1997 A probabilistic analysis of the Rocchio algorithm with TFIDF for text categorization. In *Proc. 14th Int. Conf. on Machine Learning*, pp. 143–151. San Francisco, CA: Morgan Kaufmann.

Joachims, T. 1998 Text categorization with support vector machines: learning with many relevant features. In *Proc. 10th Eur. Conf. on Machine Learning*, pp. 137–142. Springer.

Joachims, T. 1999a Making large-scale SVM learning practical In *Advances in Kernel Methods: Support Vector Learning* (ed. B. Schölkopf, C. J. C. Burges and A. J. Smola), pp. 169–184. Cambridge, MA; MIT Press.

Joachims, T. 1999b Transductive inference for text classification using support vector machines. In *Proc. 16th Int. Conf. on Machine Learning (ICML)*, pp. 200–209. San Francisco, CA: Morgan Kaufmann.

Joachims, T. 2002 *Learning to Classify Text using Support Vector Machines*. Kluwer.

Jordan, M. I. (ed.) 1999 *Learning in Graphical Models*. Cambridge, MA: MIT Press.

Jordan, M. I., Ghahramani, Z. and Saul, L. K. 1997 Hidden Markov decision trees. In *Advances in Neural Information Processing Systems* (ed. M. C. Mozer, M. I. Jordan and T. Petsche), vol. 9, pp. 501–507. Cambridge, MA: MIT Press.

Jumarie, G. 1990 *Relative information*. Springer.

Kask, K. and Dechter, R. 1999 Branch and bound with mini-bucket heuristics. In *Proc. Int. Joint Conf. on Artificial Intelligence (IJCAI99)*, pp. 426–433.

Kessler, M. 1963 Bibliographic coupling between scientific papers. *Am. Documentat.* **14**, 10–25.

Killworth, P. and Bernard, H. 1978 Reverse small world experiment. *Social Networks* **1**, 159.

Kira, K. and Rendell, L. A. 1992 A practical approach to feature selection. In *Proc. 9th Int. Conf. on Machine Learning*, pp. 249–256. San Francisco, CA: Morgan Kaufmann.

Kittler, J. 1986 Feature selection and extraction. In *Handbook of Pattern Recognition and Image Processing* (ed. T. Y. Young and K. S. Fu), ch. 3. Academic.

Kleinberg, J. 1998 Authoritative sources in a hyperlinked environment. In *Proc. 9th Ann. ACM–SIAM Symp. on Discrete Algorithms*, pp. 668–677. New York: ACM Press. (A preliminary version of this paper appeared as IBM Research Report RJ 10076, May 1997.)

Kleinberg, J. 1999 Hubs, authorities, and communities. *ACM Comput. Surv.* **31**, 5.

Kleinberg, J. 2000a Navigation in a small world. *Nature* **406**, 845.

Kleinberg, J. 2000b The small-world phenomenon: an algorithmic perspective. In *Proc. 32nd ACM Symp. on the Theory of Computing*.

Kleinberg, J. 2001 Small-world phenomena and the dynamic of information. *Advances in Neural Information Processing Systems (NIPS)*, vol. 14. Cambridge, MA: MIT Press.

Kleinberg, J. and Lawrence, S. 2001 The structure of the Web. *Science* **294**, 1849–1850.

Kleinberg, J., Kumar, R., Raghavan, P., Rajagopalan, S. and Tomkins, A. 1999 The Web as a graph: measurements, models, and methods. In *Proc. Int. Conf. on Combinatorics and Computing*. Lecture notes in Computer Science, vol. 1627. Springer.

Kohavi, R. and John, G. 1997 Wrappers for feature subset selection. *Artif. Intel.* **97**, 273–324.

Koller, D. and Sahami, M. 1997 Hierarchically classifying documents using very few words. In *Proc. 14th Int. Conf. on Machine Learning (ICML-97)*, pp. 170–178. San Francisco, CA: Morgan Kaufmann.

Koller, D. and Sahami, N. 1996 Toward optimal feature selection. In *Proc. 13th Int. Conf. on Machine Learning*, pp. 284–292.

Korte, C. and Milgram, S. 1978 Acquaintance networks between racial groups: application of the small world method. *J. Pers. Social Psych.* **15**, 101.

Koster, M. 1995 Robots in the Web: threat or treat? *ConneXions* **9**(4).

Krishnamurthy, B., Mogul, J. C. and Kristol, D. M. 1999 Key differences between HTTP/1.0 and HTTP/1.1. In *Proc. 8th Int. World-Wide Web Conf.* Elsevier.

Kruger, A., Giles, C. L., Coetzee, F., Glover, E. J., Flake, G. W., Lawrence, S. and Omlin, C. W. 2000 DEADLINER: building a new niche search engine. In *Proc. 2000 ACM–CIKM International Conf. on Information and Knowledge Management (CIKM-00)* (ed. A. Agah, J. Callan and E. Rundensteiner), pp. 272–281. New York: ACM Press.

Kullback, S. 1968 *Information theory and statistics*. New York: Dover.

Kumar, S. R., Raghavan, P., Rajagopalan, S. and Tomkins, A. 1999a Extracting large-scale knowledge bases from the Web. *Proc. 25th VLDB Conf. VLDB J.*, pp. 639–650.

Kumar, S. R., Raghavan, P., Rajagopalan, S. and Tomkins, A. 1999b Trawling the Web for emerging cyber communities. In *Proc. 8th World Wide Web Conf. Comp. Networks* **31**, 11–16.

Kumar, S. R., Raghavan, P., Rajagopalan, S., Sivakumar, D., Tomkins, A. and Upfal, E. 2000 Stochastic models for the Web graph. In *Proc. 41st IEEE Ann. Symp. on the Foundations of Computer Science*, pp. 57–65.

Kushmerick, N., Weld, D. S. and Doorenbos, R. B. 1997 Wrapper induction for information extraction. In *Proc. Int. Joint Conf. on Artificial Intelligence (IJCAI)*, pp. 729–737.

Lafferty, J., McCallum, A. and Pereira, F. 2001 Conditional random fields: probabilistic models for segmenting and labeling sequence data. In *Proc. 18th Int. Conf. on Machine Learning*, pp. 282–289. San Francisco, CA: Morgan Kaufmann.

Lam, W. and Ho, C. Y. 1998 Using a generalized instance set for automatic text categorization. In *Proc. SIGIR-98, 21st ACM Int. Conf. on Research and Development in Information Retrieval* (ed. W. B. Croft, A. Moffat, C. J. van Rijsbergen, R. Wilkinson and J. Zobel), pp. 81–89. New York: ACM Press.

Lang, K. 1995 Newsweeder: Learning to filter news *Proc. 12th Int. Conf. on Machine Learning* (ed. A. Prieditis and S. J. Russell), pp. 331–339. San Francisco, CA: Morgan Kaufmann.

Langley, P. 1994 Selection of relevant features in machine learning. In *Proc. AAAI Fall Symp. on Relevance*, pp. 140–144.

Lau, T. and Horvitz, E. 1999 Patterns of search: analyzing and modeling Web query refinement. In *Proc. 7th Int. Conf. on User Modeling*, pp. 119–128. Springer.

Lauritzen, S. L. 1996 *Graphical Models*. Oxford University Press.

Lauritzen, S. L. and Spiegelhalter, D. J. 1988 Local computations with probabilities on graphical structures and their application to expert systems. *J. R. Statist. Soc.* B **50**, 157–224.

Lawrence, S. 2001 Online or invisible? *Nature* **411**, 521.

Lawrence, S. and Giles, C. L. 1998a Context and page analysis for improved Web search. *IEEE Internet Computing* **2**, 38–46.

Lawrence, S. and Giles, C. L. 1998b Searching the World Wide Web. *Science* **280**, 98–100.

Lawrence, S. and Giles, C. L. 1999a Acccessibility of information on the Web. *Nature* **400**, 107–109.

Lawrence, S., Giles, C. L. and Bollacker, K. 1999 Digital libraries and autonomous citation indexing. *IEEE Computer* **32**, 67–71.

Leek, T. R. 1997 Information extraction using hidden Markov models. Master's thesis, University of California, San Diego.

Lempel, R. and Moran, S. 2001 SALSA: the stochastic approach for link-structure analysis. *ACM Trans. Informat. Syst.* **19**, 131–160.

Letsche, T. A. and Berry, M. W. 1997 Large-scale information retrieval with latent semantic indexing. *Information Sciences* **100**, 105–137.

Lewis, D. D. 1992 An evaluation of phrasal and clustered representations on a text categorization task. In *Proc. 15th Ann. Int. ACM SIGIR Conf. on Research and Development in Information Retrieval*, pp. 37–50. New York: ACM Press.

Lewis, D. D. 1997 *Reuters-21578 text categorization test collection*. (Documentation and data available at http://www.daviddlewis.com/resources/testcollections/reuters21578/.)

Lewis, D. D. 1998 Naive Bayes at forty: the independence assumption in information retrieval. In *Proc. 10th Eur. Conf. on Machine Learning*, pp. 4–15. Springer.

Lewis, D. D. and Catlett, J. 1994 Heterogeneous uncertainty sampling for supervised learning. In *Proc. ICML-94, 11th Int. Conf. on Machine Learning* (ed. W. W. Cohen and H. Hirsh), pp. 148–156. San Francisco, CA: Morgan Kaufmann.

Lewis, D. D. and Gale, W. A. 1994 A sequential algorithm for training text classifiers. In *Proc. 17th Ann. Int. ACM SIGIR Conf. on Research and Development in Information Retrieval*, pp. 3–12. Springer.

Lewis, D. D. and Ringuette, M. 1994 Comparison of two learning algorithms for text categorization. In *Proc. 3rd Ann. Symp. on Document Analysis and Information Retreval*, pp. 81–93.

Li, S., Montgomery, A., Srinivasan, K. and Liechty, J. L. 2002 Predicting online purchase conversion using Web path analysis. Graduate School of Industrial Administration, Carnegie Mellon University, Pittsburgh, PA. (Available at http://www.andrew.cmu.edu/~alm3/papers/purchase%20conversion.pdf.)

Li, W. 1992 Random texts exhibit Zipf's-law-like word frequency dsitribution. *IEEE Trans. Inform. Theory* **38**, 1842–1845.

Lieberman, H. 1995 Letizia: an agent that assists Web browsing. In *Proc. 14th Int. Joint Conf. on Artificial Intelligence (IJCAI-95)* (ed. C. S. Mellish), pp. 924–929. San Mateo, CA: Morgan Kaufmann.

Little, R. J. A. and Rubin, D. B. 1987 *Statistical Analysis with Missing Data*. John Wiley & Sons, Ltd/Inc.

Liu, H. and Motoda, H. 1998 *Feature Selection for Knowledge Discovery and Data Mining*. Kluwer Academic.

Lovins, J. B. 1968 Development of a stemming algorithm. *Mech. Transl. Comput. Linguistics* **11**, 22–31.

McCallum, A. and Nigam, K. 1998 A comparison of event models for naive Bayes text classification. *AAAI/ICML-98 Workshop on Learning for Text Categorization*, pp. 41–48. Menlo Park, CA: AAAI Press.

McCallum A., Freitag, D. and Pereira, F. 2000a Maximum entropy Markov models for information extraction and segmentation. In *Proc. 17th Int. Conf. on Machine Learning*, pp. 591–598. San Francisco, CA: Morgan Kaufmann.

McCallum, A., Nigam, K. and Ungar, L. H. 2000b Efficient clustering of high-dimensional data sets with application to reference matching. In *Proc. 6th ACM SIGKDD Int. Conf. on Knowledge Discovery and Data Mining*, pp. 169–178. New York: ACM Press.

McCallum, A. K., Nigam, K., Rennie, J. and Seymore, K. 2000c Automating the construction of Internet portals with machine learning. *Information Retrieval* **3**, 127–163.

McCann, K., Hastings, A. and Huxel, G. R. 1998 Weak trophic interactions and the balance of nature. *Nature* **395**, 794–798.

McClelland, J. L. and Rumelhart, D. E. 1986 *Parallel Distributed Processing: Explorations in the Microstructure of Cognition*. Cambridge, MA: MIT Press.

McEliece, R. J. 1977 *The Theory of Information and Coding*. Reading, MA: Addison-Wesley.

McEliece, R. J. and Yildirim, M. 2002 Belief propagation on partially ordered sets. In *Mathematical Systems Theory in Biology, Communications, and Finance* (ed. D. Gilliam and J. Rosenthal). Institute for Mathematics and its Applications, University of Minnesota.

McEliece, R. J., MacKay, D. J. C. and Cheng, J. F. 1997 Turbo decoding as an instance of Pearl's 'belief propagation' algorithm. *IEEE J. Select. Areas Commun.* **16**, 140–152.

MacKay, D. J. C. and Peto, L. C. B. 1995a A hierarchical Dirichlet language model. *Natural Language Engng* **1**, 1–19.

McLachlan, G. and Peel, D. 2000 *Finite Mixture Models*. John Wiley & Sons, Ltd/Inc.

Mahmoud, H. M. and Smythe, R. T. 1995 A survey of recursive trees. *Theory Prob. Math. Statist.* **51**, 1–27.

Manber, U. and Myers, G. 1990 Suffix arrays: a new method for on-line string searches. In *Proc. 1st Ann. ACM–SIAM Symp. on Discrete Algorithms*, pp. 319–327. Philadelphia, PA: Society for Industrial and Applied Mathematics.

Mandelbrot, B. 1977 *Fractals: Form, Chance, and Dimension*. New York: Freeman.

Marchiori, M. 1997 The quest for correct information on the Web: hyper search engines. In *Proc. 6th Int. World-Wide Web Conf., Santa Clara, CA. Comp. Networks* **29**, 1225–1235.

Mark, E. F. 1988 Searching for information in a hypertext medical handbook. *Commun ACM* **31**, 880–886.

Maron, M. E. 1961 Automatic indexing: an experimental inquiry. *J. ACM* **8**, 404–417.

Maslov, S. and Sneppen, K. 2002 Specificity and stability in topology of protein networks. *Science* **296**, 910–913.

Melnik, S., Raghavan, S., Yang, B. and Garcia-Molina, H. 2001 Building a distributed full-text index for the Web. *ACM Trans. Informat. Syst.* **19**, 217–241.

Mena, J. 1999 *Data Mining your Website*. Boston, MA: Digital Press.

Menczer, F. 1997 ARACHNID: adaptive retrieval agents choosing heuristic neighborhoods for information discovery. In *Proc. 14th Int. Conf. on Machine Learning*, pp. 227–235. San Francisco, CA: Morgan Kaufmann.

Menczer, F. and Belew, R. K. 2000 Adaptive retrieval agents: internalizing local context and scaling up to the Web. *Machine Learning* **39**, 203–242.

Mihail. M. and Papadimitriou, C. H. 2002 On the eigenvalue power law. In *Randomization and Approximation Techniques, Proc. 6th Int. Workshop, RANDOM 2002, Cambridge, MA, USA, 13–15 September 2002* (ed. J. D. P. Rolim and S. P. Vadhan). Lecture Notes in Computer Science, vol. 2483, pp. 254–262. Springer.

Milgram, S. 1967 The small world problem. *Psychology Today* **1**, 61.

Milo, R., Shen-Orr, S., Itzkovitz, S., Kashtan, N., Chklovskii, D. and Alon, U. 2002 Network motifs: simple building blocks of complex networks. *Science* **298**, 824–827.

Mitchell, T. 1997 *Machine Learning*. McGraw-Hill.

Mitzenmacher, M. 2002 A brief history of generative models for power law and lognormal distributions. Technical Report, Harvard University, Cambridge, MA.

Moffat, A. and Zobel, J. 1996 Self-indexing inverted files for fast text retrieval. *ACM Trans. Informat. Syst.* **14**, 349–379.

Montgomery, A. L. 2001 Applying quantitative marketing techniques to the Internet. *Interfaces* **30**, 90–108.

Mooney, R. J. and Roy, L. 2000 Content-based book recommending using learning for text categorization. In *Proc. 5th ACM Conf. on Digital Libraries*, pp. 195–204. New York: ACM Press.

Mori, S., Suen, C. and Yamamoto, K. 1992 Historical review of OCR research and development. *Proc. IEEE* **80**, 1029–1058.

Moura, E. S., Navarro, G. and Ziviani, N. 1997 Indexing compressed text. In *Proc. 4th South American Workshop on String Processing* (ed. R. Baeza-Yates), International Informatics Series, pp. 95–111. Ottawa: Carleton University Press.

Najork, M. and Wiener, J. 2001 Breadth-first search crawling yields high-quality pages. In *Proc. 10th Int. World Wide Web Conf.*, pp. 114–118. Elsevier.

Neal, R. M. 1992 Connectionist learning of belief networks. *Artif. Intel.* **56**, 71–113.

Neville-Manning, C. and Reed, T. 1996 A PostScript to plain text converter. Technical report. (Available at http://www.nzdl.org/html/prescript.html.)

Newman, M. E. J., Moore, C. and Watts, D. J. 2000 Mean-field solution of the small-world network model. *Phys. Rev. Lett.* **84**, 3201–3204.

Ng, A. Y. and Jordan, M. I. 2002 On discriminative vs generative classifiers: a comparison of logistic regression and Naive Bayes. *Advances in Neural Information Processing Systems 14. Proc. 2001 Neural Information Processing Systems (NIPS) Conference*. MIT Press.

Ng, A. Y., Zheng, A. X. and Jordan, M. I. 2001 Stable algorithms for link analysis. In *Proc. 24th Ann. Int. ACM SIGIR Conf. on Research and Development in Information Retrieval*, pp. 258–266. New York: ACM Press.

Nigam, K. and Ghani, R. 2000 Analyzing the effectiveness and applicability of co-training. In *Proc. 2000 ACM–CIKM Int. Conf. on Information and Knowledge Management (CIKM-00)* (ed. A. Agah, J. Callan and E. Rundensteiner), pp. 86–93. New York: ACM Press.

Nigam, K., McCallum A., Thrun, S. and Mitchell, T. 2000 Text classification from labeled and unlabeled documents using EM. *Machine Learning* **39**, 103–134.

Nothdurft, H. 2000 Salience from feature contrast: additivity across dimensions. *Vision Res.* **40**, 1183–1201.

Olshausen, B. A., Anderson, C. H. and Essen, D. C. V. 1993 A neurobiological model of visual attention and invariant pattern recognition based on dynamic routing of information. *J. Neurosci.* **13**, 4700–4719.

Oltvai, Z. N. and Barabási, A.-L. 2002 Life's complexity pyramid. *Science* **298**, 763–764.

O'Neill, E. T., McClain P. D. and Lavoie, B. F. 1997 A methodology for sampling the World Wide Web Annual Review of OCLC Research. (Available at http://www.oclc.org/research/publications/arr/ 1997/oneill/o%27neillar98%0213.htm.)

Page, L., Brin, S., Motwani, R. and Winograd, T. 1998 The PageRank citation ranking: bringing order to the Web. Technical report, Stanford University. (Available at http://www-db.stanford.edu/-backrub/pageranksub.ps.)

Paine, R. T. 1992 Food-web analysis through field measurements of per capita interaction strength. *Nature* **355**, 73–75.

Pandurangan, G., Raghavan, P. and Upfal, E. 2002 Using PageRank to characterize Web structure. In *Proc. 8th Ann. Int. Computing and Combinatorics Conf. (COCOON)*. Lecture Notes in Computer Science, vol. 2387, p. 330. Springer.

Papineni, K. 2001 Why inverse document frequency? *Proc. North American Association for Computational Linguistics*, pp. 25–32.

Passerini, A., Pontil, M. and Frasconi, P. 2002 From margins to probabilities in multiclass learning problems. In *Proc. 15th Eur. Conf. on Artificial Intelligence* (ed. F. van Harmelen). Frontiers in Artificial Intelligence and Applications Series. Amsterdam: IOS Press.

Pazzani, M. 1996 Searching for dependencies in Bayesian classifiers. In *Proc. 5th Int. Workshop on Artificial Intelligence and Statistics*, pp. 239–248. Springer.

Pearl, J. 1988 *Probabilistic Reasoning in Intelligent Systems*. San Mateo, CA: Morgan Kaufmann.

Pennock, D. M., Flake, G. W., Lawrence, S., Glover, E. J. and Giles, C. L. 2002 Winners don't take all: characterizing the competition for links on the Web. *Proc. Natl Acad. Sci.* **99**, 5207–5211.

Perline, R. 1996 Zipf's law, the central limit theorem, and the random division of the unit interval. *Phys. Rev. E* **54**, 220–223.

Pew Internet Project Report 2002 Search engines. (Available at http://www.pewinternet.org/reports/toc.asp?Report=64.)

Phadke, A. G. and Thorp, J. S. 1988 *Computer Relaying for Power Systems*. John Wiley & Sons, Ltd/Inc.

Philips, T. K., Towsley, D. F. and Wolf, J. K. 1990 On the diameter of a class of random graphs. *IEEE Trans. Inform. Theory* **36**, 285–288.

Pimm, S. L., Lawton, J. H. and Cohen, J. E. 1991 Food web patterns and their consequences. *Nature* **350**, 669–674.

Pittel, B. 1994 Note on the heights of random recursive trees and random *m*-ary search trees. *Random Struct. Algorithms* **5**, 337–347.

Platt, J. 1999 Fast training of support vector machines using sequential minimal optimization. In *Advances in Kernel Methods – Support Vector Learning* (ed. B. Schölkopf, C. J. C. Burges and A. J. Smola,), pp. 185–208. Cambridge, MA: MIT Press.

Popescul, A., Ungar, L. H., Pennock, D. M. and Lawrence, S. 2001 Probabilistic models for unified collaborative and content-based recommendation in sparse-data environments. In *Proc. 17th Int. Conf. on Uncertainty in Artificial Intelligence*, pp. 437–444. San Francisco, CA: Morgan Kaufmann.

Porter, M. 1980 An algorithm for suffix stripping. *Program* **14**, 130–137.

Quinlan, J. R. 1986 Induction of decision trees. *Machine Learning* **1**, 81–106.

Quinlan, J. R. 1990 Learning logical definitions from relations. *Machine Learning* **5**, 239–266.

Rafiei, D. and Mendelzon, A. 2000 What is this page known for? Computing Web page reputations. In *Proc. 9th World Wide Web Conf.*

Raggett, D., Hors, A. L. and Jacobs, I. (eds) 1999 *HTML 4.01 Specification*. W3 Consortium Recommendation. (Available at http://www.w3.org/TR/html4/.)

Raskinis, I. M. G. and Ganascia, J. 1996 Text categorization: a symbolic approach. In *Proc. 5th Ann. Symp. on Document Analysis and Information Retrieval*. New York: ACM Press.

Redner, Q. 1998 How popular is your paper? An empirical study of the citation distribution. *Eur. Phys. J.* B **4**, 131–134.

Resnick, P., Iacovou, N., Suchak, M., Bergstrom, P. and Riedl, J. 1994 GroupLens: an open architecture for collaborative filtering of netnews. In *Proc. 9th ACM Conf. on Computer-Supported Cooperative Work*, pp. 175–186. New York: ACM Press.

Ripeanu, M., Foster, I. and Iamnitchi, A. 2002 Mapping the Gnutella network: properties of large-scale peer-to-peer systems and implications for system design. *IEEE Internet Comput. J.* **6**, 99–100.

Roberts, M. J. and Mahesh, S. M. 1999 Hotmail. Technical report, Harvard University, Cambridge, MA. Case 899-185, Harvard Business School Publishing.

Robertson, S. E. 1977 The probability ranking principle in IR. *J. Doc.* **33**, 294–304. (Also reprinted in Jones and Willett (1997), pp. 281–286.)

Robertson, S. E. and Spärck Jones, K. 1976 Relevance weighting of search terms. *J. Am. Soc. Informat. Sci.* **27**, 129–146.

Robertson, S. E. and Walker, S. 1994 Some simple effective approximations to the 2-Poisson model for probabilistic weighted retrieval. In *Proc. 17th Ann. Int. ACM SIGIR Conf. on Research and Development in Information Retrieval*, pp. 232–241. Springer.

Rosenblatt, F. 1958 The perceptron: a probabilistic model for information storage and organization in the brain. *Psychol. Rev.* **65**, 386–408.

Ross, S. M. 2002 *Probability Models for Computer Science*. San Diego, CA: Adademic Press.

Russell, S. and Norvig, P. 1995 *Artificial Intelligence: A Modern Approach*. Prentice Hall.

Sahami, M., Dumais, S., Heckerman, D. and Horvitz, E. 1998 A Bayesian approach to filtering junk e-mail. *AAAI-98 Workshop on Learning for Text Categorization*, pp. 55–62.

Salton, G. 1971 *The SMART Retrieval System: Experiments in Automatic Document Processing*. Englewood Cliffs, NJ: Prentice Hall.

Salton, G. and McGill, M. J. 1983 *Introduction to modern information retrieval*. McGraw-Hill.

Salton, G., Fox, E. A. and Wu, H. 1983 Extended Boolean information retrieval. *Commun. ACM* **26**, 1022–1036.

Sarukkai, R. R. 2000 Link prediction and path analysis using Markov chains. *Comp. Networks* **33**, 377–386.

Sarwar, B. M., Karypis, G., Konstan, J. A. and Riedl, J. T. 2000 Analysis of recommender algorithms for e-commerce. In *Proc. 2nd ACM Conf. on Electronic Commerce*, pp. 158–167. New York: ACM Press.

Saul, L. and Pereira, F. 1997 Aggregate and mixed-order Markov models for statistical language processing. In *Proc. 2nd Conf. on Empirical Methods in Natural Language Processing* (ed. C. Cardie and R. Weischedel), pp. 81–89. Somerset, NJ: Association for Computational Linguistics.

Saul, L. K. and Jordan, M. I. 1996 Exploiting tractable substructures in intractable networks. In *Advances in Neural Information Processing Systems* (ed. D. S. Touretzky, M. C. Mozer and M. E Hasselmo), vol. 8, pp. 486–492. Cambridge, MA: MIT Press.

Savage, L. J. 1972 *The foundations of statistics*. New York: Dover.

Schafer, J. B., Konstan, J. A. and Riedl, J. 2001 E-commerce recommendation applications. *J. Data Mining Knowl. Discovery* **5**, 115–153.

Schapire, R. E. and Freund, Y. 2000 Boostexter: a boosting-based system for text categorization. *Machine Learning* **39**, 135–168.

Schoelkopf, B. and Smola, A. 2002 *Learning with Kernels*. Cambridge, MA: MIT Press.

Sebastiani, F. 2002 Machine learning in automated text categorization. *ACM Comput. Surv.* **34**, 1–47.

Sen, R. and Hansen, M. H. 2003 Predicting a Web user's next request based on log data. *J. Computat. Graph. Stat.* (In the press.)

Seneta, E. 1981 *Nonnegative Matrices and Markov Chains*. Springer.

Shachter, R. D. 1988 Probabilistic inference and influence diagrams. *Oper. Res.* **36**, 589–604.

Shachter, R. D., Anderson, S. K. and Szolovits, P. 1994 Global conditioning for probabilistic inference in belief networks. In *Proc. Conf. on Uncertainty in AI*, pp. 514–522. San Francisco, CA: Morgan Kaufmann.

Shahabi, C., Banaei-Kashani, F. and Faruque, J. 2001 A framework for efficient and anonymous Web usage mining based on client-side tracking. In *Proceedings of WEBKDD 2001*. Lecture Notes in Artificial Intelligence, vol. 2356, pp. 113–144. Springer.

Shannon, C. E. 1948a A mathematical theory of communication. *Bell Syst. Tech. J.* **27**, 379–423.

Shannon, C. E. 1948b A mathematical theory of communication. *Bell Syst. Tech. J.* **27**, 623–656.

Shardanand, U. and Maes, P. 1995 Social information filtering: algorithms for automating 'word of mouth'. In *Proc. Conf. on Human Factors in Computing Systems*, pp. 210–217.

Shore, J. E. and Johnson, R. W. 1980 Axiomatic derivation of the principle of maximum entropy and the principle of minimum cross-entropy. *IEEE Trans. Inform. Theory* **26**, 26–37.

Silverstein, C., Henzinger, M., Marais, H. and Moricz, M. 1998 Analysis of a very large AltaVista query log. Technical Note 1998-14, Digital System Research Center, Palo Alto, CA.

Slonim, N. and Tishby, N. 2000 Document clustering using word clusters via the information bottleneck method. In *Proc. 23rd Int. Conf. on Research and Development in Information Retrieval*, pp. 208–215. New York: ACM Press.

Slonim, N., Friedman, N. and Tishby, N. 2002 Unsupervised document classification using sequential information maximization. In *Proc. 25th Int. Conf. on Research and Development in Information Retrieval*, pp. 208–215. New York: ACM Press.

Small, H. 1973 Co-citation in the scientific literature: a new measure of the relationship between two documents. *J. Am. Soc. Inf. Sci.* **24**, 265–269.

Smyth, P., Heckerman, D. and Jordan, M. I. 1997 Probabilistic independence networks for hidden Markov probability models. *Neural Comp.* **9**, 227–267.

Soderland, S. 1999 Learning information extraction rules for semi-structured and free text. *Machine Learning* **34**, 233–272.

Sperberg-McQueen, C. and Burnard, L. (eds) 2002 *TEI P4: Guidelines for Electronic Text Encoding and Interchange*. Text Encoding Initiative Consortium. (Available at http://www.tei-c.org/.)

Spink, A., Jansen, B. J., Wolfram, D. and Saracevic, T. 2002 From e-sex to e-commerce: Web search changes. *IEEE Computer* **35**, 107–109.

Stallman, R. 1997 The right to read. *Commun. ACM* **40**, 85–87.

Sutton, R. S. and Barto, A. G. 1998 *Reinforcement Learning: An Introduction*. Cambridge, MA: MIT Press.

Tan, P. and Kumar, V. 2002 Discovery of Web robot sessions based on their navigational patterns. *Data Mining Knowl. Discov.* **6**, 9–35.

Tantrum, J., Murua, A. and Stuetzle, W. 2002 Hierarchical model-based clustering of large datasets through fractionation and refractionation. In *Proc. 8th ACM SIGKDD Int. Conf. on Knowledge Discovery and Data Mining*. New York: ACM Press.

Taskar, B., Abbeel, P. and Koller, D. 2002 Discriminative probabilistic models for relational data. In *Proc. 18th Conf. on Uncertainty in Artificial Intelligence*. San Francisco, CA: Morgan Kaufmann.

Tauscher, L. and Greenberg, S. 1997 Revisitation patterns in World Wide Web navigation. In *Proc. Conf. on Human Factors in Computing Systems CHI'97*, pp. 97–137. New York: ACM Press.

Tedeschi, B. 2000 Easier to use sites would help e-tailers close more sales. *New York Times*, 12 June 2000.

Tishby, N., Pereira, F. and Bialek, W. 1999 The information bottleneck method. In *Proc. 37th Ann. Allerton Conf. on Communication, Control, and Computing* (ed. B. Hajek and R. S. Sreenivas), pp. 368–377.

Titterington, D. M., Smith, A. F. M. and Makov, U. E. 1985 *Statistical Analysis of Finite Mixture Distributions*. John Wiley & Sons, Ltd/Inc.

Travers, J. and Milgram, S. 1969 An experimental study of the smal world problem. *Sociometry* **32**, 425.

Ungar, L. H. and Foster, D. P. 1998 Clustering methods for collaborative filtering. In *Proc. Workshop on Recommendation Systems at the 15th National Conf. on Artificial Intelligence*. Menlo Park, CA: AAAI Press.

Vapnik, V. N. 1982 *Estimation of Dependences Based on Empirical Data*. Springer.

Vapnik, V. N. 1995 *The Nature of Statistical Learning Theory*. Springer.

Vapnik, V. N. 1998 *Statistical Learning Theory*. John Wiley & Sons, Ltd/Inc.

Viterbi, A. J. 1967 Error bounds for convolutional codes and an asymptotically optimum decoding algorithm. *IEEE Trans. Inform. Theory* **13**, 260–269.

Walker, J. 2002 Links and power: the political economy of linking on the Web. In *Proc. 13th Conf. on Hypertext and Hypermedia*, pp. 72–73. New York: ACM Press.

Wall, L., Christiansen, T. and Schwartz RL. 1996 *Programming Perl*, 2nd edn. Cambridge, MA: O'Reilly & Associates.

Wasserman, S. and Faust, K. 1994 *Social Network Analysis*. Cambridge University Press.

Watts, D. J. and Strogatz, S. H. 1998 Collective dynamics of 'small-world' networks. *Nature* **393**, 440–442.

Watts, D. J., Dodds, P. S. and Newman, M. E. J. 2002 Identity and search in social networks. *Science* **296**, 1302–1305.

Weiss, S. M., Apte, C., Damerau, F. J., Johnson, D. E., Oles, F. J., Goetz, T. and Hampp, T. 1999 Maximizing text-mining performance. *IEEE Intell. Syst.* **14**, 63–69.

Weiss, Y. 2000 Correctness of local probability propagation in graphical models with loops. *Neural Comp.* **12**, 1–41.

White, H. 1970 Search parameters for the small world problem. *Social Forces* **49**, 259.

Whittaker, J. 1990 *Graphical Models in Applied Multivariate Statistics*. John Wiley & Sons, Ltd/Inc.

Wiener, E. D., Pedersen, J. O. and Weigend, A. S. 1995 A neural network approach to topic spotting. In *Proc. SDAIR-95, 4th Ann. Symp. on Document Analysis and Information Retrieval, Las Vegas, NV*, pp. 317–332.

Witten, I. H., Moffat, A. and Bell, T. C. 1999 *Managing Gigabytes: Compressing and Indexing Documents and Images*, 2nd edn. San Francisco, CA: Morgan Kaufmann.

Witten, I. H., Neville-Manning, C. and Cunningham, S. J. 1996 Building a digital library for computer science research: technical issues. In *Proc. Australasian Computer Science Conf., Melbourne, Australia*.

Wolf, J., Squillante, M., Yu, P., Sethuraman, J. and Ozsen, L. 2002 Optimal crawling strategies for Web search engines. In *Proc. 11th Int. World Wide Web Conf.*, pp. 136–147.

Xie, Y. and O'Hallaron, D. 2002 Locality in search engine queries and its implications for caching. In *Proc. IEEE Infocom 2002*, pp. 1238–1247. Piscataway, NJ: IEEE Press.

Yang, Y. 1999 An evaluation of statistical approaches to text categorization. *Information Retrieval* **1**, 69–90.

Yang, Y. and Liu, X. 1999 A re-examination of text categorization methods In *Proc. SIGIR-99, 22nd ACM Int. Conf. on Research and Development in Information Retrieval* (ed. M. A. Hearst, F. Gey and R. Tong), pp. 42–49. New York: ACM Press.

Yedidia, J., Freeman, W. T. and Weiss, Y. 2000 Generalized belief propagation. *Neural Comp.* **12**, 1–41.

York, J. 1992 Use of the Gibbs sampler in expert systems. *Artif. Intell.* **56**, 115–130.

Zamir, O. and Etzioni, O. 1998 Web document clustering: a feasibility demonstration. In *Proc. 21st Int. Conf. on Research and Development in Information Retrieval (SIGIR)*, pp. 46–54. New York: ACM Press.

Zelikovitz, S. and Hirsh, H. 2001 Using LSI for text classification in the presence of background text. In *Proc. 10th Int. ASM Conf. on Information and Knowledge Management*, pp. 113–118. New York: ACM Press.

Zhang, T. and Iyengar, V. S. 2002 Recommender systems using linear classifiers. *J. Machine Learn. Res.* **2**, 313–334.

Zhang, T. and Oles, F. J. 2000 A probability analysis on the value of unlabeled data for classification problems. In *Proc. 17th Int. Conf. on Machine Learning, Stanford, CA*, pp. 1191–1198.

Zhu, X., Yu, J. and Doyle, J. 2001 Heavy tails, generalized coding, and optimal Web layout. In *Proc. 2001 IEEE INFOCOM Conf.*, vol. 3, pp. 1617–1626. Piscataway, NJ: IEEE Press.

Zukerman, I., Albrecht, D. W. and Nicholson, A. E. 1999 Predicting user's requests on the WWW. In *Proc. UM99: 7th Int. Conf. on User Modeling*, pp. 275–284. Springer.

Index

absolute URI 35
access protocol 29
accuracy 86, 105
AdaBoost 108
agent log 41
aggregate Markov model 91
aging of documents 160, 161–7
aliases 49
aliasing 163
AltaVista 58, 202, 204, 205
America Online (AOL) 202
American Registry for Internet
  Numbers (ARIN) 160
anchor based prediction 154
anchors 32
Arachnid agents 157
aspect model 91
attributes 31, 94
authority 35, 131
automated logging 172
automated recommender systems
  212–24
  evaluating 214–5
  model-based collaborative
    filtering 218–23
  nearest-neighbor collaborative
    filtering 215–8

Backus–Naur Form (BNF) 35
bag of words representation 83, 95,
  106
base subgraph 130
Bayes' decision rule for classification
  87
Bayes' optimal separation 87
Bayes' rule 3, 18
Bayes' theorem 2, 11
Bayesian interpretation of probability
  2
Bayesian (belief) networks 13–15,
  116
Bayesian parameter estimation for
  Markov models 187–9
belief propagation (inference) 16, 92
Bellman optimality principle 156
Bernoulli distribution 96
Bernoulli likelihood model 96
Bernoulli model for the Naive Bayes
  classifier 87
best-first search strategy 149
Beta distribution 7, 8, 240
bibliographic coupling matrix 145
binary independence retrieval (BIR)
  87
Binomial distribution 22, 237–8, 240
biological networks 54
bow-tie structure 58–9
breadth-first search (BFS) 47
breakeven point 105
broad-scale networks 57
browser 29

Modeling the Internet and the Web
P. Baldi, P. Frasconi and P. Smyth
© 2003 P. Baldi, P. Frasconi and P. Smyth
ISBN: 0-470-84906-1

browsing behavior, client-side studies
    179–83
    1995–1997 180–1
    2002 181–3
    probabilistic models 184–201
    video-based analysis of Web
        usage 183
bucket 78–9
business networks 54

caching 174–6
canonicalization 49
canopies 119
Central Limit Theorem 238
centroids 20
chaining effect 118
Chi square distribution 240
CiteSeer 121, 157–8
classification 17–20
classification rules, induction of 93
classification trees 19
classifier 17
client-side data 177–9
    handling massive Web server logs
        179
cliquishness 62
clustering 20–2
    for documents 117–9
co-citation 141
co-citation matrix 145
coding theory 15
collective classification 116
common gateway interface (CGI) 49
completeness 33
compression techniques 79–80
ComScore 177–8, 212
ComScore Media Metrix 229
conditional entropy 247
conditional independence 220
conditioning 2
confinement 159–60
confusion matrix 104
connectionism 127

connectivity 236
    power-law 53–5
content-based ranking 82–6
context graphs 154–5
continuous (discrete power-law)
    distributions 22, 23, 238–40
co-occurrences 112
cookies 176–7
coordination 159
copy models 66–7
cosine coefficient 83
co-training 114–5
covariance 237
covariates 199
coverage 44, 45–9
crawler (spider; Web robot) 44
crawling
    basic 46–9
    distributed 158–60
    focused 152–8
    vs graph visiting 49
    relevance prediction 152–4
    selective 149–52
cross-entropy 84, 243, 244, 246
crossover 160
current search action 207
curse of dimensionality 99
customer data on the Web 212

data logging at Web servers 174
DEADLINER 158
Debian 89
decision-theoretic surfing model
    198–9
decision tree approximation 14
decision trees 93, 102
degree distribution 58
degrees of belief 2
dendrogram 118
dependency networks 219
depth of scoring functions 150
descriptive markup 30
dimension-reduction techniques 119
directed acyclic graph (DAG) 236

Dirichlet distribution 7, 188, 239–40, 250
Dirichlet prior 7, 12, 96, 97, 189
discrete distributions 237–8
discrete power-law (continuous) distributions 22, 23, 238–40
discriminative (decision-boundary) approach to classification 18
disordered lattice approach 63
distributed search algorithms 68–70
document clustering 116–20
    algorithm 117–9
    background and examples 116–7
    related approaches 119–20
document identifier (DID) 78–80
document similarity 83–4
Document Type Definition (DTD) 30
    frameset 32
    of HTML 31
    strict 32
    transitional 32
domain name service (DNS) 38, 49
domain name system 37–9
dominant eigenvalue 129
dominant eigenvector 129
dynamic coordination 159

e-commerce logs 212
eigengap 138
80/20 rule 20, 24
element name 31
element types 30
email 110–11
    filtering 17–18
    product recommendations 224–6
embarrassment level 169
energy function 5
entropy 243, 244–5
error function 5
error log 41
error rate 86
escape matrix 137
exchange 160

Excite search engine 202, 203, 204, 205
expectation 237
expectation maximization (EM) 5, 16
    algorithm 10–13, 21
    Naive Bayes and 112
exploratory browsing state 232
exponential distribution 22, 239, 240
exponential family 240–1
extensibility 33
eXtensible Markup Language (XML) 30
extreme value distribution 241

FASTUS system 120
feature selection 102–4
feature space 102
Fermi's model 25–7, 61
filters 102
finite-state Markov models 10, 14
firewall 160
first-in first-out (FIFO) policy 47
Fish algorithm 157
FOIL 115–6
frameset DTD 32
Freenet 73
frequentist interpretation of probability 2
freshness 167

Gamma distribution 22, 239, 240
Gaussian distribution 11, 22, 23, 238–9
Gentoo 89
geometric distribution 238,,240
GET method 39, 43, 174
Gibbs sampling 15
Gnutella 70, 73
'good citizen' search engines 173
Google search engine 44, 53, 57, 225
grace period 160
gradient descent 5
graph theory 235–7
graph visiting vs crawling 49

graphical models 13–17
greedy agglomerative clustering 120
greedy navigation algorithms 69
Gumbel distribution 241

hard focusing 153
hazard rate 161
HEAD method 41
Heaps law 24–5
Heaviside step function 164
heterogeneity of behavior 200
hidden (invisible) Web 39
hidden Markov models (HMMs) 14,
    122
hierarchical clustering 117, 118
hierarchical model 200
Hypertext Induced Topic Selection
    (HITS) 130–4
    stability of 138–9
Hotmail 225–7
hubness 131
hubs–authority bipartite graph
    pattern 71
human-generated traffic 173–4
hyper information 127
hyperlinks, text analysis and 114–6
HyperText Markup Language
    (HTML) 30–1
    body 32
    document structure 31–2
    DTD of 31
    header 32
    tokenization of 80–1
    version information 32
hypertext transfer protocol 38–9
    methods in 39

impact factor 125–6
incremental crawler 169
indegree distribution 58
index age of a document 167
indexing 77–80
inductive logic programming 93
inference 16, 92

'Informant, The' 163
information bottleneck 120
information extraction 120–2
information gain 102–3
information retrieval (IR) 77
information theory 243–51
informational goals 207, 208
Inquirus 46
intelligent crawling 157
Internet Corporation for Assigned
    Names and Numbers (ICANN) 38
Internet Engineering Task Force 34,
    44
Internet Protocol (IP) 37
    address space sampling 40–1
    addresses 37
Internet service provider (ISP) 175
Internet tomography 74
inverse document frequency (IDF) 84
inverse Gaussian distribution 198–9
inversion 78
inverted index 78
irreducible matrix 128
ISO/OSI reference model 36

Jensen's inequality 246, 247
junction tree 15

k nearest neighbours (k-NN) 93
    classifier 108
Kalman filter models 14
Karhunen–Loève transform 242
Karush–Kuhn–Tucker condition 101
kernel approaches 19, 102
Kleinberg's algorithm 132–4
K-means 12
    algorithm 20, 21
    clustering 117, 120
knowledge bases (KB) 106
kth-order Markov model 190
Kullback–Leibler distance 84, 243,
    244, 246
Kullback–Leibler divergence 84

Lagrange multiplier 11

Lagrangian 6, 11
Laplace smoothing 97
LASER 156
latent semantic analysis 88–92
latent semantic indexing (LSI)
    88–92, 112
    text documents and 89
    probabilistic 89–92
lattice perturbation models 61–3
law of total probability 2
learning directed graphical models
    from data 16–17
learning effect 201
lexical processing 80–6
likelihood 3
likelihood function 187
linear classifiers 219
link analysis 71, 126–45
    early approaches 126–8
    probabilistic 140–2
    limitations 142–5
    stability 138–9
link farm 143
links 32
Linux 89
log files 41–4
log probability score 192
luminosity 126

machine learning 77, 93, 102
macroaveraging 106
Markov blanket 103–4
Markov chain 7, 21, 57–8, 243
    mixtures of 191
Markov model 120
    aggregate 91
    Bayesian parameter estimation
        187–9
    finite-state 10, 14
    first-order 190–1
    fitting to observed page-request
        data 186–7
    hidden (HMMs) 14, 122
    kth-order 190

for page prediction 184–6
    position-dependent 193
    predicting page requests 189–93
    variable-order 193
    zero-order 9
markup 30
maximum *a posteriori* (MAP) 16
    estimation 5, 6
    framework 12
    optimization problem 8
    parameter estimation 6
maximum-distance algorithm 118
maximum likelihood 16, 187
    equations 12
    estimation 5, 6, 195–6
mean field approximation 62
mean field methods 15
mean field solution 62
mean posterior (MP) estimation 5, 9,
    16
measures of performance 104–6
Mercator 48
meta-items 217
metalanguage 30
meta-search engines 46
metasymbols 31
methods in HTTP 39
microaveraging 106
minimum description length 104
minimum-distance algorithm 118
mixture coefficients 11
mixture models 10–12, 14
ModApte split 108, 112, 113
model-based clustering 21
model-based collaborative filtering
    218–23
    conditional distribution models
        221–3
    general concepts 218–9
    joint density modeling 219–21
model-based combining of votes and
    content 223–4
Monte Carlo methods 15
mortality force 161

multinomial distribution 237–8, 240
multiple conditional model paradigm
   219
mutual information 243, 244, 246–7

Naive Bayes' classifier 18, 94–7, 102,
   105, 106
Naive Bayes' model 21, 93, 96, 119,
   220
nearest-neighbor collaborative
   filtering 215–8
   basic principles 215
   computational issues and
      clustering 218
   curse of dimensionality 217–8
   defining weights 216–7
negative binomial distribution 240
network value 227
networks and recommendations
   224–8
   diffusion model 226–8
   email-based product
      recommendations 224–6
neural networks 14, 19, 93
news filtering 110–11
news stories, classification of 107–10
NewsWeeder 110–11
next search action 207
Nielsen 177, 212
noisy OR models 14
nonnegative matrices 128–30
nonstationary Markov chain 193
normal distribution 240

one-shot coding strategy 102
one-step ahead prediction 191
one vs all coding strategy 102
ontology 106
Open Directory Project (ODP) 44
Open System Interconnect (OSI) 36
optical character recognition (OCR)
   81
optimal resource allocation 168

optimal separating hyperplane
   99–100
overall information 127
overlap analysis 45

page request prediction 199–201
PageRank 134–8, 151
   models 67–8
   power law of 57–8
   stability of 139
panel data 212
parent based prediction 153–4
Pareto–Zipf distribution (Zipf) 22, 80
partitioning 160
path 35
pattern recognition 102
Pearson correlation coefficient 104
peer-to-peer networks 70, 73
perceptrons 19, 98
period 128
Perl programming language 40, 81
Perron–Frobenius theorem 128–30
Pew Internet Project Report 44
PHITS 116, 142
PLSA 116
Poisson distribution 22, 96, 194, 238,
   240
polynomial time algorithms 71
polysemy 88
popularity 126, 150–1
Portable Document Format (PDF) 81
   tokenization 81
posterior distribution 247
PostScript 81
   tokenization of 81
power-law connectivity 53–5
power-law distribution 22–7, 52, 53
   origin of 25–7
power-law size 53
precision 86, 105
predicates 157
preferential attachment models 63–6
PreScript 81
primitive matrix 129

principal component analysis (PCA) 242
prior probability 2
priors 2, 3, 4
  distribution 247
probabilistic approach to classification 18
probabilistic decision trees 221–3
probabilistic link analysis 140–2
probabilistic Markov network model 116
probabilistic model-based clustering 117, 119, 120
probabilistic ranking principle (PRP) 86–8
probabilistic retrieval 86–8
probability 1–3
procedural markup 30
Prolog 115
protocols 29, 36–41
prototypes 20
proxy server 174–6
Pstotext 81
purchase-oriented state 232

quadratic programming 101
quality of ranking 44
quasi-random graphs 74
query 35

radial basis function (RBF) kernel 108
random effects model 200
random graphs 63, 236–7
random recursive trees 65
random sparse graphs 74
random tree 65
random uniform graph (random exponential graph) 72
recall 86, 105
recency 44–5, 160, 180
  measures of 167
  synchronization policies and 167–9

recursive query 38
reference models 36–7
referrer log 41
referrer page 174
reinforcement learning 155–7
relational Bayesian networks 116
relational learning 115–6
relative entropy (cross-entropy; Kullback–Leibler distance) 84, 243, 244, 246
relative URI 36
relevance 250–1
Rényi's entropies 250
Réseaux IP Européens (RIPE) 160
resolution 38
resource 29
resource identifiers 29, 31, 33–6
retrieval and evaluation measures 85–6
retrieval status value (RSV) 88
rich-get-richer model 63–4
Riemann's zeta function 23, 195
robot-generated traffic 173
robot program 41
Robots Exclusion Protocol 48
robustness 71–3
ROC curves (Receiver Operating Characteristic) 105
run-length modeling 193–4

SALSA (Stochastic Approach for Link Structure Analysis) 140–2
SavvySearch 46
scale-free networks 57
scale-free properties (80/20 rule) 20, 24
scaling property of power laws 53
Search Engine Persuasion (SEP) 142–3
search engines 44–9
  empirical studies of search behavior 202–6
  models for search strategies 207–8

querying, modeling and
    understanding 201–8
search query logs
    general characteristics of 203–5
    power-law characteristics 205–6
second-generation Web searching
    tools 126
security 73
seeds 46
self-scaling 24
self-similarity 24
semantic linkage 154
semi-Markov model 193
separating hyperplane 98
server 29
server access log 41
server log files 41
server-side data 174–7
server transfer log 41
session length modeling 194–8
Shannon's information 247–8
Shark algorithm 157
sigmoidal belief networks 14
simple Markov chain models 14
simulated annealing 5
single-link clustering 118
single-scale networks 57
singular value decomposition (SVD)
    88, 89, 217–8, 241–2
six degrees of separation 55, 69
slack variables 101
sliding-window based classification
    121
small-world networks 55–7, 65
    classes 57
social networks 54
soft focusing 154
source anchor 32
spam 17–8, 145, 224, 225–6
sparseness 52, 83
specificity 105
spherical K-means algorithm 119
spider 44
spider program 41

spreading activation mechanism 225
stability 138–9
    of HITS 138–9
    of PageRank 139
Standard Generalized Markup
    Language (SGML) 30–1
static coordination 159
stationary distribution of the Markov
    chain 130
status code 41
stemming 82
stop words, removal of 82
strict DTD 32
strongly connected component (SCC)
    58
structural risk minimization 97
subgraph patterns and communities
    70–1
supervised learning 20
    with unlabeled data 111–4
support vector classifiers 97–102
support vector machines (SVMs) 19,
    93, 97–102
    classifier 106
    transductive 113–4
support vectors 101
surprise 249
    computing 249–50
    learning and 250
symbolic rule learning 104
synonymy 88

tags 31
target anchor 32
TCP/IP 35, 36–7
telecommunication networks 54
tempered EM (TEM) 92
term frequencies (TF) 84
terms 78
text categorization 77, 93–114
text conflation 82
Text REtrieval Conference (TREC)
    86
textual information 127

TF–IDF 118
    weight 84
Tightly Knit Community (TKC)
    Effect 141
time interval 207
tokenization 80–1
topic drift problem 134
topic locality 153
top-level domain (TLD) 38
topological sort 127
transductive support vector machines
    113–4
transitional DTD 32
Transmission Control Protocol (TCP)
    37
transportation networks 54
traps 49
triangulated moral graph 15

unconditional marginal dependence
    220
uniform random graphs 73
Uniform Resource Identifier (URI)
    32, 33–6
Uniform Resource Locator (URL)
    34, 48–9
Uniform Resource Name (URN) 34
unsupervised learning 20
user agent 38
user search action 207

variance 237
variable-order Markov model 193
variational methods 15
vector-space model 82–3
visibility 126
visible Web 46
Viterbi algorithm 122

Vivisimo search engine 117, 203, 205
vocabulary 78
vocabulary reduction 82
vulnerability 71–3

Web data and measurement 172–9
Web documents 30–2
Web dynamics 160–9
Web graphs
    applications 68–73
    Internet and 51–3
    reason for studying 52
Web-KB system 120–1
Web pages
    analysis for purchase prediction
        228–32
    classification of 106–7
    growth 60–1
Web robot 44
Web server logs, identifying
    individual users from 176–7
Web user agents 29
WebFountain 169
Weblogs 145
WebWatcher 157
Weibull distribution 16–6, 240
weighted graph 74
within-site lock-in effect 201
word-of-mouth concept 224–6
wrappers 102
www.amazon.com 179, 211
www.google.com 179

XHTML 32

zero-order Markov model 9
Zipf distribution 22, 80
Zipf's Law 22, 24–5